15.00

WORLD FISH FARMING:
CULTIVATION AND ECONOMICS

WORLD FISH FARMING: CULTIVATION AND ECONOMICS

E. Evan Brown

Professor
University of Georgia

THE AVI PUBLISHING COMPANY, INC.
Westport, Connecticut

Library of Congress Cataloging in Publication Data

Brown, E.   Evan.
    World fish farming.

    Includes index.
    1. Aquaculture. 2. Aquaculture — Economic aspects.
I. Title.
SH135.B76   338.3'71'3   77-17211
ISBN 0-87055-234-1

Printed in the United States of America

Preface

Previously there was little or no economic and technical information by countries or by species on commercial aquaculture. This book was written to fill this gap existing in the literature. I have tried to present in an easy-to-read style detailed information about 28 countries, including all major producing countries. Between 5.5 and 6.0 million metric tons of such production of finfish, shrimp and crayfish are covered, which accounts for about 90% of the world's total.

The People's Republic of China and the USSR, as well as the continents of Europe, Asia, North America and Oceania are represented. Because of only minimal reported cultured fish production in Africa and South America, these continents were omitted. Also, according to information received from New Zealand, there is no culturing of fish in that country.

Ninety-three species of cultured fish, 7 species of shrimp and prawns, and 6 species of crayfish are discussed. The manuscript includes 100 tables for the serious reader to peruse, 41 illustrations and 91 explanatory photographs. Data are presented in both the metric and American systems.

Cultural data include feeding practices, foods, and types of production facilities for freshwater, brackish and marine production. Pond, raceway, fiord, cage, net and other culturing methods are presented. New techniques, such as the submergible net cage of Japan, are also presented and fully illustrated. For some countries, the intensity and density of production are included. A number of farm examples are presented when and where possible.

Economic data, which are difficult to obtain, are presented when possible for production systems by species. Producer prices, marketing margins and consumer prices appear. Data on imports and exports are presented when possible. An attempt has been made to present the most recent and pertinent information available.

For some countries, information is presented by species for restocking of natural waters.

Because of considerable interest in and good profit potentials for operators of recreation or fee fishing facilities, this area is included.

When possible, the outlook for the cultured fish industry of the individual countries is also presented.

I have tried to cover a wide range of subjects for private and public aquaculturalists, as well as for all those who have an interest in fish culture, whether as a large-scale enterprise or as a hobby, for importers or exporters, fisheries economists, students, as well as those individuals who wish to learn more about an increasingly important worldwide food source.

I am pleased to acknowledge the assistance given me in this undertaking by the 10 contributing chapter authors given in the list of contributors. Acknowledgment is also made of the contributions of the following individuals associated with the Department of Fisheries, Food and Agriculture Organization, Rome, Italy: Mr. W.P. Appleyard, Fisheries Industries Development Service; Mr. Jean-Louis Gaudet, Secretary to EIFAC; Mr. Michael N. Mistakidis, Senior Fisheries Resources Office; and Dr. Ziad H. Shehadeh, Fisheries Resources and Environment Division. I would also like to recognize the special assistance of Mr. Wayne Goble for preparation of illustrations and Mrs. Mary T. Prather, for diligence and assistance in typing the manuscript.

E. EVAN BROWN

January 1977

Contributors

ARAI, SHIGERU, Ph.D., Freshwater Fisheries Research Agency, Tokyo, Japan

ATKINSON, CLINTON E., Fisheries Consultant and Advisor, Seattle, Washington, USA

CHEN, HSI-HUANG, Ph.D., Senior Specialist, Joint Commission on Rural Reconstruction, Taipei, Taiwan

HOMMA, AKIRO, Research Division, Fisheries Agency, Kasumigaseki, Tokyo, Japan

MAC CRIMMON, HUGH R., Ph.D., Professor of Zoology, University of Guelph, Guelph, Canada

MAC LEAN, J.L., Department of Primary Industry, Fisheries Division, Canberra, A.C.T.

NISHIMURA, S., Fisheries Economist, Freshwater Fisheries Research Agency, Tokyo, Japan

PARK, KOO-BYONG, Ph.D., Department of Fisheries Business and Administration, Pusan Fisheries College, Pusan, Korea

SHIBUKAWA, HIROSHI, Research Division, Fisheries Agency, Kasu   -seki, Tokyo, Japan

WOHLFARTH, G.W., Ph.D., Fish and Aquacultural Research Stat   Institute of Animal Science, Agricultural Research Organization, Dor   l

Contents

Dedication

This book is dedicated to a man who made it
possible and who collaborated on the Japan Chapter.

Mr. S. Nishimura

A man who can move mountains.

Introduction To Fish Farming

Fish farming is very old. The Romans were able to maintain and raise fish in brackish water along the Italian coast. They probably learned methods of primitive fish farming from the Etruscans, who in turn learned it from the Phoenicians. Any exact date of development would be pure conjecture. However, in the Mediterranean area, fish farming dates back several thousand years. Egyptian bas-reliefs have scenes depicting fish raised in ponds. In China and the Indo-Pacific regions fish culture is very old. The origin of aquaculture or fish farming in China is commonly attributed to Wen Fang, the founder of the Chou Dynasty. The last ruler of the Shang Dynasty had Wen Fang confined to an estate in Honan Province (1135-1122 B.C.). There, Wen Fang built one or more ponds, filled them with fish and made the first recorded references to their behavior and growth. Much of the early skill in the rearing of fish is said to have originated with him and those working with him. For the next several centuries a number of early Chinese writings appeared that were related to fish rearing. In 460 B.C., Fan Li wrote his *Fish Culture Classic*, which described in detail the results of numerous experiments made by Fan Li and others. It was in this era that the keeping of carp for pleasure, as described by Wen Fang, changed to the rearing of carp for food. The size of ponds expanded and the ventures were exceptionally profitable. (See Chapters 11 and 12 for Italy and the People's Republic of China.)

TODAY'S TECHNOLOGY

From the early beginnings fish farming has expanded in area, methods and technology. The modern-day industry can only be described as highly variable.

Species cultured today are raised in fresh (both warm and cold), brackish and marine waters. Species include catadromous fish, such as the eel, which live their lives in fresh or brackish waters and go to the sea for spawning. Anadromous fish, such as salmon, live their lives in marine waters and ascend rivers to go to spawning areas. Other species include marine fish which spend their entire life cycles in salt water, but may be raised in brackish waters. Freshwater fish can sometimes be raised in brackish waters nearly as well as in fresh waters.

Culturing systems have expanded from the earlier ponds to include flowing water systems (called raceways) and enclosure systems ranging from rafts and cages, both floating and submerged, to closing off and farming fiords. Even the traditional pond system has been modified by varying the amount of water added, which ranges from simple replacement to water currents so rapid that they verge on being considered flowing water systems. The various systems may contain forage as well as predator fish. The systems may be both extensive and intensive. These two terms are not well defined and may mean different things to different people. For example, a managed pond may yield only a few kilograms of fish per hectare. By adjusting for pH, adding natural or artificial fertilizers and relying only on natural foods, yields may increase to 7 metric tons (MT) per hectare (ha) (3.1 short tons (ST) per acre) as can be found in Israel. This may still be considered extensive fish farming by some individuals, who differentiate on the basis that intensive production requires the feeding of trash fish or artificial foods such as pellets, or who are accustomed to very large yields per unit of area which may be found only in flowing water systems.

Considerable variation may also be found in the sources of and supplies of either eggs, fry or fingerlings. With some species, the eggs, fry and fingerlings for stocking can only be obtained from wild stocks. In other cases, fish for stocking can be obtained through raising and spawning adult fish. In the latter example, the modern practice of injecting hormones is often used to obtain spawning as well as to adjust photoperiods.

The development of artificial foods such as sinking or flowing pellets within the past 2 or 3 decades has stimulated the production of some species. However the cost of such feeds, coupled with feed conversion, mortality and other costs has sometimes placed these fish in the "luxury" class. Further advances in production of these fish may well be limited by the ability to secure necessary feed ingredients, such as fish meal, and by the price of such ingredients.

Sea ranching, which may be called a variety of names in different parts of the world, is creating considerable interest. This method involves releasing hatchery-reared animals of various sizes into marine waters for rearing and the subsequent recapture of the adult fish upon their return to the point of release. This method involves political and public decisions which may restrict further development in some countries and areas.

GOALS AND ORGANIZATION

The goals of the author have been to present in easy-to-read fashion, on a country-by-country basis, the present state of the aquacultural arts, and to answer some of the questions which hopefully have been aroused by the preceding section. An attempt has also been made to present today's technology, using different culturing techniques. Coupled with these goals is one of presenting economic data (extremely difficult to obtain and constantly changing) to the reader so that differences that exist between culturing systems, species and countries may be more thoroughly understood. The most recent cost-relationships, marketing margins, and other economic data have been included when possible. Sometimes these data become useless after only a brief time as the result of one of more changing relationships. However, the data presented do serve as a benchmark for future comparisons.

To allow for maximum understanding by as many people as possible from various countries, both the metric and British (called American in this book, since the British have converted to metric) systems of measurement are used and presented. Hopefully, this helps rather than confuses the reader.

This book is organized by continents and by countries. It begins with North America, including the USA and Canada. It then proceeds to Europe where countries are presented from northern Europe through southern Europe followed by eastern Europe. Asia is then presented, again proceeding from northern Asia (Asian USSR) to southern Asia. Then comes Oceania. The book concludes with a brief outlook by the author.

Africa and Central and South America are not presented. Essentially, these continents were omitted because of limited production. For example, Africa in 1975 was estimated to produce only 2.7% and Central and South America only 0.7 of 1% of the world's cultured fish (Pillay 1976).

REFERENCE

PILLAY, T.V.R. 1975. The State of Aquaculture. Paper presented at the FAO Tech. Conf. on Aquaculture, Kyoto, Japan, 26 May - 2 June, 1976.

Chapter 1

United States of America

COLD FRESHWATER CULTURED FISH

Rainbow Trout (Salmo gairdneri)

Rainbow trout is the only species of cold freshwater fish cultured commercially in the United States. It is cultured in 45 of the contiguous 48 states (Fig. 1.1). Only Texas, Louisiana and Mississippi do not culture trout. In the past two years new techniques of growing-out trout in the colder winter months in what is normally warm water is changing this situation. In 1975 some trout were grown-out in catfish ponds in Mississippi but only on a pilot basis. Hence, depending on profitability, trout may soon be raised in all 50 states.

The number of trout producers by types is unknown. It may be reasonable to assume a minimum of 250 and a maximum of 500. The leading states in numbers of commercial growers are: Idaho, Wisconsin, Colorado, Michigan and Pennsylvania. To this number must be added 1000 to 2000 operators of trout fee fish-out ponds. In 1974 there were more than 100 of these in North Carolina alone.

The trout industry in the United States has specialists, each dealing with one phase of production. These can be classified as:

(1) Egg producers who raise fish to sexual maturity (usually 2 to 3 years) and spawn them for 3 or 4 years before they are sold for either processing or fee-fishing stock. Fertilized eggs may be sold as eyed eggs or incubated and hatched. Fry and fingerlings are sometimes sold.

(2) Fingerling producers who either raise their own fry or purchase them in sizes varying from 2.5 to 12.5 cm (1 to 5 in.) and sell them at sizes ranging from 10 to 20 cm (4 to 8 in.). These fish are sold alive to individuals who raise them to marketable sizes for human consumption, or to individuals who operate fee fish-out facilities.

(3) Market fish producers who usually buy eyed eggs to be incu-

FIG. 1.1.   GEOGRAPHIC DISTRIBUTION OF RAINBOW TROUT FARMS IN THE UNITED STATES

FIG. 1.2.   RAINBOW TROUT FEE FISHING, GEORGIA

bated and hatched. Fry are then raised to market sizes of 25 to 35 cm (10 to 14 in.). At that time they are transferred alive to processing plants, which may or may not be on the premises. Some of these producers may buy fingerlings for growing-out. Depending on water temperatures, a marketable rainbow trout may take from 10 to 18 months to mature from hatching.

(4) Grow-out producers who may raise fish for another producer. The grow-out operator raises 10 to 15 cm (4 to 6 in.) fingerlings to market size without ever taking title to the fish. He is paid on the basis of weight gained. Time required to do this is usually 4 to 6 months. This depends on water temperatures, size of fingerlings stocked, size of marketable fish desired and management.

(5) Fee fish-out operators who usually buy catchable-size fish which can be caught within a short period of time by fishermen. Sometimes these operators buy fingerlings and raise them for 4 to 6 months and then transfer them to the fish-out facility.

(6) Live-haulers who haul live fish from one farm to another, or to a processing plant or fee-fishing facility. Some of these haulers may transport live fish 750 to 3000 km (465 to 1860 mi.).

(7) Processors who may or may not have a fish production facility. They receive live fish for processing into fresh and frozen items (Table 1.1).

TABLE 1.1
DRESS-OUT FIGURES FOR A 28 CM (11.3 IN.) RAINBOW TROUT

Item	Market Weights (oz)	(g)	Dressing Loss (% of round weight)
Round weight	9.25	263	0
Dressed weight	7.4	210	20
Boned weight	6.45	183	31.3
Fillet weight	5.0	142	46

Feeds.—In general trout rations contain:

Protein	32 to 55%
Carbohydrate	9 to 12%
Fat	5 to 14%
Fiber	15 to 17%

Essential trace minerals and vitamins

The wide range in ingredients is due to different types of diets for particular sizes of fish. For example, fry have higher protein requirements than fingerlings. Prepared feeds are sold either in bulk or in 22.7 kg (50 lb) bags. Rations are in the form of dry crumbles or pellets. There are more than 15 trout feed manufacturing companies. Each has its own formulations.

History.—The business of raising trout for market began in the 1870s in the northeastern United States. The main produce of these early commercial trout farms was adult brook trout (*Salvelinus fontinalis*) which were sold on the fresh fish markets of large eastern cities. The eggs stripped from the ripe females were a secondary consideration as a saleable item. In the early 1900s brook trout farming became important. Fish farms increased in number and size by constructing raceways. Food for the fish was obtained from packinghouse products such as spleen, liver, heart and lungs, which were ground together with frozen bottom fish and fed as a thick paste.

Rainbow trout were officially introduced into the eastern United States in the 1880s. Their introduction into commercial fish farming did not occur until after 1900. At that time they were raised primarily as a hobby-type operation by private fishing clubs. Limited production of brown trout (*Salmo trutta*) and cutthroat trout (*Salmo clarki*) has occurred at various times. Essentially, recent production has centered on rainbow trout.

From the early 1920s until after World War II, private trout hatchery production developed very slowly because of the availability of sport-caught fish and low market demand. From the late 1940s until the present the trout industry has grown rapidly. The most rapid change has been in processed fish destined for food use.

Production data for cultured fish (both live and processed) sold annually in the United States are very difficult to obtain. Data quoted in trade and scientific journals are often misleading and conflicting. There are no official state or federal production statistics.

Present Status.—Even with the lack of corroborative production data it is safe to say that the trout industry is growing. Evidence of this comes from feed manufacturers who state that they are selling more trout feed than ever before. Similar statements come from egg producers. This increased production is occurring from enlargement of older farms, new farms and fee fish-out ponds. Production of processed trout has also been expanding.

Klontz and King (1975) state that the main trout production states are Idaho, California, Wisconsin, Michigan, Colorado and Pennsylvania. The author adds North Carolina to this listing.

Data obtained by Klontz and King (1975) indicate a total production of 12,272 MT (27 million lb) by private growers. Of this volume, about 9100 MT (20 million lb) was probably processed and
about 3172 MT (7 million lb) was marketed through fee fish-out
ponds.

Of the processed volume about 90% was contributed by one state
(Idaho). The remaining 10% was contributed by all other states combined which market their trout largely through fee fish-out facilities
or by limited sales of fresh fish to local restaurants. For example,
Colorado is considered a major trout-producing state, yet a reliable
estimate by Hagan (1972) of Colorado State University indicates
that production is probably less than 114 MT (250,000 lb).

The primary reason for the overwhelming importance of Idaho
as the major trout-producing state is adequate water of proper
temperature. This happy event is due to a geological accident referred to as the Southern Idaho Aquifer. The water originates in
the mountains and enters the vast lava plain in southern Idaho.
This lava plain extends about 240 km (150 mi.) from east to west and
about 80 to 120 km (50 to 75 mi.) from north to south. This rock is
extremely porous and the mountain runoff is absorbed. Water
emerges from this aquifer from the side of a deep fissure cut by the
Snake River. In essence underground rivers emerge to the surface
at the canyon wall. The water is 14.4°C (58°F) and temperatures

FIG. 1.3.   LARGE RAINBOW TROUT FARM WITH FEED BEING DISPENSED BY FEEDING
CART SUSPENDED FROM OVERHEAD RAILS, IDAHO

fluctuate only by about 0.5°C (1°F). Water flow through a single hatchery may be as much as 920 liters per sec (14,583 gal. per min).

In Colorado there are no private producers with more than 56 liters per sec (897 gal. per min) and average flow is probably less than one-third of this figure. In other major trout-producing states the same situation exists. Surface water is often unreliable; it may be too warm in summer or have reduced flows in winter because of freezing. Springs are usually small and too far apart for an individual to have a large volume of production. Most of these individuals are only part-time fish farmers. Even with small volumes, which in many cases do not exceed 5 MT (11,020 lb), some of these producers fare well. This is done by selling fresh dressed fish to local restaurants at prices above the frozen price or by sales through fee fish-out ponds.

The Idaho Fish and Game Department issued 72 permits to raise fish commercially in 1974. The majority of these permit holders were located along the Snake River.

The commercial food fish industry in Idaho is complex. No two individual facilities are alike from the viewpoint of raceway design, water utilization, feeding practices, fish density per unit of water volume, or flow and fish husbandry methods. In general a raceway segment is 10 times longer than it is wide and is 1 m (36 in.) deep. The industry in Idaho consists of: (1) egg producers who sell eggs both within and outside the state; (2) growers; (3) contract grow-out operators who furnish management, labor and facilities (with contract grow-out operators the owner retains title and provides feed and technical assistance); (4) processors, who are integrated with a fish-raising facility; (5) fee fish-out operators; (6) live-haulers; and (7) feed manufacturers.

Management.—*Egg Incubation.*—After spawning, eggs are acclimated to water temperature at the hatchery. This is usually 13° to 14°C (56° to 58°F). After acclimation the eggs are placed in shallow trays in a single layer in baskets many layers deep. Trays of eggs are stacked either in deep troughs with baffles to provide upwelling currents of water through each stock, or in Heath incubators where water flows through each tray in an upwelling direction, beginning with the top tray and progressing to the bottom tray. The baskets are either put into deep troughs or into incubation boxes in which the water flow is upwelling at a rate that gently rolls the eggs. The time required for hatching is 21 to 23 days in 13° to 14°C (56° to 58°F) water. Hatching takes place ten months out of the year. Only in July and August are eggs unavailable. In 1973 the Idaho trout industry eyed an estimated 165,320,000 salmonoid eggs of which 50,470,000 were incubated for hatching. Hatchability was 81.6%.

FIG. 1.4.   RAINBOW TROUT HATCHING TRAYS IN USE

Fry.—The emergent yolk-sac fry are removed from incubator trays or baskets and put into troughs. When nearly all the yolk has been absorbed, they are started on feed dispensed by hand or automatic feeders. Several times a day they are fed an amount based upon the feed formulation of 7 to 9% of body weight, dependent upon water temperature. This management phase lasts for 3 to 5 weeks, depending upon facilities. Of the 41,180,000 fry hatched in 1973, 6,590,000 died from various causes. The survivors were from 2.5 to 7.5 cm long (1 to 3 in.).

Fingerlings.—In this phase fish are in outside ponds or raceways and are growing rapidly, sometimes over 2.5 cm (1 in.) per month. They are fed 4 to 5% of their body weight. During this phase, as the density of fish dictates by weight per cubic unit of water, they are distributed to other ponds or raceways. On some farms they are separated by size groups as they are distributed to new ponds. In 1973, 41,180,000 fry developed into 34,590,000 fingerlings. These fingerlings were from 7.5 to 15 cm (3 to 6 in.).

Stockers.—During this management phase the fish are transported to either farm pond (grow-out) operations or fee-fishing establishments, or remain on the farm until they are of processable size. Inventory and grading by size is done at monthly intervals. Since the growth rate has slowed, the level of feeding drops to 1 to 3% of body weight per day, depending on water temperature. In 1973,

FIG. 1.5.   FRY REARING TROUGHS WITH AUTOMATIC FEEDERS FOR RAINBOW TROUT

34,590,000 fingerlings developed into 30,300,000 stockers.

Marketables.—These are fish that are 30 cm (12 in.) long or more. They are shipped from rearing ponds to processing plants. During 1973, 26,910,000 fish were sold live or processed. Hence, the survivability after hatching was over 65%.

Production and Marketing.—There is no recent data available concerning production costs. Farmers in Idaho interviewed by the author in 1976 indicated that farm prices were near production costs. Farm prices for 280 to 390 g (10 to 14 oz) processable fish in May 1976 ranged between $1.43 and $1.54 per kg ($0.65 to $0.80 per lb). After processing, dressed fish having undergone an approximate 20% weight loss sold for $3.30 per kg ($1.50 per lb). Boned fish sold for $3.74 per kg ($1.70 per lb). Both dressed and boned fish prices were for frozen, packaged fish, FOB processing plant. Processed fish are usually sold by the processing plants to fish wholesalers whose marketing margin is about 20%. Hence, wholesale price was $3.96 per kg ($1.80 per lb) for frozen, packaged dressed fish for sale to restaurants and retail stores, and about $4.48 per kg ($2.04 per lb) for boned fish. Retail store markup varies between 10 and 25% of wholesale price. Thus, retail store prices for dressed fish would vary from $4.36 to $4.95 per kg ($1.98 to $2.25 per lb). Prices of boned trout would vary from $4.92 to $5.60 per kg ($2.25 to $2.55 per lb).

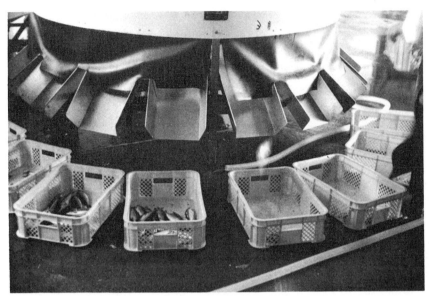

FIG. 1.6.   SIZING DRESSED RAINBOW TROUT WITH AUTOMATIC MACHINE, IDAHO

Market sizes for the restaurant trade range between 225 and 336 g (8 to 12 oz) per dressed fish. Supermarkets prefer two 200-250 g fish in each package. This keeps the unit price down per 454 g (1 lb package) and is thought to increase retail sales. About 85% of Idaho's processed volume is sold frozen and 15% is sold as fresh, iced trout.

Fee Fishing.—No one has any accurate idea of the number of trout fee fish-out operations in the United States. The author hazards a guess that there are between 1000 and 2000. In North Carolina alone there are over 100. These operators may buy fish already of catchable size (250 g or ½ lb) as large as fish that have been used for spawning, or they may buy fingerlings and raise them up to catchable size. These operators generally buy locally-produced live trout from nearby small-scale producers. These producers are insulated from the competition of large-scale commercial producers in Idaho by several factors. To have live fish delivered from Idaho, hundreds or perhaps thousands of miles, is expensive. To lower costs Idaho haulers may not deliver less than 1400 to 1800 kg (3000 to 4000 lb) in one load. Small-scale fish-out operators may be willing and able to purchase only a few hundred kilograms or pounds at one time because of water restrictions. Hence, a long distance hauler may have to coordinate sales for 6 to 10 outlets to dispose of one load. This is not profitable or feasible for haulers. The local producer can and does make deliveries of small lots at irregular intervals. The fee fish-out operator may secure from $3.30 to $4.40 per kg ($1.50 to $2.00 per lb) for fish caught at his pond. Thus, he is willing and able to pay more for his fish than the $1.43 to $1.54 per kg ($0.65 to $0.70 per lb) that Idaho producers secure from processing plants.

Fee fish-out facilities vary widely in volume sold annually and size. Some may sell only a few hundred kilograms while others may sell many tons. Usually the catch-out price is a fixed amount per kilogram or pound of fish caught. The operator supplies fishing tackle and bait, dresses the fish, and packs them in ice for a small additional charge. These facilities are usually open to the public on weekends and in the main tourist season, which varies by location. Either earthen ponds or concrete impoundments are used. The earthen ponds bear closest resemblance to a natural setting and are more popular with fishermen.

An example of popularity of such operations is the use of one fish-out facility in Georgia by over 3000 fishermen one year, from which the operator sold more than 7300 kg (16,000 lb) of fish and grossed over $25,000. There are undoubtedly many other larger operations in the United States.

Total Supply.—Estimated cultured production volume of approximately 12,272 MT (27 million lb) of domestic production must be adjusted by deducting exports and adding imports to arrive at total supply. There are practically no exports of U.S. trout other than 50 to 60 MT annually to Canada. The major, if not only, country from which imports are made is Japan. Imports of dressed frozen trout from Japan between 1967 and 1974 averaged 1136 MT (1252 ST) annually.[1] This would be equivalent to 1363 MT (1502 ST) of trout on a live weight basis. Thus, total supply is about 13,635 MT or 15,026 ST (30,052,000 lb). About 90% of total supply is produced domestically and 10% is imported.

Outlook.—The future of trout farming in the United States must be discussed from two viewpoints: (1) the large-scale commercial processing segment, which is essentially in Idaho, and (2) the small-scale part-time enterprise which sells fish to one or two specialized trout restaurants and/or sells its output through fee fish-out facilities.

The outlook for the large-scale commercial processing segment varies from one of extreme optimism to one of cautious conservatism. At present many producers feel that they are producing near cost and, unless production becomes more profitable, there will be little

---

[1] Export data obtained from customs data, Tokyo, Japan.

or no expansion. The majority of knowledgeable individuals believe that there are sufficient unused water sources in Idaho to increase production by 20% within the next 10 years if producer prices increase faster than costs. Some producers express the view that average production costs can be slightly reduced under present conditions by better disease control. In the event production increases or remains stationary, the number of managements will decline, while average production per management increases.

The outlook for the small-scale part-time trout farmer is good. Interest in and public demand for trout fishing is projected to increase rapidly. This means that market prices for fish producers will remain good or increase. Existing producers are unlikely to discontinue production while there are still sufficient water resources in the states of Wisconsin, Michigan, Missouri, Arkansas, Minnesota, Pennsylvania and North Carolina for some expansion by new producers.

WARMWATER CULTURED FISH

Catfish (*Ictalurus punctatus*)

Private warmwater fish culture in the United States probably began in the late 1920s and early 1930s. During this period a few individuals began raising minnows to supply the growing demand for fish bait for sports fishing.

Shortly after World War II, with the boom in farm pond and reservoir construction, the demand for minnows increased. By the late 1940s a dozen or more private operators were successfully producing bait fish. By 1953 the number had increased enormously and farmers had begun raising buffalo fish (*Ictiobus*), bass (*Micropterus*) and crappies (*Pomoxis*). Many of these early attempts failed because of inexperience, improper construction of ponds or because low-value species were stocked.

The Southeastern Fish Cultural Laboratory was established at Marion, Alabama in 1959; the Fish Farming Experiment Station at Stuttgart, Arkansas in 1960; and the Fish Farming Development Center, Rowher, Arkansas in 1963. These stations began research mainly in catfishes (*Ictalurus*) and bait fishes.

By 1961 the fish farming industry was again expanding. This time the expansion was based on greater experience and early results of research; especially with catfish, which have proved to be the most desirable and profitable warmwater species.

In 1963 about 6830 ha (16,870 acres) were devoted to warmwater

fish culture, 960 ha (2370 acres) of which were used in raising catfish. By 1969 the estimated totals were 27,530 ha and 16,194 ha (68,-
000 and 40,000 acres) respectively. This comparison dramatically at-
tests the explosive growth of the industry and particularly catfish
production. The 1969 crop of catfish had a wholesale value of more
than $33 million and generated an estimated $75 million retail business.

During these earlier days fish farming was a welcome addition in
the delta areas of Arkansas and Mississippi. The field crops were
rice and soybeans grown in rotation. Because of crop controls,
some farmers had idle land that could not be planted to cotton, rice
or soybeans. The fish returned a higher margin of profit than did
some of the less productive crop acreage. In addition, fish in the
rotation added nitrogen for field crops and aided in weed control,
which resulted in higher crop yields following the fish crop.

The farm-raised or cultured catfish industry, until about 1968,
was based almost entirely on the market for live fish for stocking
farm ponds, other fish farming operations, pay lakes and reservoirs. Although these markets are still relatively important and
still account for the majority of total sales, there has been a big
shift toward production of catfish for the food market. With this
shift has come the realization that marketing is at least as important as production efficiency and that independent restaurants,
grocery chains and fast food restaurants each have their own needs
for portion control, custom preparation and packaging, and contin-
uity of supply.

The channel catfish (*Ictalurus punctatus*) is the species most often raised on commercial warmwater fish farms. It adapts readily
to pond conditions as well as raceways, cages and tanks; accepts
artificial feeds; and tolerates crowded conditions associated with
intensive culture. Techniques for propagation have been developed
so that large and adequate numbers of fingerlings can be produced.

Channel catfish are esteemed for the high quality of their meat,
especially in southern states where the fish is a traditional delicacy. Dressed catfish retains a high quality when packed in ice
and sold through traditional fish marketing channels. Frozen fish
can be stored for extended periods without undue loss of quality
if adequately packaged and refrigerated.

These attributes form the background for the expanding catfish
industry. This expansion is described in the following sections.

Cultural Methods.—*Brood Stock.*—Large catfish usually spawn
earlier than smaller ones and produce more eggs. Culturists prefer 1 to 5 kg (2 to 11 lb) brood fish, although stunted catfish can be-

come sexually mature when they weigh only 340 g (0.75 lb). Fish larger than 5 kg (11 lb) are hard to handle. Channel catfish brood stock can be raised and reliably spawned in three years.

Brood stock should be selected prior to feeding. The female ready to spawn should have a well rounded abdomen with the fullness extending past the pelvis to the genital orifice. The ovaries should be palpable and soft, and the genitals swollen and reddish. Less care needs to be used in selecting males. Males with prominent secondary sexual characteristics, such as a heavily muscled head wider than the body, dark pigmentation under the jaw, and large protruding genital papilla, usually have well-developed testes. Such males may be used successfully for as many as three times, whereas males with poor secondary sexual characteristics may be capable of fertilizing only one spawn.

Pairing.—Successful pairing of fish, which is an essential part of pen and aquarium spawning, depends on the skill of the culturist in sexing and selecting the fish. Sex determination is slightly less important in pond spawning.

Channel catfish fight during the spawning season. Their bites are often deep enough to break the skin, and the resulting wounds may become infected. Fish sometimes die from severe bites. For this reason, special care must be taken to pair fish properly in pens and aquariums. If the female in particular is not ready to spawn the male will fight with her, and in the confinement of an aquarium may inflict enough injuries in 15 or 20 min to kill her. In spawning pens, the condition of the female is not quite as important, since the pen is usually large enough for her to escape from the attacking male. Even so, it is not uncommon to see bite marks on the female in a pen. When fish reproduce in ponds they pair when they are ready to spawn, and fighting is not a serious problem.

Although most culturists prefer or require that the male be slightly larger than the female, biologists who have observed hundreds of pairs spawned in aquariums under different experimental conditions have concluded t.. + males and females of similar size should be used. If the male is considerably larger than the female, spawning is usually successful. If the female is much the larger, however, she usually attacks the male and may not mate with him.

Care and Handling.—Immediately after the spawning season, brood stock may be placed in a pond at the rate of 150 per 0.40 ha (150 per acre). Although the fish are easily frightened, 90 to 95% of the fish held in small ponds will learn to come for feed within 1 week. If this same number of fish is held in a 0.40 ha (1 acre) or larger pond, 25 or 30% may never come for feed, and consequently will be

Courtesy of Fish Farming Experimental Station, Stuttgart, Arkansas

FIG. 1.7.   CATFISH SPAWN

in poor condition and undesirable as brood stock during the next spawning season. One recommended practice is that of dividing the fish among several ponds to prevent destruction of the entire brood stock by a possible epizootic. Some hatcheries distribute the adult catfish throughout all available ponds during the summer. The scattered fish then need not be fed because sufficient natural food is available. When the ponds are drained in fall and winter, the catfish from several ponds are returned to a single pond and fed until time for spawning. The feeding of brood stock is important because diet quality and quantity largely govern the number and size of eggs, spawning time and general health.

Brood fish that weigh 1 to 1.5 kg (2 to 3 lb) should be stocked at the rate of 300 to 400 kg per 0.40 ha (660 to 880 lb per acre) when additional growth is desired. Larger fish should be stocked at the rate of 800 kg per 0.40 ha (800 lb per acre) if further growth is not wanted. The feeding rate and diet depend partly upon water temperature. Brood fish should be fed 2 to 3% of their body weight on each of 3 or 4 days a week when water temperature is above 13°C (55°F). Since brood stock channel catfish feed sparingly, even when the water temperature is as low as 7°C (45°F), they should be fed only on the warmest days of cold weather periods. When it is very cold, catfish feed better on meat or diets high in animal protein than they do on cereal feeds. Meat diets can be readily utilized by the fish. It is generally accepted that fresh or frozen meat or fish should be included in the brood stock diet.

Number of Eggs Laid.—Females weighing 0.5 to 1.8 kg (1 to 4 lb) and in good condition produce about 4000 eggs per 0.45 kg (1 lb) of body weight; larger fish usually yield about 3000 per 0.45 kg (1 lb). Fish in poor condition produce fewer eggs.

Egg Development and Incubation.—The number of days required for eggs to reach various developmental stages varies according to temperature. Channel catfish spawn at 21° to 29°C (70° to 85°F). The optimum temperature is about 27°C (80°F). The incubation period ranges from 10 days at 21°C (70°F) to 5 days at 29°C (85°F). At incubation temperatures above 29°C (85°F) many deformed fry are produced.

The male channel catfish assumes a position over the eggs after spawning is finished and cares for the eggs during the incubation period. Although the female aerates the eggs during spawning, she is driven away along with other intruders after the male takes possession. The male generally faces in the same direction with his pelvic fins working alternately in a continuous beat. Occasionally he circles away from the eggs and returns. The most striking activity of the male is the vigorous shaking of his body as he presses and packs the eggs with the side of his pelvic fins in a manner that moves the entire egg mass. Apparently this act helps aerate the developing eggs, especially those deep within the mass, but it may also serve to move the embryos within the eggs.

Many fish farmers who produce fingerlings for sale prefer to collect spawns and hatch them in artificial hatching systems. Good incubation and hatching are obtained at some hatcheries by the paddle wheel method (described later), which simulates the male's agitation of the eggs.

After the eggs hatch, the fry accumulate on the bottom and remain there for about two days before coming to the surface. At this time, the yolk is greatly reduced and the skin pigment is visible. By the third day the fry start to feed and swim actively.

Pond Method of Spawning.—In early attempts to induce spawning, tiles, beer kegs, nail kegs, or boxes were partly embedded in the bank of a pond about 0.6 to 1 m (2 or 3 ft) below the water surface. After the brood fish placed in the pond had spawned, the newly hatched fish were removed from the containers and transferred to a clean pond. In later years the egg masses were removed from the pond for incubation and continuous-motion paddles were used to agitate the water and the eggs. This system for hatching catfish eggs is still in use.

The pond method remains essentially unchanged. Brooders are placed in small, usually shallow ponds, ranging up to about 2 m (6 to 7 ft) deep. Equal numbers of males and females are placed in the pond at a stocking rate of 24 to 150 fish per 0.4 ha (1 acre).

Forty-liter (10 gal.) milk cans and small drums are popular spawning containers. Ordinarily it is not necessary to provide a spawning receptacle for each pair of fish, since not all fish spawn at the same time. Most culturists allow 2 to 3 receptacles for every 4 pairs of fish. The cans or drums are usually placed with the open end toward the center of the pond. Fish have spawned in containers in water as shallow as 15 cm (6 in.) and as deep as 1.5 m (5 ft). The receptacles can be most easily checked in water no deeper than arm's length.

Frequency of examination of spawning containers depends on the number of brood fish in the pond and the rate at which spawning is progressing. Caution should be used because an attacking male can bite severely. In checking a container the culturist gently raises it to the surface. If this is done quietly and carefully the male is not disturbed. If the water is not clear the container may be slowly tilted and partly emptied until the bottom can be checked for eggs or fry.

Spawns may be handled in different ways by the fish farmer. In the pond method, he may either remove eggs or fry or leave them in the spawning receptacle. Removal of the eggs has several advantages: it minimizes the spread of diseases and parasites from adults to young; provides protection from predation; and may increase the percentage hatched. The main reason for removing fry is to improve control of stocking rates, although it also protects them from predation. Fry or eggs are often removed when spawns are produced for stocking other ponds. Large-scale producers commonly use special brood ponds.

If eggs and fry are left in the pond, the brood stock should be removed with a large-mesh seine. Periodic seining with a small-mesh seine provides information about numbers and growth of fingerlings.

Advantages of the Pond Method.—The pond method is inexpensive because it requires minimal facilities of spawning containers and a pond, and does not place demands on the farmer for critically selecting, sexing and pairing his brood stock. The fish in the pond continue to feed and develop until they pair and spawn. If the brood fish are of marginal quality, the pond method is more likely to produce spawns than are the other methods.

Pen Method of Spawning.—Pens about 3 m long and 1.5 m wide

(10 ft long and 5 ft wide) are commonly used by federal and state hatcheries and by a few private hatcheries. The pens are constructed of wood, wire fencing or concrete blocks. They may be enclosed on four sides, or the bank of the pond may be used as one side. The sides should be embedded in the pond bottom and should extend at least 30 cm (12 in.) above the water surface to prevent the escape of the fish. Water in the pen should be 0.6 to 1 m deep (2 to 3 ft).

Location of the spawning receptacle in the pen is not critical, but the opening generally faces the center of the pond and the receptacle should be staked down. Forty-liter (10 gal.) milk cans, 45 kg (100 lb) grease drums, and earthenware crocks are popular spawning, containers. After spawning, eggs or fry and parent fish may be removed and a new pair placed in the pen. Alternatively, the female may be removed as soon as an egg mass is found, and the male then allowed to hatch the eggs.

Advantages of the Pen Method.—The pen method has several advantages: (1) it provides close control over the time of spawning, which can be delayed by separating females from males; (2) it offers the advantage of pairing selected individuals; (3) it facilitates removal of spent fish to a separate pond where they can be given special care; (4) it protects the spawning pair from intruding fish; and (5) it allows the use of hormones.

To succeed with the pen method, the culturist must know his fish well enough to be able to pair the right fish at the right time.

Aquarium Method of Spawning.—The aquarium method provides still greater control than the pen. A pair of fish is placed in an aquarium with running water and induced to spawn by the injection of hormones. The method capitalizes on limited facilities, use of hormones and expert brood fish selection. It is an intensive type of culture in which many pairs can be successively spawned in a single aquarium during the breeding season, since eggs are immediately removed to a mechanical hatching trough. The technique is used in federal, state and a few private hatcheries.

In this method only well-developed females nearly ready to spawn should be used. Males need not be injected with hormones, but should be about the same size as the females with which they are paired. If the male attacks the female, he should be removed until after the female has been given 1 to 3 hormone injections. He may then be placed with the female again. Males may be left to attend the eggs in the aquarium; or, preferably, the eggs will be removed to a mechanical hatching trough.

Spawning may be induced in catfish by injecting the female with pituitary material from carp, buffalo fish, flathead catfish or chan-

nel catfish. Potency of the pituitaries differs little among these species and is not affected by the date of collection. The total amount
of acetone-dried pituitary material required varies widely. However, most females require about 6 mg per 454 g (1 lb) — that is, 3 injections of 2 mg per 454 g of body weight at 24 to 48 hr intervals.
Most fish begin spawning within 16 to 24 hr after the injection.

Human chorionic gonadotropin has been used successfully at a
dosage of about 800 international units (IU) per 454 g. A single injection is usually sufficient.

Fish spawned by the hormone method are not particularly disturbed by people moving around the area.

Advantages of the Aquarium Method.—The aquarium method
has several advantages. (1) Spawn can be obtained at a convenient
time. The hormone injections eliminate such environmental vari-
ables as spawning areas, light, temperature and other climatic conditions. (2) The spawning period can be altered within reasonable
limits and total spawn-taking time can be reduced. (3) Fish that
will not spawn naturally can sometimes be induced to spawn. (4)
Culture ponds can be stocked with fry of uniform age and size. (5)
Disease transmission from brood stock to offspring, as well as predation by adults, is minimized.

Controlling Spawning Time.—The date and length of the spawning season for channel catfish varies from year to year and among
localities. In various natural waters the season may begin as early
as April and end as late as August.

Encouraging Spawning.—In June and early July fish in pens occasionally spawn for a few days and then completely stop. Raising
the water level rapidly 5 to 8 cm (2 to 3 in.) will sometimes cause spawning to resume immediately.

Some farmers inject brood females with human chorionic gonadotropin before transferring them to the spawning pond. Others
have advanced the spawning time about two weeks by taking advantage of the warmer water of small, shallow brood ponds.

Spawning can be delayed 20 to 30 days by keeping the sexes separated. It may also be delayed by holding the fish at water temperatures of 17° to 18°C (65° to 66°F) during May, June and July.

Hatching Eggs.—For egg incubation, temperatures below 18°C
and above 29°C (65° and 85°F) should be avoided. Temperatures from
26° to 28°C (78°and 82°F) are considered optimal. In this range the eggs
will hatch in about six days.

Color is an important index of the condition and stage of catfish
eggs. Under proper conditions the yellow eggs turn pink as the em-
bryo develops and establishes its blood supply. Unfertilized or dead

eggs turn white and enlarge.

All hatching devices must provide sufficient agitation to supply the entire egg mass with oxygenated water of a suitable temperature. When eggs are hatched in troughs, they are agitated with paddles driven by an electric motor or a water wheel. The agitation should be sufficient to move the whole spawn, but not enough to throw eggs out of the holding baskets. If well water is used it must be aerated and of suitable temperature and quality. For example, water with a high iron content is not considered desirable. Gravity-flow water should be used if available because this system is not likely to fail.

Trough hatching systems may be constructed from a variety of materials. Aluminum is commonly used, but wood or steel serves equally well. The shaft is fitted with a pulley at one end and is belt-driven by an electric motor (frequently ½ hp) at a preferred speed of 30 rpm. Combinations of pulley sizes or a variable-speed gear box may be used to deliver the desired speed of rotation.

Spawn baskets are constructed of 0.25 in. hardware cloth. Each basket is divided into four equal sections and fitted with wire hooks so that it can be hung on the sides of the trough with the top edge 2.5 cm (1 in.) above the water. A flow of about 10 liters (2.5 gal.) per min of well-aerated water should be provided.

It is important that spawns in each section of the trough be of the same age, because the mixing of spawns of different ages may prevent the use of prophylactic treatments.

A flush treatment of malachite green at about 2 ppm may be introduced at the head of the trough once or twice a day if needed to control fungus. Fungi grow on dead eggs and spread to living ones, eventually destroying the whole spawn. This chemical is not to be applied within 24 hr of hatching. If fry are present it will kill them.

Rearing Fry.—To remove fry from the hatching trough the culturist simply siphons them from the trough with a hose into a washtub or pail. To remove fry from a pond spawning receptacle he first removes the male, usually by frightening him away, then lifts the spawning container to the surface and carefully pours out part of the water. The remaining water with the fry can then be emptied into a floating tub that contains an inch or two of water. Fry should be counted and then moved to either a rearing trough or a pond.

Trough Method.—Rearing troughs may be made of wood, metal, fiberglass or plastic. Typical troughs are 2.5 to 3.0 m long, 30 cm deep and 20 to 50 cm wide (8 to 10 ft long, 1 ft deep, and 8 to 20 in. wide). Each trough must be supplied with running water and equipped with a drain and a standpipe. The fry from one or two spawns are put into each trough. A flow of about 20 liters (5 gal.) of fresh oxygenated water per min is sufficient. Standpipes are screened so the fry are not washed over the standpipe and down the drain.

Fry begin to feed shortly after the yolk sac is absorbed and the fish begin to develop a grayish color. This usually occurs at three days of age. It is mandatory that suitable feed be available at this time. Channel catfish fry eat a variety of feeds. Feeding frequencies and particle size are important considerations. Young fry should be fed every 2 to 4 hr around the clock for the first week. Thereafter the fry should be fed about four times a day. Diets for channel catfish are now available commercially in the USA. Fry may be raised to the fingerling stage in troughs or may be moved to a rearing pond at any time.

Pond Method.—Although the area of rearing ponds for channel catfish varies from 1000 m² to 2 ha (0.1 to 5 acres) and larger, it is usually about 0.4 ha (1 acre).

Predatory insects are often a problem in the pond culture of fry. If a pond is not filled until immediately before it is stocked with fry, establishment of the insects is prevented. If water has been standing in the pond for several days, or if surface water is used, it should be treated with a nonresidual insecticide 2 to 3 days before the pond is stocked. The operator should use extreme care because insecticides are dangerous to man.

Fry can either be released directly into the open pond or held for the first few days in floating cages made of screen or in a wooden frame with a screen bottom. They are then protected and can be fed during this vulnerable period. If a pond has a basin, fry may be placed in it and the rest of the pond kept dry. As the fry get larger the pond is gradually filled.

Fry are stocked at the rate of 50,000 to 250,000 per 0.40 ha (1 acre), depending on the size of fingerlings sought at the end of the growing season. Young fish are fed daily along most of the shoreline at a rate of about 4 to 5% of their body weight at each feeding.

Rearing Fingerlings.—Channel catfish fingerlings are reared in either ponds or troughs. In ponds low-cost pelleted fish feed may be used, but in troughs a more expensive, balanced feed is required. In the two environments different methods and techniques are used to stock, feed and harvest the fish. Some farmers prefer to start the fry in rearing troughs and transfer the fingerlings to ponds after they are actively feeding. The choice of method depends on facilities and labor available, and on the number and size of fingerlings to be produced.

Regardless of the rearing method selected, attention should be given to the water supply. In the trough rearing method the incoming water should range between 24° and 29°C (75° to 85°F) and contain not less than 6 ppm of dissolved oxygen. Factors such as pH, hardness and dissolved iron content influence production of fish in troughs that are supplied with well water. When water from ponds is used for trough culture, these conditions may be disregarded. Then, however, a Saran screen or sand filter between pond and trough is desirable. In pond rearing a major problem is fry-eating insects and fish. Predatory fish can be controlled by filling the pond with fish-free well water or by using a fine-mesh screen to filter water from other sources. Fish-eating insects can be controlled by treating weekly with oil or kerosene until the fish are about 3.8 cm (1.5 in.) long.

Trough culture of fingerling catfish begins with yolk-sac fry from the hatching trough or from spawning containers in the ponds. It is a good practice if time and facilities permit, because it gives the culturist complete control of the small fish. When fry about 2 cm (0.75 in.) long are stocked in rearing ponds, 60 to 90% can be harvested the following fall.

Techniques for feeding catfish fry are extremely important, particularly when the fry are learning to feed. For about the first 3 or 4 days after hatching, fry subsist on yolk and remain on the bottom of the pond or trough. After they have absorbed the yolk sac they become known as "swim-up" fry. When they are seen swimming along the sides and surface of the trough in search of food they must be fed at once. Fry that do not learn to feed during the first few days after absorption of the yolk sac will die.

In pond culture either sac fry, swim-up fry or feeding fingerlings may be stocked. Stocking rates vary, depending on the fingerling size desired. If the fish are to be harvested at 5 to 10 cm (2 to 4 in.) lengths after about 120 days, they are stocked at the rate of 100,000 to 150,000 per 0.4 ha (1 acre). If 20 cm (8 in.) fish are desired, the stocking rate should be reduced to 14,000 to 20,000 per 0.4 ha. A combination of these methods may be used. The fish can be stocked at the maximum rate initially and then partially harvested for sale or transfer to other ponds when they reach the 5 to 10 cm (2 to 4 in.) size.

When all the fingerling catfish in a pond are to be harvested, as many as 75% of the fish may be removed by seining the feeding areas before the pond is drained. In summer it is best to harvest the fish early in the morning while water and air are cool. Care should be taken to avoid excessive muddying of the water. If possible, sites with a firm obstacle-free bottom should be selected as feeding and seining areas. After the seining is completed the pond is drained to recover the rest of the fish.

Rearing Fingerlings to Market Size.—A large healthy fingerling, a good environment and a conscientious feeding program are necessary for a profitable food-fish production program. If market-size fish are to be produced in one growing season, fingerlings 15 to 20 cm (6 to 8 in.) long or longer must be stocked. Such fingerlings will weigh at least 454 g (1 lb) after about 210 days if properly cared for.

Time of stocking is not as critical as some believe. A pond should be stocked whenever it is ready to receive fish. A 5 cm (2 in.) fish stocked in July will be only 24 to 25 cm (9 to 10 in.) long by November and must be reared to market size the next summer.

The poorest stocking months are December and January because of the low water temperature. Fish feed least at this time of the year and sometimes do not resume feeding readily after they are moved. The fish that die after stocking at this time of year may never be seen.

Fish should be fed during winter, but the feeding rate should be reduced as the water cools. Self-feeders are useful under such conditions. At low water temperatures fish move slowly and do not seek out feed as they do when the water is warm. They also consume less food at each feeding and digest it much more slowly.

It is very important that fingerlings start feeding immediately after they are stocked in a pond. Well-fed healthy fish are more resistant than others to parasites, disease and predators, and reach marketable size sooner. Fish soon learn to feed as food is provided along the entire edge of the pond on the day after they are stocked.

Once the fish start eating a good feeding program should be initiated and followed. Food allowances are 3 to 6% per day of the estimated weight of fish in the pond; rates are lower (1.5 to 2%) during unusually hot or cold periods. Feed is offered in the early morning and late afternoon in summer but only in late afternoon during the cooler seasons. If sinking pellets are used it is desirable to scatter them along the shallow pod margin where feeding activity can be observed. Floating feeds may be scattered over the entire surface of the pond. Feeding activity is a good index of the well-being of the fish; rapid and vigorous consumption of the food suggests good environmental conditions and good health.

Catfish are being raised successfully in water from many sources. Well water is best, but other uncontaminated supplies such as clean streams or springs are acceptable if they are free of fish and disease organisms.

Oxygen depletion is the greatest fish farming problem. Most oxygen depletion kills are preceded by a phytoplankton die-off and decay. This situation is aggravated by the decomposition of uneaten fish feeds and fecal waste. When excessively thick blooms of phytoplankton (algae) occur, it is desirable to add fresh water to the pond. Feed should be reduced in amount or withheld entirely until the condition has improved.

Catfish culture is also influenced by aquatic vegetation. Although rooted aquatic plants and filamentous algae are not as troublesome in pond rearing of catfish as they are in some other forms of fish culture, they should be removed if they appear. Manual removal and some chemical controls are feasible.

When treating a pond with chemicals the culturist should be aware that the chemical may be toxic to the fish, and that killing too much vegetation at one time can result in an oxygen-depletion mortality. Ponds should be carefully checked for low oxygen each day for 7 to 10 days after applications of herbicides.

Production.—In 1973 Dr. C. A. Oravetz, National Marine Fisheries Service, reported that 22,118 ha (54,633 acres) of water was engaged in commercial catfish production. Of this, 12,122 ha (29,942 acres) or 55% was harvested. A total of 22,640,500 kg (49,860,000 lb) were produced. This production was 96.3% from ponds and 3.7% from raceways, cages and tanks. Output per hectare of water in ponds harvested was 1798 kg (1602 lb per acre). Total value of the harvested crop to producers was $26,188,154. Estimated retail value was in the neighborhood of $60 million. The largest producers were:

State	Ha	Acres
Mississippi	10,572	26,112
Arkansas	3723	9197
Alabama	1932	4773
Louisiana	1717	4241
Texas	968	2391
Georgia	651	1607
Tennessee	559	1381
Kansas	462	1141

Smaller amounts of fish production were found in at least 12 other states. Nearly one-half of all commercial production in raceways, cages and tanks was in Georgia.

In 1976 there were probably between 1400 and 1800 food-size catfish producers and 500 to 600 fingerling producers. Of the food-size producers about 88% used ponds, about 7% used cages and about 5% used raceways. Only a few farms were using tank culture.

A typical food-size fish producer buys his fingerlings from a specialized producer. The specialized fingerling producer keeps brood fish, from which he obtains spawns, and raises the fry to fingerling sizes. The size of fingerlings stocked by food-size fish producers (grow-out operations) ranges from 6 to 22 cm (2 to 9 in.), depending on when he stocks the fish and when he hopes to harvest at 454 g (1 lb) size. In pond culture most fingerlings stocked are between 13 and 20 cm (5 to 8 in.). They are usually stocked in March, April or May. These sizes reach market weights in the fall. The raceway and cage producers buy slightly larger fingerlings. Producers who plan to harvest in late spring or early summer of the following year buy smaller fingerlings.

The primary producing states in 1976 were essentially in the southeastern United States, an area blessed with ample water and a long grow-out season with warm conditions. Exceptions to this were California and Idaho (Fig. 1.8). California produces fish in certain areas because the market price structure is conducive to higher cost grow-out operations. In Idaho warmwater springs and wells are used for year-round growth. The fish are marketed on the west coast.

Double Cropping.—One rather interesting development taking place is the raising of channel catfish and rainbow trout in the same facility at different times of the year. Trout fingerlings are stocked in the fall of the year when water temperatures drop below 21°C

FIG. 1.8.   PRIMARY CATFISH-PRODUCING STATES

(70°F). They are then fed out until the spring of the year when water temperatures increase above 21° or 22°C (70° or 72°F). By stocking the proper size fingerlings the fish can reach market sizes in 120 days or fewer. After the trout are harvested, channel catfish fingerlings are stocked in the warm water. They are then fed out to the fall of the year when it is time to restock rainbow trout again. By stocking the proper sizes of catfish fingerlings the fish reach market sizes of 454 g (1 lb) during an 8 month season. This development, which was begun in Georgia by the author and two co-workers in raceways, is slowly being developed commercially.

The major advantage to double-cropping is the sharing of fixed expenses by two crops of fish. This lowers the production costs of both species. Net returns can be increased by 100% or more using this technique.

Feeds and Feeding.—The objective of any animal husbandry is to convert relatively low-cost feedstuffs into high value, high quality protein. Fish farming is no exception.

Catfish are desirable for recreation and food. They may be raised in still water, flowing water raceways, tanks, troughs, pens or cages. They may be fed wet or dry feeds prepared as meals, sinking pellets, floating pellets, blocks or crumbles. When stocked at high population densities in a restricted area, catfish soon exceed the production limit of natural foods and must depend on artificial feeds for growth.

In pond tests, channel catfish fingerlings stocked at 3705 per ha (1500 per acre) showed an average net gain of more than 454 g (1 lb) in 6 months, with a feed conversion of 1.3 or 1 unit of gain per 1.3 units of feed.

A general formulation guide is:

	(%)
Crude protein, more than	30
Digestible protein, more than	25
Fish meal protein, more than	35
Crude fat, more than	6
Crude fiber, more than	8
Crude fiber, less than	20

An economical minimum level of fish meal in the diet for ponds is 5 to 6%.

Feeds for raceways and cage culture, as well as other intensive production techniques, should contain more high quality protein and more vitamins and minerals than those used in pond culture. This, of necessity, requires a more expensive feed.

FIG. 1.9.   CATFISH CAGE CULTURE, ARKANSAS

The ingredients and amounts to make a ton of feed are given in Table 1.2. This feed formula has given excellent results at the Fish Farming Experimental Station, Stuttgart, Arkansas.

The Bureau of Commercial Fisheries recommends that feed be provided at the rate of 3% of body weight per day, 6 or 7 days a week. The feed should be distributed around the pond edge in water 0.6 to 1.3 m deep (2 to 4 ft). Feeding should be checked periodically to see

TABLE 1.2
INGREDIENTS AND AMOUNT TO MAKE A TON OF FEED FOR CHANNEL CATFISH

Item	Pounds	Kilograms	%
Fish meal	300	150	15.0
Bloodmeal	100	50	5.0
Feather meal	100	50	5.0
Rice bran	700	350	35.0
Distillers' solubles	100	50	5.0
Rice milldust	200	100	10.0
Soybean meal	400	200	20.0
Dehydrated alfalfa	70	35	3.5
Mineralized salt	20	10	1.0
Vitamin premix	10	5	0.5

Source: Anon. (no date).

that the food is being consumed. Checking can be done by using submerged feeding tables that can be raised and lowered. If after several hours feed remains on the table, the amount fed should be reduced.

According to the Fish Farming Experiment Station, Stuttgart, Arkansas the following feeding rates have proved to be satisfactory. Feeding at the rate of 3% of body weight per day is recommended when the water temperature is 21° to 29°C (70° to 85°F). At water temperature 27°C (85°F) or above, only the feed that is consumed in a 10 min period should be fed. When the water temperature is 15.5° to 21°C (60° to 70°F), feeding should be at the rate of 2% of body weight per day. At temperatures of 7° to 15.5°C (45° to 60°F) it should be 1% of body weight. At temperatures below 7°C (45°F) no feed should be given.

Channel catfish are excellent converters of feed. Research reports from several experiment stations show conversion rates of 2:1 under actual field conditions. Under controlled experimental conditions conversions as low as 1.25 units of feed to 1 unit of gain have been achieved. It is not at all unrealistic to expect conversion rates of 2:1 for most fish farming operations.

Harvesting.—Harvesting catfish raised in tanks or cages is relatively easy; they are lifted out with a dip net, loaded into transfer containers, weighed and placed in the truck for transportation.

Courtesy of U.S. Bureau of Commercial Fisheries, Kelso, Arkansas

FIG. 1.10.   BAILING CATFISH FROM LIVE CAR, ARKANSAS

Courtesy of U.S. Bureau of Commercial Fisheries, Kelso, Arkansas

FIG. 1.11.   MECHANICAL LOADING OF HAULING TRUCK
FROM CATFISH LIVE CARS, ARKANSAS

Raceways can be harvested using a seine. Since the raceway segments are usually not over 6 m wide (20 ft), a seine about twice this
length is used. The seine is dragged the length of the raceway segment, crowding the fish against the concrete headwall where they
are dipped out. A second alternative is to drain the segment. With
this system a fish trap is placed below the headwall and the fish are
trapped as they leave the raceway segment.

Pond harvesting is much more complex. The USA has some of the
largest, if not the largest, fish culturing ponds in the world. A few
ponds are as large as 65 ha (160 acres). Ponds of 16 ha (40 acres) are
commonplace.

In current practice extensive preparations are usually required
to ready ponds for harvesting the fish. Draining a pond and prepar-
ing for final harvesting operations by means of pumps or ditches require several days. During draining, some of the fish may die from
being concentrated in a small volume of water with inadequate oxygen. Also, valuable amounts of water are wasted. Coordinating
production with market demands is difficult. For example, to harvest and transport to market over 20 MT per day from one pond may
not be possible. Hence, for ponds larger than 8 ha (20 acres) the volume of fish in the ponds exceeds the capacity of the market. Since
each truckload is about 5 MT (5.5 ST), an 8 ha (20 acre) pond re-

quires the movement of 4 truckloads daily. This is not always possible. Harvesting the fish often must be limited to periods of the year when it does not conflict with other activities on the farm.

The smaller ponds below 16 ha (40 acres) are usually harvested using a haul seine. Ponds of about 16 ha (40 acres) and larger have necessitated innovative harvesting techniques. A mechanical haul seine is commonly used.

Because larger ponds require such a long and heavy seine, a boat or barge is used to set the net around the pond. The seine is retrieved by using a mechanized wire-cable puller or a rope-line puller. The mechanized puller winds the cable or rope onto a double drum winch. As the fish are crowded into a harvesting area, they are dipped out by dip nets or crowded into a fish loading conveyor. When more than 8 ha (20 acres) are to be harvested, even these mechanized aids are not the sole answer due to the inability to sell over 20 MT (22 ST) per day. Thus, the fisheries experiment stations at Stuttgart and Kelso, Arkansas have developed a new technique.

This new technique consists of having a "fish holding bag" or live car in the middle of the chute formed by the seine net. As the fish are crowded they swim into the live car. Each live car holds about 5 MT (5.5 ST) of fish. When a live car is filled it is detached and an empty live car attached (Fig. 1.13). The filled live car is then floated

Courtesy of Fish Farming Experimental Station, Kelso, Arkansas

FIG. 1.12.   THREE LIVE CARS FILLED WITH CATFISH DURING SEINING OF LARGE POND

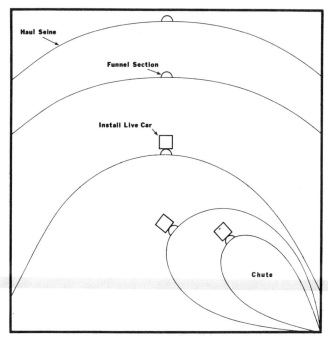

FIG. 1.13.   SCHEMATIC DRAWING OF HAUL SEINING PROCEDURE USED WITH
LIVE CAR FOR HARVESTING CATFISH

to deeper water to give them more water room and to get them out
of the muddy harvesting water. The colder the water, the longer the
fish can be held in the live car. If the fish become distressed aeration can be started, or the fish can be released back into the pond in
case of an emergency.

When the pond is harvested and the fish are in the live cars, each
live car can be positioned near the bank and a mechanized brailer
is used to dip the fish out onto a sorting table positioned in the water.
The small fish of less than 340 g (0.75 lb) can be released back into
the pond or transported to another finishing pond. The fish between
340 and 1135 g (0.75 to 2.5 lb) are loaded into the awaiting trucks
for transport to the processing plant or other market. The larger fish
over 1135 g (2.5 lb) are sorted out for use as brood fish or are sold
separately for fish fillets.

As many as 90% of the fish in a pond can be captured in a single
haul. The remaining fish are permitted to stay in the pond, which is
then restocked with fingerlings.

Some farmers with large ponds may not use haul seines at all.
These farmers use trapping devices which may harvest several tons
per day. Over a period of several weeks as many as 80 to 90% of the

fish may be harvested, and trapping ceases for that year. Use of this system permits a large pond to be harvested slowly so that the market is not depressed by too large a quantity of fish being harvested at one time. Several commercial trapping devices have been invented which, when baited with food, serve as efficient harvesters.

Utilization.—There is no detailed information on how cultured fish are utilized. Dr. C. A. Oravetz of the National Marine Fisheries Service did a study of utilization in 1973. The study was only partial and not designed to yield anything more than some idea of utilization after harvesting. Only about 25 to 30% of total production in the USA was included with data collected in only 8 producing states. The study indicated that only about 32% of production went to processors, 24% to local retailers for retail sale, 13% to fee fish-out pay lakes and 31% to live-haulers. A follow-up of fish sold to live-haulers was not made. Since we know that the live-haulers resold the fish, it is logical to assume that these fish were resold for stocking fee fish-out lakes and farm ponds, with some going to retailers located some distance from the producing area. Thus, it can be assumed that in total perhaps 32% went to processors, 26% to retailers, 33% to fee fish-out lakes and perhaps 10% to restocking farm ponds for family fishing.

In the heavy production areas of Mississippi and Arkansas larger amounts went for processing, while in some states with limited production, such as California, perhaps two-thirds ended up in fee fish-out lakes. This hypothesis was also borne out by prices received by the producers in 1973. The lowest prices of about $1.14 per kg ($0.52 per lb) were in Arkansas, a state of heavy production. At the other extreme was California with an average producer price of $2.09 per kg ($0.95 per lb).

Live-haulers.—Live-haulers buy a significant part of total cultured production. They, in turn, sell the live unprocessed fish to farmers for stocking farm ponds, pay lakes and other outlets. An unknown quantity of live domestic wild catfish also are purchased by live-haulers. Hence, the activity of live-haulers is of considerable importance, although little is known about them. In fact, accurate estimates of the number of live-haulers is not available. Information gathered for 9 out of 19 catfish-producing states indicates that there were more than 100 licensed live-haulers in the 9 states. The other states failed to reply or did not license live-haulers.

Fee Fishing.—Live-haulers and some catfish producers have good market outlets for catchable-size fish for stocking fee fish-out ponds. In the main production areas and in other southeastern states as well as the midwest there are thousands of fee fish-out ponds operated by individuals. Perhaps one-third of all cultured catfish are sold for fee fishing.

The ponds are smaller than those used for grow-out operations. Normally they are not more than twice the distance that a fisherman can throw a line with a fishing rod. This enables the fishermen to fish the entire pond without the use of a boat. All tree stumps and debris are removed during construction. Ponds are earthen and are usually supplied with fresh water by a spring or by surface run-off. If a small stream is used as a water source the water goes through a screen to remove unwanted species. Channel catfish are the main species stocked.

Fish of different sizes are stocked. The minimum size is about 454 g (1 lb). Fish as large as 16 kg (35 lb) have been caught out of some ponds. The operator usually furnishes the bait, which may range from goldfish, earthworms, crawfish, chicken livers, blood bait, dough bait, cotton soap balls and fatback to grapes.

Most catfish fee fish-out operators charge a fee for each kilogram or pound of fish caught. If no fish are caught, there is little or no fee charged.

The amount paid for fish caught varies widely. It may range from $1.43 to $2.20 per kg ($0.65 to $1.00 per lb).

Some fee fish-out operations have two ponds. One pond raises fingerlings and one is for the catching area. Other operators may have several ponds and buy only catchable-size fish. The fish are stocked in fertilized ponds. Ponds are fertilized at the rate of about 111 kg of 8-8-2 fertilizer per ha (100 lb per acre). This is applied twice a year. It increases the natural carrying capacity of the pond and permits a lower feeding rate of commercial catfish feeds.

Net returns vary widely, ranging from a few dollars to $10,000 per year. The profit depends on number of customers, competition, availability of wild fishing waters, survival of fish stocked and prices charged per kilogram or pound of fish caught.

Processing.—Processed catfish for the 7-year period 1969-75 are shown in Table 1.3. The table indicates the heavy harvest season in January-February, and March in each year. It also shows the increased volume of fish being processed for food use. Processed production in round weight (live weight) was only 1,455,000 kg (3,210,-000 lb) in 1969. It increased dramatically through 1973 when 8,969,-000 kg (19,731,000 lb) were processed. As a result of high fish food prices in 1973 and 1974 caused by the worldwide shortage of fish meal (due to the poor Peruvian catch and relaxed restrictions on rice production in the United States), fish acreage was cut back. Since the processed market is the residual market after live fish are

TABLE 1.3

PROCESSED PRODUCTION OF FARM-RAISED CATFISH, 1969-75
(ROUND WEIGHT IN THOUSANDS)

Month	1969 (lb)	1969 (kg)	1970 (lb)	1970 (kg)	1971 (lb)	1971 (kg)	1972 (lb)	1972 (kg)	1973 (lb)	1973 (kg)	1974 (lb)	1974 (kg)	1975 (lb)	1975 (kg)
January	169	77	689	313	926	421	1203	547	2128	967	1267	576	1644	747
February	439	200	1092	496	1154	525	1520	691	2257	1026	1418	644	1729	786
March	434	197	420	236	1386	630	2134	970	2244	1020	1734	788	1504	684
April	194	88	177	81	897	408	1487	676	1388	631	1355	616	1011	459
May	23	11	97	44	487	221	1531	696	1259	572	1395	634	790	359
June	52	23	149	68	556	253	1365	620	1324	602	1436	653	1481	673
July	104	47	221	101	716	435	1180	536	1646	748	1303	592	1426	648
August	193	88	241	110	918	417	1638	745	1773	806	1541	701	1369	622
September	273	124	348	158	1008	458	1483	674	1642	746	1277	581	1339	608
October	349	158	723	329	1673	760	1673	760	1690	768	1530	695	1402	637
November	421	191	715	325	1097	499	1781	810	1249	568	1324	602	1100	500
December	550	250	769	350	1006	457	1221	555	1129	513	1364	620	1325	602
Total	3201	1455	5741	2610	11,257	5117	18,331	8332	19,731	8969	16,944	7702	16,140	7336

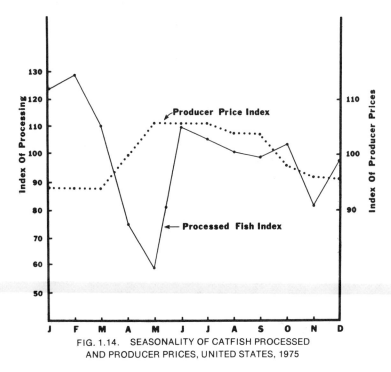

FIG. 1.14.   SEASONALITY OF CATFISH PROCESSED
AND PRODUCER PRICES, UNITED STATES, 1975

sold for recreational fishing and stocking farm ponds, the volume
going to processing plants in 1974 and 1975 declined. However, producers in the main cultural area of Mississippi and Arkansas believe
that production will increase in 1976 as a result of increasing fish
prices and relative declines in rice, cotton and soybean prices.

In 1975 a total of 7,336,495 kg (16,140,289 lb) of live-weight fish
were processed for the food market (Table 1.4). Volume of fish processed monthly varied considerably. The peak months for processing were January and February, which are also the two months when
labor requirements for harvesting do not conflict with other farm
work. In March, harvesting and processing decline until the low
point is reached in May (the busiest months for farm labor). In June
harvesting resumes, and from then until December continues on a
more or less regular basis (Fig. 1.14).

During January, February and March, when harvesting and
processing are above the yearly norm, producer prices for fish are
at the yearly low (Fig. 1.14). When harvesting and processing decline in May, June and July, producer fish prices rise and stay relatively high until the fall of the year. Then both processing and producer prices decline as processors adjust frozen inventories prior to

TABLE 1.4

VOLUME OF CATFISH PROCESSED, PRODUCER PRICES, FOB PLANT PRICES FOR ICED AND FROZEN CATFISH, 1975

Month	Live Weight of Fish Processed		Prices Paid Producer at Plant		Avg Price Paid Producer		FOB Plant Ice Pack Price		FOB Plant Frozen Price	
	(kg)	(lb)	($ per kg)	($ per lb)	($ per kg)	($ per lb)	($ per kg)	($ per lb)	($ per kg)	($ per lb)
January	756,487	1,664,271	0.95-1.10	0.43-0.50	1.01	0.46	2.09-2.42	0.95-1.10	2.29-2.79	1.04-1.27
February	786,032	1,729,270	0.99-1.10	0.45-0.50	1.01	0.46	2.24-2.42	1.02-1.10	2.22-2.75	1.01-1.25
March	638,698	1,504,135	0.99-1.10	0.45-0.50	1.01	0.46	2.29-2.66	1.04-1.21	2.38-2.75	1.08-1.25
April	459,445	1,010,779	0.99-1.17	0.45-0.53	1.08	0.49	2.29-2.62	1.04-1.19	2.35-2.62	1.07-1.19
May	359,254	790,358	1.10-1.21	0.50-0.55	1.14	0.52	2.35-2.71	1.07-1.23	2.40-2.86	1.09-1.30
June	672,977	1,480,550	1.10-1.19	0.50-0.54	1.14	0.52	2.42-2.57	1.10-1.17	2.53-2.66	1.15-1.21
July	648,150	1,425,931	1.10-1.21	0.50-0.55	1.14	0.52	2.42-2.60	1.10-1.18	2.53-2.97	1.15-1.35
August	622,173	1,386,781	1.10-1.21	0.50-0.55	1.12	0.51	2.49-2.71	1.10-1.23	2.42-2.97	1.10-1.35
September	608,469	1,338,632	1.03-1.21	0.47-0.55	1.12	0.51	2.42-2.71	1.10-1.23	2.38-2.71	1.08-1.23
October	637,285	1,402,026	0.99-1.10	0.45-0.50	1.06	0.48	2.24-2.60	1.02-1.18	2.53-2.97	1.15-1.35
November	500,126	1,100,278	0.95-1.10	0.43-0.50	1.03	0.47	2.40-2.60	1.09-1.18	2.53-2.97	1.15-1.35
December	602,399	1,325,278	0.95-1.10	0.43-0.50	1.03	0.47	2.38-2.60	1.08-1.18	2.53-2.97	1.15-1.35
Totals or Averages	7,336,495	16,140,289	0.95-1.21	0.43-0.55	1.08	0.49	2.09-2.71	0.95-1.23	2.22-2.97	1.01-1.35

Source: Monthly data from National Marine Fisheries Service, Little Rock, Arkansas.

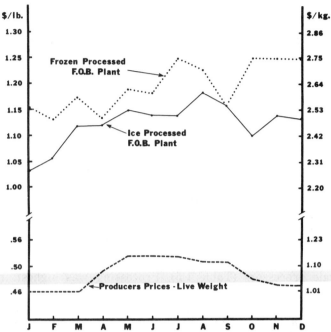

FIG. 1.15.   PRODUCER PRICES DELIVERED TO PLANT, FROZEN AND ICED PROCESSED
CATFISH FOB PLANT, UNITED STATES, 1975

Courtesy of Fish Farming Experimental Station, Stuttgart, Arkansas

FIG. 1.16.   CATFISH HAULING TRUCK WITH LOAD GOING TO MARKET, ARKANSAS

the big buying splurge of fish in January, February and March.

Processor prices FOB plant bear a resemblance to producer prices (Fig. 1.15). Both processed iced and frozen catfish were lowest when producer prices were lowest; they reached a high during the summer when producer prices did. During October, November and December, when producer prices declined so did processed iced fish prices. However, frozen processed fish prices were at their high.

Rather interesting from the viewpoint of Asians and Europeans is the difference between dressed, iced fish prices compared to frozen. In the USA frozen fish command a premium over fresh iced fish, while in Europe and Asia frozen fish prices are generally lower.

Marketing.—There are about 12 large-scale processors of catfish in the United States. In addition a considerable number of producers process all or part of their own fish for sale to fish dealers, restaurants and individuals.

In 1975, 7,336,000 kg (16,140,000 lb) of fish were processed by the large-scale processors (Table 1.5). Processing was highest in January, February and March, reached a low in May, and stabilized during the remainder of the year. Sales by the processors follow the same ebb and flow of volumes. Table 1.5 shows the volume of fish processed monthly by national processors, the inventory on hand at the end of each month and changes in inventory and monthly net sales. Inventory usually consists of about two weeks of sales. The table shows the range of prices for dressed fish either in ice packs or frozen FOB plant. The frozen product sells for slightly more than the ice-packed fish. The table also shows the ranges and averages of prices paid producers for fish delivered to the plant. Prices paid to producers in 1975 ranged from $1.01 per kg ($0.46 per lb) in January, February and March to $1.14 per kg ($0.52 per lb) in May, June and July.

It should be realized that processing volumes and producer and processor prices shown in Table 1.5 do not include sales by farm processors who maintain their own small processing facilities, or sales by seafood dealers or restaurants who may buy live catfish and do their own processing.

Between 90 and 95% of sales are pan-ready dressed fish. Fish steaks and fillets account for the remainder. This emphasis on pan-ready dressed fish may indicate the strong desires of producers to sell a 454 g (1 lb) fish rather than to raise larger, older fish for the steak and fillet market when feeding efficiency declines.

Fish dealers and restaurants are the primary markets for processors. Minor but growing volumes are being sold to supermarkets

TABLE 1.5

PRODUCTION-INVENTORY-SALES DATA OF CHANNEL CATFISH, 1975 (NATIONAL PROCESSORS)

	Round Weight Processed[1]	Cumulative Production[1]	Ending Inventory	Change in Inventory	Net Sales	FOB Plant Ice Pack Price	FOB Plant Frozen Price	Prices Paid[2]	Weighted Avg of Price Paid
	Kilograms					$ per kg			
Jan	756,487	756,487	339,264[3]	44,985	482,197	2.09 - 2.42	2.29 - 2.79	0.95 - 1.10	1.01
Feb	786,032	1,542,519	319,765	19,499	491,115	2.24 - 2.42	2.22 - 2.75	0.99 - 1.10	1.01
Mar	683,698	2,226,216	239,159[4]	80,606[3]	549,859	2.29 - 2.66	2.38 - 2.75	0.99 - 1.10	1.01
Apr	459,445	2,685,661	134,211	104,948	381,827	2.29 - 2.62	2.35 - 2.62	0.99 - 1.17	1.08
May	359,254	3,044,915	100,205	34,006	251,679	2.35 - 2.71	2.40 - 2.86	1.10 - 1.21	1.14
Jun	672,977	3,717,892	121,728	21,523	366,558	2.42 - 2.57	2.53 - 2.66	1.10 - 1.19	1.14
Jul	648,150	4,366,043	114,051[3]	7,677	399,681	2.42 - 2.60	2.53 - 2.97	1.10 - 1.21	1.14
Aug	622,173	4,988,216	135,478	21,426	351,878	2.49 - 2.71	2.42 - 2.97	1.10 - 1.21	1.12
Sep	608,469	5,596,685	147,436[4]	11,958	362,114	2.42 - 2.71	2.38 - 2.71	1.03 - 1.21	1.12
Oct	637,285	6,233,970	166,504	19,068	364,061	2.24 - 2.60	2.53 - 2.97	0.99 - 1.10	1.06
Nov	500,126	6,734,096	150,409	16,095	318,310	2.40 - 2.60	2.53 - 2.97	0.95 - 1.10	1.03
Dec	602,399	7,336,495	162,646	12,242	349,352	2.38 - 2.60	2.53 - 2.97	0.95 - 1.10	1.03

TABLE 1.5 (Continued)

	Round Weight Processed[1]	Cumulative Production[1]	Ending Inventory	Change in Inventory	Net Sales	FOB Plant Ice Pack Price	FOB Plant Frozen Price	Prices Paid[2]	Weighted Avg of Price Paid
			Pounds				$ per lb		
Jan	1,664,271	1,664,271	746,380[3]	98,968	1,060,834	0.95 - 1.10	1.04 - 1.27	0.43 - 0.50	0.46
Feb	1,729,270	3,393,541	703,483	42,897	1,080,453	1.02 - 1.10	1.01 - 1.25	0.45 - 0.50	0.46
Mar	1,504,135	4,897,676	526,150[4]	177,333[3]	1,209,690	1.04 - 1.21	1.08 - 1.25	0.45 - 0.50	0.46
Apr	1,010,779	5,908,455	295,264	230,886	840,019	1.04 - 1.19	1.07 - 1.19	0.45 - 0.53	0.49
May	790,358	6,698,813	220,451	74,813	553,693	1.07 - 1.23	1.09 - 1.30	0.50 - 0.55	0.52
Jun	1,480,550	8,179,363	267,802	47,351	806,428	1.10 - 1.17	1.15 - 1.21	0.50 - 0.54	0.52
Jul	1,425,931	9,605,294	250,913[3]	16,889	879,298	1.10 - 1.18	1.15 - 1.35	0.50 - 0.55	0.52
Aug	1,368,781	10,974,075	298,051	47,138	774,132	1.13 - 1.23	1.10 - 1.35	0.50 - 0.55	0.51
Sep	1,338,632	12,312,707	324,359[4]	26,308	796,651	1.10 - 1.23	1.08 - 1.23	0.47 - 0.55	0.51
Oct	1,402,026	13,714,733	366,309	41,950	800,935	1.02 - 1.18	1.15 - 1.35	0.45 - 0.50	0.48
Nov	1,100,278	14,815,011	330,900	35,409	700,282	1.09 - 1.18	1.15 - 1.35	0.43 - 0.50	0.47
Dec	1,325,278	16,140,289	357,822	26,933	768,575	1.08 - 1.18	1.15 - 1.35	0.43 - 0.50	0.47

Source: (Anon. 1976A).

[1] Total live weight of fish delivered for processing.

[2] Prices paid to farmer, harvested, at plant site.

[3] Higher than normal because of carryover of one or more plants from December, 1974.

[4] Inventory adjustment.

and grocery stores. It is estimated that between 75 and 85% of processed fish sold is frozen while the remainder is iced. Most processed fish weigh between 225 and 454 g (8 to 16 oz). Sales are primarily in the southeastern states and the midwest. People in these areas are accustomed to eating catfish because of the past availability of wild fish. This custom has carried over to processed cultured fish.

Imports and Exports.—There are virtually no exports of either live or dressed catfish from the United States. However, there are considerable imports.

In 1969, the first year for which records were maintained separately for catfish imports, there were 1,710,000 kg (3,762,000 lb) of dressed catfish imported. Since the dress-out percentage is approximately 60%, this volume represented 2,856,000 kg (6,283,000 lb) of live or round weight (Table 1.6). Since imports were not recorded in January, 1969, the total imports may have been 6 to 10% higher than this volume. Since that time catfish imports have increased considerably. After a slight drop in 1971, imports continued to increase. In 1975 imports were 190% higher than in 1969, when dressed imports totaled 4,958,000 kg (10,907,000 lb). On a live weight basis this represented 8,279,000 kg (18,215,000 lb), for an average yearly increase of 32%. Between 80 and 90% of the imports are from Brazil, with lesser amounts from other Central and South American countries

The major attraction of imported catfish is the relatively low

TABLE 1.6

PROCESSED CATFISH IMPORTS ENTERING USA, 1969-75[1]

Month	1969	1970	1971	1972	1973	1974	1975
January	No Record	420	378	321	543	947	1496
February	285	341	85	743	307	290	641
March	367	413	190	290	632	454	663
April	200	277	58	394	737	670	1421
May	344	584	184	216	434	561	749
June	364	208	123	277	523	1671	782
July	255	475	209	358	335	486	997
August	344	423	338	678	473	795	1090
September	518	358	321	361	593	187	332
October	274	437	121	503	1068	937	979
November	531	644	264	450	396	789	606
December	280	221	932	370	572	657	1151
Total lb dressed	3762	4801	3203	4871	6613	8444	10,907
Total lb round weight	6283	8018	5349	8135	11,044	14,101	18,215
Total kg dressed	1710	2182	1456	2214	3006	3838	4958
Total kg round weight	2856	3645	2431	3698	5020	6410	8279

[1] Figures given indicate thousands of pounds.

price. Wild imported catfish commonly retail at prices as much as 40% below the domestic cultured product. For example, it is commonplace for the imported catfish to be priced at $0.99 to $1.29 per 454 g (1 lb) in a supermarket, while domestically cultured fish in adjacent stores are priced at $1.59 to $1.79. Estimated retail value of imported catfish sold in 1975 was about $5 million.

Wild Catch of Catfish and Bullheads.—Catfish and bullheads are native to nearly all fresh waters east of the Rocky Mountains. The largest catch occurs in the southeastern United States, particularly in the Mississippi River and its tributaries. Whether this is because of larger native populations, more intensive fishing pressures, or both is not known. States having the largest landings are: (1) Florida, (2) Louisiana, (3) Arkansas, (4) North Carolina, (5) Minnesota, (6) Tennessee, (7) Kentucky, (8) Mississippi, and (9) Virginia.

From 1969 through 1974 the catch was relatively stable, varying from a low of 15,088,000 kg (33,193,000 lb) in 1969 to a high of 17,892,-000 kg (39,363,000 lb) in 1971 (Table 1.7). The value per kilogram or pound caught was stable from 1969 through 1972 at $0.51 to $0.53 per kg ($0.23 to $0.24 per lb), but in 1973 and 1974 the value per unit increased rapidly. In 1974 the unit value was $0.86 per kg ($0.392 per lb).

Because of the increased values per unit of catch, the total value of all landings increased from $7,635,000 in 1969 to $14,062,000 in 1974.

It should be emphasized that this catch was obtained only by commercial fishermen and does not include the catch of sports fishermen. Hence, the total number of landings of wild catfish and bullheads is much larger than these impressive data indicate.

Total Supply.—In 1975 an estimated 20,909,000 kg (46,000,000 lb) of catfish were produced by fish farmers on a live weight basis. Lit-

TABLE 1.7

WILD CATCH OF CATFISH AND BULLHEADS BY WEIGHT AND VALUE, 1969-74

Year	Catch or Landings (000 lb)	(000 kg)	Value ($ 000)	Avg Value Per Unit ($ per lb)	($ per kg)
1969	33,193	15,088	7,635	0.23	0.506
1970	35,217	16,008	8,241	0.234	0.515
1971	39,363	17,892	9,138	0.232	0.51
1972	36,568	16,622	8,788	0.24	0.528
1973	36,576	16,625	10,512	0.287	0.631
1974[1]	35,420	16,100	14,062	0.392	0.862

Source: National Marine Fisheries Service, Little Rock, Arkansas.
[1] Estimated from information concerning catch by major regions.

tle or no catfish were exported. Imports amounted to 8,279,000 kg (18,215,000 lb) and the wild catch was 16,591,000 kg (36,500,000 lb). This wild catch is understated because it includes only the catch of commercial fishermen and does not include the sizeable catch of millions of sports fishermen. Hence, the total supply, estimated conservatively, was 45,779,000 kg (100,715,000 lb). Total supply was 46% cultured, 18% imported and 36% wild catch.

Outlook.—The outlook for catfish in the United States is for increased supplies. The wild catch is likely to keep declining slowly as the number of commercial fishermen decline. Culturing is predicted to increase, but at a much slower rate than in the past explosive 15 years. During the next 10 years cultured production may only increase 15 to 20%. This will be largely due to the impact of foreign imports. The market for food fish is by no means saturated, but as the market has widened imports of foreign wild catfish have increased since 1969 at an annual rate of 24%. Unless this torrent of imports is checked or modified by quotas or tariffs, or purchased by some other country, the cheaper imports are likely to continue to have a restrictive influence on domestic production.

American Eel (*Anguilla rostrata*)

The United States is not an eel-consuming country. Each year between 2000 and 3000 MT (4,408,000 to 6,612,000 lb) of wild fish are caught. The major market for these are countries in western Europe, with a few tons sold to Japan. Domestic consumption is probably less than 100 MT (110 ST). The fisherman's price in 1974 was between $0.44 and $0.55 per kg ($0.20 to $0.25 per lb). The major catching areas for larger and adult eel are in Virginia and Maryland. Limited quantities of elvers are caught in South Carolina for export to Japan. Within the past 12 months (1975-76) limited commercial culturing has started. This production is destined for Japan. Total cultured production is probably not over 50 MT (55 ST).

Prawns (Freshwater) (*Macrobrachium rosenbergii*)

Experimental work has been under way in the United States for about 8 to 10 years on producing freshwater prawns. Most of the work is still experimental and concentrated in South Carolina, Florida, Georgia, Texas, Puerto Rico and Hawaii. Several farms are reported to be producing prawns commercially. In 1975 output of 3 farms in Puerto Rico and Hawaii was estimated at 67 MT (146,-872 lb). No information was received by the author on costs of production, selling prices or profitability. However, it is known that the prawns are sold for more per unit than the marine wild species.

Crayfish

Crayfish are versatile animals and are found on nearly every continent except Africa and Antarctica. About 500 species are found throughout the world and over 250 species and subspecies are found in North America.

Biology.—Life for most crayfish begins in a hole in the ground known as a burrow. Some of these may be only a few centimeters deep, serving as a temporary home, while others may be more than 1 m (39 in.) deep. Most burrows have a mud chimney or are capped with a mud plug. The time of mating varies by species and region. A fairly typical mating may occur in May or June. At this time the female will dig a burrow. The male deposits sperm in an external receptacle on the female. The sperm is held until late summer or early fall until the female lays her eggs. As the eggs are laid they are fertilized by the sperm. The fertilized eggs are held to the female's body by a sticky substance. The number of eggs laid varies greatly by species. Some may have as few as 10 and others up to 700.

The eggs appear as a bunch of "grapes" on the underside of the female's tail and she is said to be "in berry." The eggs hatch in 1 to

FIG. 1.17.   ADULT CRAYFISH

2 weeks. Once hatched, the young stay attached for a week or two while they undergo two molts. Later the young leave the female and they burrow and forage for themselves. Crayfish eat a variety of foods, usually plant material and small organisms.

Crayfishing.—Crayfish can be caught in almost every body of water in North America. In California the species *Pacifastacus klamathensis* is the principal crayfish caught. In Oregon, one of the leading states of production, fishing is carried out in the slower moving streams of the flat agricultural valleys. In the state of Washington some crayfishing occurs. Other than the Pacific states, commercial catching in the East is largely in Louisiana and Mississippi. In these 2 states there are at least 29 species. The two dominant commercial ones are the red swamp, *Procambarus clarki*, and the white river crayfish, *Procambarus blandingi*.

Crayfish Farming.—Most of the crayfish farming area is found in Louisiana and until a few years ago was confined to the southern part of the state. In this area over 4.5 million kg (10 million lb), valued at over $5 million, is harvested annually, and about 35 crayfish processing plants are licensed.

Crayfish farming, as such, began over 25 years ago, possibly by accident. It is thought that a rice farmer flooded his rice field one fall, following the harvest of rice, to provide duck hunting. The next spring the duck pond was teeming with crayfish. Duck hunters were transformed into crayfishermen and harvested the unexpected crop of crayfish. Today crayfish farming has changed little from this accidental beginning.

Crayfish are currently being farmed in three types of ponds: rice-field ponds, wooded ponds and open ponds. In rice-field ponds crayfish are rotated with the rice. The general procedure followed by the rice farmers is to remove water from the rice field about two weeks before harvesting. This permits drying of the field to facilitate harvest. When drying begins, crayfish burrow. The second growth of rice and grasses, along with rice straw, provides food for the crayfish.

Wooded areas are also used, but make poor crayfish ponds. Dense growths of trees and shrubs hinder harvest. Wooded ponds usually have poor wind circulation, resulting in oxygen depletion. Also, water in wooded ponds is often acid, resulting in a low pH and a low total hardness. Neither condition is conducive to good crayfish production. Despite these drawbacks wooded areas are sometimes used for crayfish ponds because the land is idle and owners feel some production is better than none.

Open ponds are often constructed solely for crayfish farming. The

Courtesy of James W. Avault, Jr.

FIG. 1.18.   CRAYFISH BURROWS, LOUISIANA

procedure for farming these ponds is generally the same as for farming wooded and rice-field ponds. Crayfish are stocked in the ponds in late May or June. Brood stock, usually bought from a dealer, is stocked at rates of 23 to 56 kg per ha (25 to 50 lb per acre), depending on the amount of vegetation and the number of native crayfish present. Once stocked, the crayfish burrow. In July the ponds are drained, mainly as a means of predator control, since racoons and wading birds may make serious inroads on the crayfish. When young crayfish are found in the burrows, generally in September or October, the ponds are flooded to release them. Once the ponds fill the young crayfish forage on native plants. If the winter is mild crayfish can be harvested the same year. Some crayfishermen may harvest the first crop as early as November 25. Typically it is spring of the next year before the main crop is ready.

Crayfish cultured in ponds are harvested in the same manner as in the wild. Both lift nets and funnel traps of wire are used. Wire baskets or cages are baited with fish or fish heads. A trapper usually handles 12 to 25 traps per ha (5 to 10 per acre), depending on accessibility. Lift nets are fished in a similar manner. Most crayfish farmers employ professional crayfishermen who are paid one-half the going market price for live crayfish.

Ponds with good production might yield from 225 to 900 kg per ha (200 to 800 lb per acre) of crayfish. Some reports tell of up to 1136 kg per ha (1000 lb per acre). The early crop of farm-raised crayfish brings the best price, and crayfish may sell for over $1.32 per kg ($0.60 per lb). Later, when the "wild" crop comes in, the price drops. By May the season has peaked and by June the harvest is nearly complete.

The future of the crayfish industry looks promising and the number of ponds devoted to crayfish farming is increasing rapidly. In Louisiana over 4850 ha (12,000 acres) were devoted to crayfish culture in 1969. By 1971 the area had increased to nearly 16,200 ha (nearly 40,000 acres).

Levees can be constructed with rice-field equipment or with conventional farm equipment. By setting up a pump for filling ponds from a nearby stream or sinking a well, the farmer has his water. No feed is used and harvesting presents no real problem since trapping rights are usually leased. Profits for the pond owner range from $150 to $300 per ha ($50 to $100 per acre).

Processing.—Upon arrival at a typical processing plant, crayfish are stored in a cooler at 3°C (38°F). The next day the live crayfish are removed from the cooler and thoroughly washed. After washing, the clean crayfish are put into stainless steel tanks for blanching. The crayfish are blanched for 5 min and no spices or additives are used. The blanching does three things: (1) kills the crayfish, (2) kills the bacteria in crayfish, and (3) facilitates manual peeling.

The blanched crayfish are then transported into the processing room where the meat is separated from the shell manually. As many as 50 persons may be employed for this job.

There are various methods used for packaging. A common package is a 454 g (1 lb) bag. It is kept under refrigeration and has a shelf-life of about five days.

MARINE CULTURE

Salmon

Salmon are cultured for food in three states in the USA. These are Washington and Oregon on the West Coast and Maine on the East Coast. In 1975 there were two farms in Washington specializing in fingerling production only, three farms producing fingerlings and pan-size cultured salmon, and three farms producing only pan-size salmon. Two of these eight farms were also engaged in sea

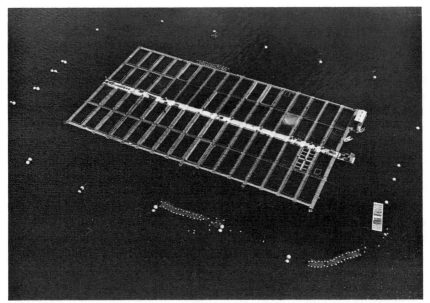

Courtesy of Mr. Jon M. Lindbergh, Domsea Farms, Inc.

FIG. 1.19. FLOATING NET CAGE CULTURE OF SALMON, WASHINGTON

ranching. Two farms were in Maine and one was under construction.

Production of pan-size salmon in 1976 for Washington and Oregon was estimated at 681,000 kg (1.5 million lb). Production in Maine was estimated at 50,000 kg (110,000 lb). Since most of the production is centered in Washington and Oregon, the discussion presented here is related only to this area.

Salmon culture in Washington and Oregon has had many obstacles to overcome. The initial culturing was begun by the Bureau of Commercial Fisheries in 1967 in floating net pens. The predecessor of the Bureau of Commercial Fisheries is the National Marine Fisheries Service. In 1970 early attempts at private commercial culturing were initiated. By 1973-74 production was about 350 MT (385 ST). Two years later, in the 1975-76 season, production was estimated at 681 MT (750 ST).

In Washington private growers are not allowed to spawn fish. Eyed eggs are purchased from public agencies. In Oregon private spawning of chum salmon (*Oncorhynchus keta*) was first permitted in 1971. In 1973 coho (*O. kisutch*) and chinook (*O. tshawytscha*) salmon were also added to the list of species that could be privately spawned.

Culturing has evolved primarily into culturing only coho salmon in Washington, Oregon and Maine. Only small numbers of chinook and chum are cultured. Interest has centered on coho because of its resistance to disease and willingness to accept pelleted dry feed. This is fortunate because coho are one of the most desired species and sell for a relatively high price.

Unlike trout, coho salmon eggs are obtained between October and December and hatch in January. Hatching takes about 60 days in 9° to 13°C (48° to 55°F) water. The fry are maintained in heated water between 10° and 15°C (50° to 60°F). The fry are raised in vertical silos or troughs in recirculated fresh water. Liquid oxygen is sometimes used for oxygenation. This permits ten times the fry and fingerling density of non-oxygenated water. After 4 to 5 months the fingerlings transform into smolts which weigh between 15 and 25 g (0.5 to 1 oz). In June or July the smolts are acclimated to salt water and transferred to saltwater floating pens, or ponds on land into which salt water is pumped. There is no standard size of pens used. They vary from 3 m square (9 by 9 ft) to more than 7 m square (21 by 21 ft). Ponds are usually 6.1 m wide by 24.4 m long and 1.5 m deep (20 ft by 80 ft by 5 ft deep) with flowing water and mechanical aeration. By December of the same year the smolts reach pan size of 380 g (14 oz), and the larger ones are sold. About one-half are dressed and one-half are boned. The dressed weight is about 80% of live weight while the boned fish weight is about 70% of live weight. Production cost is about $2.20 per kg ($1.00 per lb). About 50% of the marketed fish are

FIG. 1.20.   SILOS FOR REARING SALMON FRY TO SMOLTS, WASHINGTON

sold directly to restaurants and hotels and about 50% to supermarkets. Supermarket prices range from $4.38 to $6.58 per kg ($1.99 to $2.99 per lb). Some of the fish are marketed fresh on ice while others are frozen.

The cultured fish are fed the Abernathy diet during production. During the last four weeks pelagic red crab is fed to pink up the flesh.

Cultured production is less than 1% of the wild catch. However, there is no competition between wild and cultured fish. The cultured fish are individual portions or pan-size and appeal to the gourmet market. Some 1 kg (2.2 lb) fish are cultured and marketed. These take up to 18 months in the 10° to 13°C (50° to 60°F) water. Most of the cultured fish are sold between December 1 and April 1 when the wild catching season is closed. However, some cultured fish are sold in every month of the year.

Feeds.—The basic feed (Fowler and Burroughs 1971) fed to cultured salmon is based on the Abernathy dry diet (Table 1.8). It is formulated to contain approximately 45% protein with about 3200 cal per kg of diet. The protein calories range from 50 to 55% of the total. The vitamin mixture added is shown in Table 1.9. Although the formula is designed for a dry feed it can be made into a moist diet by the addition of water. Water is added at the rate of 25% or less, which produces a dough-like mixture. A binder, carboxymethylcellulose, is also added at the rate of 2% to prevent the finished pellets from disintegrating too rapidly during feeding. After preparation the moist pellets must be quick-frozen and kept in frozen storage. They are usually thawed before feeding.

Recommended particle sizes are shown in Table 1.10. Particle size plays a major role in the acceptance or rejection of the diet, and is more critical in the dry than in the moist diet. Too large a dry particle will be repeatedly rejected by the fish, especially when they first start to feed or when they are being shifted to a larger pellet.

Feeding of the small fish should be as frequent as possible. This leaves some waste but a conversion ratio of 2:1 can still be expected. Later the feed conversion ratio will range from 1.0 to 1.5.

Table 1.11 was developed for chinook salmon. For other species slightly more or less feed may be required. The feeding program can be used from first feeding to adult sizes. In practice, pelagic red crab is ground and fed to pink up the flesh.

Total Supply.—Total supply of salmon is difficult to present accurately because of the composition of exports and imports. Many of these are processed salmon which do not permit conversion into round weight equivalents.

In 1975 there were 91,466 MT of commercial catch (201,591,000 lb). To this must be added 5678 MT of imports (12,515,000 lb) and 725 MT (1.6 million lb) of cultured salmon. From this amount of 97,869 MT (215,706,000 lb) must be subtracted 21,882 MT (48,229,000 lb) of exports (National Marine Fisheries Service 1976). Hence the total supply, exclusive of salmon caught by recreational fishermen, was 75,987 MT (167,477,000 lb). Cultured production was slightly less than 1% of total supply.

TABLE 1.8

FORMULA FOR THE ABERNATHY DRY DIET

Ingredient	%	Type
Fish carcass meal[1]	44.5	Salmon, dogfish, hake, herring or turbot
Dried whey produce	17.0	Not less than 15% protein (foremost or equal)
Wheat germ meal	16.5	Not less than 25% protein and 8% lipid
Cottonseed meal	15.0	Not less than 50% protein
Soybean oil	6.0	Fully refined soybean oil (National Soybean Processors Association Code) with 0.01% BHA and 0.01% BHT added
Vitamin supplement	1.0	See Table 17.9

[1] To have protein content of more than 70%, lipid less than 12%, water less than 7%, and a TBA value of less than 40.

TABLE 1.9

ABERNATHY VITAMIN SUPPLEMENT

Ingredient	Amount (g)
Thiamine mononitrate	0.15
Riboflavin	0.69
Pyridoxine hydrochloride	0.30
Niacin	4.77
d-Pantothenic acid	0.68
Inositol	13.65
Biotin	0.03
Folic acid	0.10
DL Alpha-tocopherol acetate (10,500 I.U.)	10.50
Ascorbic acid	25.50
Carrier[1]	397.23
Total	453.60

[1] May be wheat middlings or cottonseed meal sized to pass through a U.S. Sieve No. 30.

TABLE 1.10

RECOMMENDED PARTICLE SIZES OF THE ABERNATHY DIET
AS CORRELATED WITH FISH SIZE

Granule or Pellet Size	Fish Size No. Fish per 454 g (1 lb)
Starter granule[1]	More than 800
0.8 mm (2/64 in.) granule[2]	800 to 500
1.2 mm (3/64 in.) granule	500 to 200
1.6 mm (4/64 in.) granule	200 to 100
2.4 mm (6/64 in.) granule	100 to 80
2.4 mm (6/64 in.) pellet	80 to 50
3.2 mm (8/64 in.) pellet	Less than 50

[1] Composed of 95% of the basic Abernathy formula as shown in Table 1.8 plus 5% additional soybean oil.
[2] Composed of 98% of the basic formula plus 2% additional soybean oil.

Sea Ranching.—Sea ranching is another culturing technique. However, instead of feeding the fish throughout the production process, the smolts are released into the sea. The mature fish return after 2 to 5 years to the point of release. Returning fish may be as high as 6%. With this technique production cost may be only one-third that of fish fed out in pens or ponds.

Sea ranching by individuals or corporations is an extension of public restocking efforts which have been in process for years. The only difference is that the individual harvests all of the returning fish that come back to the privately owned point of release and rearing. Sea ranching of salmon is legal only in Oregon, California and Alaska. It is illegal in Washington. Efforts were being made in the 1976 Washington state legislature to legalize this technique. There are two farms engaged in sea ranching in Washington. Both of these are Indian enterprises which the state cannot regulate.

Outlook.—The future expansion of the cultured salmon industry will be greatly influenced by the legality of sea ranching, the number of permits granted for pond and pen culture, and market prices for cultured salmon. Hence, any projections could be greatly modified by these considerations. Estimates by individuals directly concerned with culturing indicate that by 1980 production could increase from the present 731,000 kg (1.6 million lb) to 6,879,000 kg (15 million lb). Sea ranching, which is in its infancy, could have equally startling gains in production.

Shrimp

There is only one commercial marine shrimp culturing farm in

TABLE 1.11

FEEDING CHART FOR THE ABERNATHY DRY DIET (AMOUNT OF FOOD PER DAY EXPRESSED
AS % OF BODY WEIGHT)

Avg Water Temperature (°F)	(°C)	Number of Fish per lb (454 g)						
		2500-1000	1000-300	300-150	150-90	90-40	40-10	10-Under
40	4.4	2.6	2.2	2.0	1.7	1.3	0.8	0.6
41	5.0	2.7	2.3	2.0	1.8	1.4	0.9	0.6
42	5.6	2.8	2.4	2.2	1.9	1.4	0.9	0.7
43	6.1	3.0	2.5	2.2	2.0	1.5	1.0	0.7
44	6.7	3.1	2.6	2.4	2.1	1.6	1.0	0.7
45	7.2	3.2	2.8	2.5	2.2	1.7	1.0	0.8
46	7.8	3.4	2.9	2.6	2.3	1.8	1.1	0.8
47	8.3	3.6	3.1	2.7	2.4	1.8	1.2	0.8
48	8.9	3.7	3.2	2.9	2.5	1.9	1.2	0.9
49	9.4	3.8	3.4	3.0	2.6	2.0	1.3	0.9
50	10.0	4.0	3.6	3.2	2.8	2.1	1.4	1.0
51	10.6	4.2	3.7	3.3	2.9	2.2	1.4	1.0
52	11.1	4.5	3.9	3.5	3.1	2.4	1.5	1.0
53	11.7	4.7	4.1	3.6	3.2	2.4	1.6	1.1
54	12.2	4.9	4.3	3.8	3.4	2.5	1.7	1.2
55	12.8	5.1	4.5	4.0	3.6	2.7	1.7	1.2
56	13.3	5.3	4.6	4.2	3.7	2.8	1.8	1.2
57	13.9	5.6	4.8	4.4	3.9	2.9	1.9	1.3
58	14.4	5.8	5.1	4.6	4.1	3.1	2.1	1.4
59	15.0	6.0	5.3	4.8	4.3	3.2	2.1	1.5
60	15.6	6.5	5.6	5.0	4.5	3.4	2.2	1.6

the United States (Bente 1975). This farm is known as Marifarms, Inc., Panama City, Florida. Experimental production was begun in 1968. Farming is extensive compared to the intensive techniques used in Japan. The operation began drawing upon the hatching and rearing experience of Japanese scientists working with kuruma shrimp. However, instead of the farm containing only a few hectares of water, it contains a 1012 ha embayment (2500 acres) with tidal exchange of water leased from the State of Florida, and two 121 ha (300 acre) marine lakes with pumped water exchange constructed on uplands leased from a private owner.

Both types of farms have bottoms which are typically soft and sandy and in which benthic organisms abound. It was intended that the shrimp feed partially on naturally occurring marine life, which would provide sufficient growth stimulants and nutrients so they could be fed a cheaper diet than is customary in Japan. Feeding has evolved from feeding a slurry of ground up trash fish to feeding a dry pellet. In general the pellets contain 30% fish meal, 24% shrimp meal or crab meal or combinations, 20% soy meal, 20% meat scraps, 2% whey, 2% vitamin mix and 2% binder. In 1974, 1000 MT (2.2 million lb) of these pellets were used to produce 65 million shrimp weighing 375,000 kg (825,000 lb).

In 1974 the two 121 ha (300 acre) marine ponds harvested produced 136,000 and 91,000 kg (300,000 and 200,000 lb), respectively. The conversion ratio using the pellet feed was 1.8 and 1.9 units fed per unit of shrimp harvested. Indications are that conversion ratios may go as low as 1.4 in raising shrimp averaging 9 g (50 per lb) each, heads-on basis.

Techniques of gathering gravid females, hatching, and rearing shrimp to the proper size for stocking the grow-out areas are similar to those used in Japan (See Chap. 21).

In farming tidal waters the baby shrimp were initially planted in circular pens enclosed by fine mesh nylon nets having openings about like that of window screening. A chain fastened to one edge of the net sealed off the bottom, while a line of cork floats at the other edge held the net to the surface. Since small shrimp move about much as do suspended particles in riled water, in rough weather the shrimp would sometimes be washed out with waves breaking over the top of the pens. Though still on the farm, these liberated shrimp were in an area where predators had not been removed and where feed was not added. Therefore, survival was low and growth somewhat slower until the shrimp grew large enough to move about in search of food.

It was found to be most productive to retain baby shrimp behind pen nets or within the ponds until they were at least 0.3 g in size.

Shrimp behave very differently in large growing areas than in small experimental ponds. White shrimp (*Penaeus setiferus*) migrate in schools, whereas brown (*P. aztecus*) and pink (*P. duorarum*) shrimp do not. Shrimp also move to find natural feeding grounds. As shrimp grow larger they seek deeper water, but frequently move back into warm, shallow waters to feed on natural foods that seem to abound where more sunlight reaches the bottoms. They do not generally seek to escape until they are young adults, at which time they would normally move out to sea with the tides. This tendency occurs just prior to and during harvest periods. At such times Marifarms attaches a specially designed supplementary collar of floats to the top of its nets in order to stop the shrimp from swimming over the top during rough weather. Marifarms uses some 24 km (15 mi.) of nets.

Multiple Cropping.—Marifarms has learned enough about the availability of mother shrimp and the hatching and growing patterns to make it practical to grow two consecutive crops of shrimp in the very same areas. The first crop is brown shrimp. These are hatched in February and March, planted in the starting areas in April and May, moved progressively through the nursery and into the grow-out areas, and harvested from June to August. Some pink shrimp may be included in this crop.

The hatchery meanwhile shifts to white shrimp, which are hatched in the May and June period. These move progressively through the same areas, arriving in the grow-out areas after the harvest of brown shrimp has been completed. The white shrimp are then harvested in the October-December period.

While it is theoretically possible to run a third crop of pink shrimp through the system — hatching them in the July-September period, holding them in juvenile form through winter when it is too cold for growth to continue, and subsequently continuing growth in the spring and harvesting them together with the brown crop during the next summer — tests show that wintering-over operations should have fairly deep holding areas so that the water doesn't cool during cold snaps as much as it does in shallower areas. Survival, particularly in the shallow pond areas, is too low to justify such efforts. When water temperature drops below 10°C (50°F) survival falls off markedly. At 4°C (40°F) the effect is lethal. Operations further south ought to make a third crop feasible.

Harvesting.—In shrimping at sea the catch is manipulated by hand. Sometimes as little as 10% of what is caught is shrimp. First of all, the trash fish and other marine life are sorted out by hand and thrown overboard. Then, while trawling continues, the crew members squeeze the heads off the shrimp by hand, place the tails in wire baskets, and finally store these with ice below deck. When a load is accumulated the trawler returns to port to unload and stock up on provisions, and then returns to fishing at sea. Ships stay out a week or more at a time.

By contrast, Marifarms uses similar trawlers, but since the farmed waters contain little but shrimp, there is no need to sort the catch by hand on deck. The catch is dropped on deck, spread out and layered with shaved ice. During trawling the catch is generally kept covered by a canvas. After a few hours of trawling, the boat has a load of several thousand pounds on its stern deck. Then the boat returns to a dock at shore on the farm where the catch is unloaded by a vacuum lift with the aid of a heavy stream of water.

In the procedure used to harvest the farmed marine lakes, the shrimp, as they are caught, are put into a transfer net on the special skiff used for the trawling. Every hour or so the skiff stocks at a docking point so that the transfer net can be lifted out and immediately emptied into a tank of ice water mounted on a trailer. Every few hours this is hauled to a vacuum lift for final unloading.

The shrimp are discharged from the lift into a large agitated tank of ice water from which they are removed by a metal chain-link-type conveyor belt. The shrimp on the conveyor pass across an inspection table where debris and the relatively few fish, crabs, etc., are removed by hand. (The crabs are accumulated and sold as a by-product. Crab yield in 1974 was about 9090 kg [20,000 lb] live weight.) At the end of the conveyor the shrimp are weighed into standard fish boxes holding 45 kg (100 lb) each. They are kept chilled by placing a layer of ice on the bottom, in the middle and on the top of each box of shrimp. These boxes are loaded into a tractor trailer. Loads of 9090 to 13,636 kg (20,000 to 30,000 lb) are hauled off to nearby processing plants within a day of being caught.

The shrimp are delivered untouched by human hands. Consequently the bacteria count is low. The shrimp do not give off a strong fishy smell when cooked. They have a sweet taste, devoid of the iodine-like flavor often encountered in other shrimp. Thus, the quality of the cultivated shrimp produced by Marifarms is reported to be superior to the standard product of the marketplace.

Total Supply.—During the latest year for which data could be obtained, total domestic landings were 25,289 MT (55,736,000 lb). In order to supply the tremendous demand for shrimp in the United States, an additional 102,497 MT (225,904,000 lb) live weight equivalent were imported. Assuming that these figures are representative of 1974, we can add the 375 MT (825,000 lb) of cultured shrimp. Hence the total supply was about 128,161 MT, or 282,465,000 lb. This was composed of 19.7% domestic catch, 80.0% imported and 0.3% cultured. The 1974 fisherman's price for heads-on shrimp was $1.69 per kg ($0.77 per lb). This price increased to $2.86 per kg ($1.30 per lb) in 1975.

Outlook.—It is extremely doubtful that there will be any future development of farms like Marifarms, Inc., in the United States. In general, sportsmen, fishermen, shrimpers and public officials will be reluctant to lease large bay areas to a single individual or corporation. While it is possible to develop pond enclosures on land and pump marine waters, it takes tremendous capital to do so. Possible investors have developed a wait-and-see attitude and are awaiting word that Marifarms is highly successful commercially before making such investments. Hence, in the foreseeable future cultured shrimp is likely to be produced only by Marifarms. From 375 MT in 1974 (825,000 lb) officials of the company expect to reach capacity of 2273 MT (5 million lb) based on 2 crops a year raised during an 8 month period.

RESTOCKING

All states except Delaware and Mississippi operated at least one state hatchery during 1973. The total number of state fish hatcheries was 425, of which 297 produced coldwater fishes, 73 produced warmwater fishes, 29 produced walleye (*Stizostedion vitreum vitreum*) and northern pike (*Esox lucius*), and 26 produced various combinations.

The average number of installations per state was 8.5. Washington had the most with 66 individual hatcheries, of which 57 produced anadromous salmonoids. Oregon was next with 31, with a strong emphasis on Pacific salmon and steelhead. The nationwide system of state hatcheries is growing slowly with 37 additional units scheduled for operation by 1978. This represents an 8.8% increase in 4 years.

There were 90 national hatcheries operated by the U.S. Fish and Wildlife Service in 1973. Of these, 34 reared trout exclusively and 22 reared trout and other types of fish. Only six raised warmwater fishes exclusively. However, 35 produced them in combination with other species.

Coldwater Fishes

In 1973 the total for all states of coldwater fishes stocked was

178,859,000, weighing 8,373,182 kg (18,421,000 lb). The distribution of these fish by states is shown in Fig. 1.21. The proportions by species are shown in Table 1.12.

The largest programs were in the western United States where 5 states exceeded 454,545 kg (1 million lb). These were California, Colorado, Washington, Oregon and Idaho.

National federal hatcheries produced 36,515,000 coldwater fish weighing 1,971,818 kg (4,338,000 lb). The species mix was much the same for national as compared to state hatcheries (Table 1.13).

Salmonoid Eggs.—All the states together produced 425 million salmonoid eggs in connection with inland fishery operations. Of that number 278 million came from domestic broodstock. An additional 102 million came from wild fish, 37 million eggs from private breeders and about 8 million from federal hatcheries. Sixty percent of all salmonoid eggs were rainbow trout.

The federal hatcheries produced 87 million salmonoid eggs. Seventy-three percent of them were rainbow trout.

Fingerlings.—Fishery managers use trout fingerlings as a farmer uses seed, stocking them annually in lakes where natural reproduction is not adequate. Such stocking often produces satisfactory angling, particularly when fishing pressure is light. Stocking fingerlings in streams with established fish populations is generally not satisfactory.

FIG. 1.21.   TOTAL POUNDAGES OF SALMONOIDS PRODUCED FOR INLAND WATERS
BY EACH STATE

TABLE 1.12

TOTAL STATE PRODUCTION OF COLDWATER FISHES FOR INLAND WATERS[1]

(ALL QUANTITIES ARE IN THOUSANDS)

Species	Fingerlings[2] (no.)	(lb)	Large (no.)	(lb)	Total (no.)	% of Total	Total (lb)	(kg)	% of Total
Rainbow trout (Salmo gairdneri)	66,099	1,569	39,683	12,575	105,782	59	14,144	6,429	77
Brown trout (Salmo trutta)	5,490	185	6,712	1,740	12,202	7	1,925	875	10.5
Brook trout (Salvelinus fontinalis)	7,443	128	4,114	1,110	11,557	6	1,228	558	7
Cutthroat trout (Salmo clarkii)	5,670	121	1,315	499	6,985	4	620	282	3
Lake trout (Salvelinus namaycush)	1,451	34	954	31	2,405	1	65	30	0.3
Kokanee or sockeye (Oncorhynchus nerka)	24,186	fry			24,186	14			
Coho salmon (Oncorhynchus kisutch)			6,995[3]	255	6,995	4	255	116	
Other			8,747	184	8,747	5	184	84	
Total	110,339	2,037	68,520	16,384	178,859	100	18,421	8,374	98.8

[1] Includes Great Lakes.
[2] Mainly fingerlings, but includes all sizes below about 17 cm (7 in.) in length except fry. Kokanee are the only fry included.
[3] All sizes of coho salmon are included in the large category for convenience. They are generally released as smolts at whatever size.

TABLE 1.13

TOTAL FEDERAL PRODUCTION OF COLDWATER FISHES FOR INLAND WATERS, INCLUDING GREAT LAKES[1]

(ALL QUANTITIES ARE IN THOUSANDS)

Species	Fingerlings		Large			Total[2]		
	(no.)	(lb)	(no.)	(lb.)	% of Total	(no.)	% of Total	(lb)	(kg)	% of Total
Rainbow trout (*Salmo gairdneri*)	12,941	327	12,936	3,268	25,877	70.9	3,595	1,634	82.9
Brown trout (*Salmo trutta*)	1,320	20	842	178	2,162	5.9	198	90	4.6
Brook trout (*Salvelinus fontinalis*)	1,307	25	834	201	2,141	5.9	226	103	5.2
Cutthroat trout (*Salmo clarkii*)	881	23	623	80	1,504	4.1	103	47	2.4
Lake trout (*Salvelinus namaycush*)	2,009	69	2,122	109	4,131	11.3	178	81	4.1
Other	209	9	491	29	700	1.9	38	17	0.9
Total	18,667	473	17,848	3,865	36,515	100.0	4,338	1,972	100.1

[1] Data are from Table 1, p. 5, of "Propagation and Distribution of Fishes from National Fish Hatcheries for the Fiscal Year 1973," Fish Distribution Report 8, U.S. Fish and Wildlife Service, 1974.

[2] Totals differ slightly from those in the table cited above because the numbers and weights of eggs and fry are not included, and because steelhead are omitted.

State hatcheries produced 86 million fingerling trout of the 5 principal species. The average size was 11 g (42 per lb) and about 10 cm (4 in.) long (Table 1.14). Rainbow trout accounted for 77% of the total number.

In general, federal production of salmonoid fingerlings is not large. A total of only about 18 million were released (Table 1.14). A large proportion of these went to the Great Lakes and to a few large western reservoirs. Seventy percent of these fingerlings were trout. In size the fingerlings released were nearly identical to those of the state hatcheries.

Large Trout.—Large trout are generally stocked in heavily fished waters where natural crops of wild fish cannot withstand fishing pressure, and in cool reservoirs deficient in spawning and nursery areas. They are increasingly being stocked in waters suitable for trout during cool weather only to provide fishing near big cities. These fish generally create a put-and-take operation in which they are caught by anglers within a few weeks.

In 1973 the state hatcheries stocked 53 million large trout weighing about 7,273,000 kg (16 million lb). The average weight per fish was 138 g (3.3 fish per lb). This means the fish averaged 25 cm (10 in.) in length. Three-fourths of the trout were rainbow (Table 1.15).

Federal hatcheries released 17,357,000 large trout, which were slightly smaller in size than those released by states. Rainbows accounted for 74% of the federal large trout.

Other Coldwater Fishes.—Kokanee (or sockeye) (*Oncorhynchus nerka*) and coho salmon (*Oncorhynchus kisutch*) are also reared in a few states. About 24 million kokanee and 7 million coho were reared in state hatcheries. Federal production for release in inland waters was less than one million kokanee fingerlings.

Warmwater Fishes

The three primary warmwater species cultured in state and federal hatcheries are largemouth bass (*Micropterus salmoides*), bluegills (*Lepomis macrochirus*) and channel catfish (*Ictalurus punctatus*). These 3 account for over 80% of total production.

State and federal hatcheries produced roughly 100 million warmwater fish in 1973. These were 6.5 million fry, 90.6 million fingerlings and 3.6 million large fish. States produced 44% of the total and federal hatcheries 56%.

Warmwater fisheries are generally stocked as fingerlings to establish breeding populations in new or reclaimed waters. Fry are

TABLE 1.14

STATE AND FEDERAL PRODUCTION OF FINGERLING TROUT COMPARED (ALL QUANTITIES ARE IN THOUSANDS)

Species	State					Federal					Total		
	(no.)	% of State Fingerlings	(lb)	No. per Lb	Grams Each	(no.)	% of Federal Fingerlings	(lb)	No. per Lb	Grams Each	(no.)	(lb)	(kg)
Rainbow trout (Salmo gairdneri)	66,099	77	1,569	42	11	12,941	70	327	40	11	79,040	1,896	862
Brown trout (Salmo trutta)	5,490	6	185	30	15	1,320	7	20	66	7	6,810	205	93
Brook trout (Salvelinus fontinalis)	7,443	8.5	128	58	8	1,307	7	25	52	9	8,750	153	70
Cutthroat trout (Salmo clarkii)	5,670	6.5	121	47	10	881	5	23	38	12	6,551	144	65
Lake trout (Salvelinus namaycush)	1,451	2	34	43	11	2,009	11	68	30	15	3,460	102	46
Total	86,153	100.0	2,037	42	11	18,458	100	463	40	11	104,611	2,500	1,136

TABLE 1.15

STATE AND FEDERAL PRODUCTION OF LARGE TROUT COMPARED (ALL QUANTITIES ARE IN THOUSANDS)

Species	State						Federal					Total	
	(no.)	% of All Large State Trout	(lb)	No. per lb	Grams Each	(no.)	% of All Large Federal Trout	(lb)	No. per lb	Grams Each	(no.)	(lb)	(kg)
Rainbow trout (Salmo gairdneri)	39,683	75	12,575	3.2	142	12,936	74	3,268	4.0	114	52,619	15,843	7,201
Brown trout (Salmo trutta)	6,712	13	1,740	3.8	119	842	5	178	4.7	96	7,554	1,918	872
Brook trout (Salvelinus fontinalis)	4,114	8	1,100	3.7	123	834	5	201	4.2	108	4,948	1301	591
Cutthroat trout (Salmo clarkii)	1,315	2	499	2.6	175	623	4	80	7.8	58	1,938	579	263
Lake trout (Salvelinus namaycush)	954	2	31	31.0	15	2,122	12	109	19.5	23	3,076	140	64
Total	52,778	100	15,945	3.3	138	17,357	100	3,836	4.5	101	70,135	19,781	8,991

not generally satisfactory and large fish cost too much to rear for routine stocking, except channel catfish. Hence over 90% of warmwater fish are stocked as fingerlings.

Other Fishes.—Among the other fishes are striped bass (*Roccus saxatilis*), which are stocked in large reservoirs because they are big enough and aggressive enough to utilize forage fishes effectively. They add an element of excitement to reservoir fisheries because they attain large sizes and are popular with fishermen.

This species cannot usually reproduce in reservoirs because the eggs and larvae require long stretches of flowing water. Hence continued restocking is necessary.

In 1973 some six million striped bass fingerlings were reared in state and federal hatcheries.

Walleye (*Stizostedion vitreum*), northern pike (*Esox lucius*) and muskellunge (*Esox masquinongy*) are raised in some states. In 1973, 9 states raised 7 million walleye fingerlings and 882 million fry. The federal hatcheries contributed 3 million fingerlings and 75 million fry. States also raised 78 million northern pike fingerlings and fry, and federal hatcheries 11 million. About two million muskellunge fingerlings and fry are reared by the states.

Marine Restocking.—The culture of Pacific salmon and steelhead trout is localized along the Pacific coast with a small satellite program in New England. Seven species are involved. The operation is very large, involving production of more than 300 million fish annually. State hatcheries produce about 80% and federal hatcheries about 20%. There are 89 state and 12 federal installations. The numbers of fish produced are shown in Table 1.16. On the East Coast, New Hampshire, Massachusetts and Rhode Island produced some anadromous coho salmon. A total of 360,000 fry and fingerlings were raised.

In addition, the U. S. Fish and Wildlife Service and the New England states are now in the process of developing a substantial Atlantic salmon program. However, during fiscal year 1973 total smolt production amounted to only 225,000 fish weighing 11,364 kg (25,-000 lb). The states involved are Connecticut, Maine, Massachusetts, New Hampshire and Vermont.

Conclusions to Restocking.—In 1973 the state hatcheries reared 179 million coldwater fish and the federal hatcheries reared 37 million, for a total of 216 million. An additional 100 million warmwater fish were produced. Striped bass accounted for six million. The midrange fishes, including walleye, northern pike and muskellunge, accounted for 967 million. Pacific salmon, steelhead and Atlantic salmon consisted of 300 million fish. The total of all fish reared and

TABLE 1.16

STATE AND FEDERAL HATCHERY PRODUCTION OF FINGERLING PACIFIC SALMON AND STEELHEAD TROUT, 1973

Species	State Totals					Federal Totals				
	(no.)	% of Total	(lb)	(kg)	% of Total	(no.)	% of Total	(lb)	(kg)	% of Total
Steelhead trout (Salmo irideus)	19,471,913	7.7	1,996,963	907,710	24.2	7,047,945	10.4	503,972	229,078	32.1
Coho salmon (Oncorhynchus kisutch)	65,456,481	25.8	2,865,957[1]	1,302,708	34.7	10,033,050	14.7	333,414	151,552	21.2
Chinook salmon (Oncorhynchus tshawytscha)	163,082,646	64.4	3,384,247[1]	1,538,294	40.9	47,076,049	69.1	725,725	329,875	46.2
Sockeye salmon (Oncorhynchus nerka)	2,496	—	32	15	—	—	—	—	—	—
Chum salmon (Oncorhynchus keta)	5,245,410	2.1	14,714	6,688	0.2	3,966,013	5.8	7,513	3,415	0.5
Pink salmon (Oncorhynchus gorbuscha)	50,000	—	1,351	614	—	—	—	—	—	—
Cherry salmon (not identified)	29,045	—	2,292[1]	1,042	—	—	—	—	—	—
Total	253,337,991	100.0	8,265,556	3,757,071	100.0	68,123,057	100.0	1,570,624	713,920	100.0

[1] Washington Department of Fisheries did not report weight allocations within the different size classes for this species. The proportion of total weight allocated to the fingerling and large size categories for this agency is based on averages for the other states. Hence, poundage totals within the size classes for this species are not exact.

released by all governmental agencies amounted to 1589 million.

Other than those fish already reported in this section, such as Pacific and Atlantic salmon and steelhead, the culture of marine and estuarine fishes is minimal.

SPECIAL ACKNOWLEDGEMENTS

DR. J.W. ANDREWS, Skidaway Institute of Oceanography, Savannah, GA

DR. JAMES W. AVAULT, JR., Fisheries Division, School of Forestry and Wildlife Management, Louisiana State University, Baton Rouge, LA

DR. JAMES W. AYERS, National Marine Fisheries Service, U. S. Department of Commerce, Little Rock, AR

MR. ROBERT BALKOVIC, National Marine Fisheries Service, Seattle, WA

MR. JOHN BISHOP, National Marine Fisheries Service, Seattle, WA

MR. PORTER BRIGGS, General Manager and Publisher, The Commercial Fish Farmer and Aquaculture News, Little Rock, AR

DR. J.L. CHESNESS, Agricultural Engineering Department, University of Georgia, Athens, GA

CLEAR SPRINGS TROUT COMPANY, Buhl, ID

CRYSTAL SPRINGS RANCH, Buhl, ID

DR. LAUREN DONALDSON, University of Washington, Seattle, WA

DR. JOHN R. DONALDSON, Oregon Aqua-Foods, Inc., Newport, OR

DR. HARRY DUPREE, Fish Farming Experimental Station, Stuttgart, AR

DR. JAMES E. ELLIS, Fish Farming Experimental Station, Stuttgart, AR

MR. ROBERT ERKINS, Fisheries Consultant, Box 108, Bliss, ID

DR. ABDULLAH ERSOZ, Department of Agricultural Economics, University of Georgia, Athens, GA

DR. JOHN GLUDE, National Marine Fisheries Service, Seattle, WA

DR. J.B. GRATZEK, College of Veterinary Medicine, University of Georgia, Athens, GA

DR. DONALD GREENLAND, Fish Farming Experimental Station, Stuttgart, AR

DR. T.K. HILL, Department of Entomology and Fisheries, University of Georgia, Tifton, GA

MR. TERRY HUDDLESTON, Department of Fisheries and Wildlife, Moscow, ID

JONES AND SANDY LIVESTOCK COMPANY, Hagerman, ID

DR. GEORGE W. KLONTZ, College of Forestry, University of Idaho, Moscow, ID

DR. RAY KRAATZ, National Marine Fisheries Service, Seattle, WA

MR. JON LINDBERG, Dómsea Farms, Incorporated, A subsidiary of Union Carbide, Seattle, WA

DR. C.E. MADEWELL, Tennessee Valley Authority, Muscle Shoals, AL

MARIFARMS, INC., Panama City, FL

MARINE PROTEIN CORPORATION, Magic Springs Trout Farm, Hagerman, ID

DR. MAYO MARTIN, Fish Farming Experimental Station, Stuttgart, AR
DR. E.W. McCOY, Department of Agricultural Economics and Rural Sociology, Auburn University, Auburn, AL
DR. FRED P. MEYER, Fish Farming Experimental Station, Stuttgart, AR
DR. JACK MILLER, Animal Science Department, University of Georgia, Athens, GA
MR. MAURICE MOORE, Editor, The Commercial Fish Farmer and Aquaculture News, Little Rock, AR
DR. W.R. MORRISON, Department of Agricultural Economics, University of Arkansas, Fayetteville, AR
DR. TONY NOVOTRY, National Marine Fisheries Service, Seattle, WA
DR. C.A. ORAVETZ, National Marine Fisheries Service, Little Rock, AR
RANGEN'S TROUT RESEARCH LABORATORY, Hagerman, ID
DR. RICHARD C. RAULERSON, National Marine Fisheries Service, U. S. Department of Commerce, St. Petersburg, FL
DR. W.A. ROGERS, Department of Fisheries and Allied Aquaculture, Auburn University, Auburn, AL
DR. KERMIT E. SNEED, Fish Farming Experimental Station, Stuttgart, AR
DR. M.E. "PETE" THORNTON, Managing Editor, Aquaculture and the Fish Farmer, Little Rock, AR
THOUSAND SPRINGS TROUT FARMS, INC., Buhl, ID
DR. J.E. WALDROP, Department of Agricultural Economics, Mississippi State University, State College, MS
WHITE WATER TROUT FARM, Hagerman, ID

REFERENCES

ANON. (no date). Catfish Feeding and Growth. Bur. Sport Fish. and Wildlife. U. S. Dep. of the Interior.
ANON. 1969A. Rearing 20,000 trout in a silo. Am. Fish Farmer 1, No. 1, 19.
ANON. 1969B. World's largest trout farm. Am. Fish Farmer 1, No. 1, 6-9.
ANON. 1971A. Farming Pacific salmon in the sea: from the "Womb to the Tomb." Fish Farming Ind. 2, No. 5, 6-9.
ANON. 1971B. Fisheries statistics of the United States, 1971. Natl. Marine Fish. Serv. Stat. Dig. 65.
ANON. 1972A. A statistical reporting system for the catfish farming industry, methodology and 1970 results. U. S. Dep. Comm. Econ. Dev. Admin. Tech. Assist. Proj. 99-6-09044-2. In cooperation with the Ind. Res. and Exten. Cen. and Dep. of Agric. Econ. and Rural Sociology, Univ. of Arkansas (Dec., 1972).
ANON. 1972B. Bob Erkins talks trout marketing. Fish Farming Ind. 3, No. 2, 20-24.
ANON. 1974A. Oregon salmon farmers plan to turn profit corner in '74. Fish Farming Ind. 5, No. 2.
ANON. 1974B. Production Survey of Catfish Farming. Natl. Marine Fish. Serv., Dep. of Comm., Little Rock, Arkansas.
ANON. 1975A. Georgia landings. Monthly Reports January through December, Nat. Oceanic Atmos. Admin. Natl. Marine Fish. Serv., Washington, D. C.
ANON. 1975B. Various articles in Commer. Fish Farmer and Aquaculture News 2, No. 1.

ANON. 1976A. Farm-Raised Catfish Processors Report. Natl. Marine Fish. Serv., Dep. of Comm., Little Rock, Arkansas.

ANON. 1976B. Fisheries of the United States, 1975. U. S. Natl. Marine Fish. Curr. Fish. Stat. *6900*, 11, 40, 50.

ANON. 1976C. Idaho's water-rich Magic Valley: heart of the food trout industry. Commer. Fish Farmer and Aquaculture News 2, No. 4, 4-8.

ANON. 1976D. New Hampshire cohos. SFI Bull. *273*. Sports Fishing Inst., Washington, D. C.

ANON. 1976E. Processed Catfish Imports. Natl. Marine Fish. Serv., Dep. of Comm., Little Rock, Arkansas.

ANON. 1976F. Processed Production Farm-Raised Catfish. Natl. Marine Fish. Serv., Dep. of Comm., Little Rock, Arkansas.

ANON. 1976G. Trout production for recreation: peaks, valleys and a bright future. Commer. Fish Farmer and Aquaculture News 2, No. 4, 9-13.

AVAULT, J.W., JR. 1973A. Crawfish farming gains attention. Fish Farming Ind. *4*, No. 1, 28.

AVAULT, J.W., JR. 1973B. Crawfish Farming in the United States. Center for Wetland Resources, Louisiana State Univ., Baton Rouge, Louisiana.

AVAULT, J.W., JR., DE LA BRETONNE, L., JR., and JASPERS, E.J. 1975. Culture of crawfish—Louisiana's crustacean king. Am. Fish Farmer 1, No. 10, 8-14, 27.

BENTE, F., JR. 1975. Pioneering the new industry of shrimp farming. The Symposium on Marine Chemistry in the Coastal Environment, the 169th Am. Chem. Soc. Meeting, Philadelphia, Pa.

BROWN, E.E., HILL, T.K., and CHESNESS, J.L. 1974. Rainbow trout and channel catfish—A double-cropping system. Ga. Agric. Exp. Stn. Res. Rep. *196*.

BROWN, E.E., LaPLANTE, M.G., and COVEY, L.H. 1969. A synopsis of catfish farming. Ga. Agric. Exp. Stn. Res. Bull. *69*.

COON, L., LARSEN, A., and ELLIS, J.E. 1968. Mechanized haul seine for use in farm ponds. Fish. Ind. Res. *4*, No. 2.

DE LA BRETONNE, L., JR., and AVAULT, J.W., JR. 1971. Liming increases crawfish production. Louisiana Agric. *15*, No. 10.

DONALDSON, J.R. 1976. Salmon aquaculture in Oregon. Proc. Conf. on Salmon Aquaculture and the Alaskan Fishing Community, Cordova, Alaska, Sea Grant Report *76-2*, 281-288.

FINKBEINER, K. 1974. Trout farm featured at popular theme park. Aquaculture and the Fish Farmer *1*, No. 3, 4-6, 11.

FOWLER, L.G., and BURROWS, R.E. 1971. The Abernathy salmon diet. Progressive Fish-Culturist *33*, No. 2, 67-75.

GOODWIN, H.L., and HANSON, J.A. (1975). The Aquaculture of Freshwater Prawns. Oceanic Institute, Waimanalo, Hawaii.

GORANSON, E. 1973. Salmon farming gains toehold on Oregon coast. Fish Farming Ind. *4*, No. 2, 8-11.

GREENLAND, D., GILL, R., and HALL, J. 1971. Live cars for use in catfish industry. Comm. Fish. Review *33*, No. 7-8, 44-53.

HAGAN, H.K. 1972. Colorado trout dilemma: big demand, limited production. Fish Farming Ind. *3*, No. 5, 8-10.

HEFFERMAN, B.E. 1974. Mt. Lassen trout farms look to 100,000,000 eggs a year. Fish Farming Ind. *5*, No. 4, 6-8, 29.

HOLEMO, F.J., BROWN, E.E., and HUDSON, H. 1973. Trout fee fishing. Italian J. Fish Culture *8*, No. 3, 73-75. (Italian)

HUNER, J.V., and AVAULT, J.W., JR.   1974.   Crawfish for bait.   Aquaculture and the Fish Farmer 1, No. 2, 12-17.

HUNER, J.V., and AVAULT, J.W.   1976.   Management implications for time of fall pond flooding in Louisiana crawfish farming with emphasis on production of bait crawfish.   Paper presented at National Fish Culture Workshop, Springfield, Missouri.

JASPERS, E., and AVAULT, J.W., JR.   1969   Environmental conditions in burrows and ponds of the red swamp crawfish Procambarus clarki (Girard), near Baton Rouge, Louisiana.   Proc. 23rd Ann. Conf. Southeastern Assoc. of Game and Fish Commissioners, Mobile, Ala.

KINNAMAN, P.V.   1975.   Weyerhauser: The salmon growing company.   Aquaculture and the Fish Farmer 2, No. 4, 4-7, 18.

KLONTZ, G.W., and KING, J.G.   1975.   Aquaculture in Idaho and Nationwide.   Idaho Dep. of Water Resources, Boise, Idaho.

MALOY, C.R.   1966.   Status of fish culture in the North American region.   Proc. FAO World Symp. on Warm-water Fish Culture, Rome, Italy. FAO Fish Rep. 44, Vol. 11 (May), 18-25.

MEYER, F.P., SNEED, K.E., and ESCHMEYER, P.T.   1973.   Second report to the fish farmers.   Resource Public. 113, U. S. Fish and Wildlife Service, Bureau of Sport, Fisheries and Wildlife, Washington, D.C.

NAEF, F.E.   1972.   Salmon culture.   Am. Fish Farmer and World Aquaculture News 3, No. 4, 12.

NYEGAARD, C.W.   1973.   Coho salmon farming in Puget Sound.   Wash. State Univ., Coop. Ext. Serv., Ext. Bull. 647.

SHANG, Y.C.   1974.   Economic feasibility of fresh water prawn farming in Hawaii.   Econ. Res. Center, Univ. of Hawaii, Honolulu, Hawaii.

WAGNER, L.C.   1973.   An evaluation of the market for pan-sized salmon.   Natl. Marine Fish. Serv., Market Res. and Serv. Div., Research Contract No. N-208-0344-72N.

Chapter 2

Canada

Dr. Hugh R. MacCrimmon

Aquaculture in Canada began in the Province of Quebec where the Atlantic salmon *(Salmo salar)* and the speckled trout *(Salvelinus fontinalis)* were propagated commercially by 1857. With establishment of the Newcastle Hatchery in Ontario by the Canadian government in 1866, a system of salmonoid production was established which strongly influenced both private and government aquacultural developments in Canada, if not North America, over the century which followed. Similarly, the construction of a series of major jar hatcheries along the Great Lakes, beginning in 1875, heralded the advent of a culture method which was generally adopted in North America for the culture of coregonid, percoid and, more recently, esocid and other fishes. The pond culture of centrachid fishes, begun by government agencies in 1900 for the production of black bass *(Micropterus)* for live release, is presently limited to modest government and private operations, largely in the Great Lakes region. While fisheries research and the management of fisheries resources has been a priority consideration in Canada for many years, a concerted effort towards the application of fisheries science to the commercial production of food fishes and marine invertebrate organisms in confinement is of comparatively recent origin.

Under terms of the British North America Act of 1867, overall responsibility for the Canadian freshwater and marine fisheries lies with the federal government, but specific aspects of education, research and management are matters of provincial concern. Consequently aquaculture in Canada has evolved within both federal and provincial frameworks. In 1972 there were some 21 federal and 31 provincial government hatcheries, 163 commercial hatcheries, and an additional 113 licensed domestic operations using sloughs situated in the prairie provinces of western Canada. Not included in these figures is an unknown number of small private hatcheries, principally in eastern Canada, which are being operated more or less exclusively by fishing clubs for stocking their own waters.

Federal and provincial governments have continued to maintain a rather traditional approach to the artificial propagation of fish, both in terms of species cultured and the almost exclusive use of the product for live release within sport and commercial fisheries. Small numbers are used for aquacultural and other aspects of fisheries research in government and university research laboratories. Production from government hatcheries in 1972 approached 369 million fish with an aggregate weight exceeding 500 MT (551 ST) of which salmonoid fishes made up only 8.8% by number and 86.4% by weight. Some 94.3% of the production by number and 85.3% by weight came from provincial facilities spread across the continent. Of the total production of salmon, trout and char in Canada by government and commercial hatcheries, private enterprises were responsible for about 28% by number and 57% by weight (Fig. 2.1). Other species reared in federal and provincial establishments for live release included 265 million whitefish and cisco *(Coregonidae)*, 265 million pickerel and perch *(Percidae)*, 29 million pike and maskinonge *(Esocidae)*, and 180,000 black bass *(Centrarchidae)*.

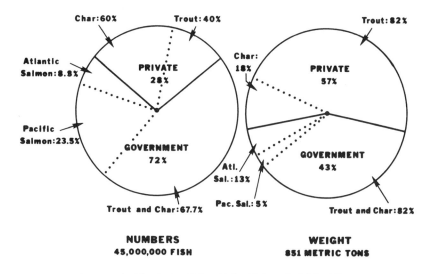

FIG. 2.1.   AQUACULTURAL PRODUCTION OF SALMONOIDS, 1972

Commercial fish farmers in the same year (1972), excluding experimental harvests of stocked rainbow trout from prairie pothole lakes, marketed nearly 13 million fish weighing an estimated 485 MT (534 ST), largely rainbow and speckled (brook) trout (Fig. 2.2). Of the salmonoids produced some 70% by number and 31% by weight were sold for live release, principally into private waters across Canada. Brook trout made up some 65% by number and 61% by weight. The balance was: rainbow trout *(Salmo gairdneri),* except for comparatively few lake trout *(Salvelinus fontinalis);* brown trout *(Salmo trutta);* cutthroat trout *(Salmo clarki);* Atlantic salmon *(Salmo salar);* and splake *(Salvelinus fontinalis × namaycush).* Collectively these made up less than 1% of the total production of salmonoids marketed. The principal market for commercially produced live fish lies in the Provinces of Ontario and Quebec where there is a strong demand for trout and char to stock private ponds and stream systems, and in the prairie provinces where considerable numbers of trout are released in privately owned pothole lakes for subsequent commercial recovery and domestic sale.

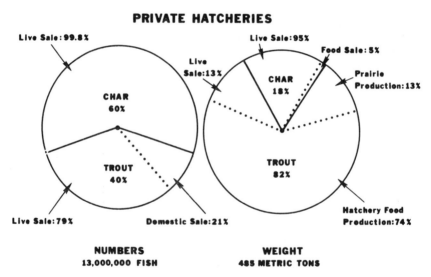

FIG. 2.2. COMMERCIAL PRODUCTION OF SALMONOIDS, 1972

The commercial farming of freshwater salmonoids for human production in Canada is of comparatively recent origin because of a traditional local respect for the speckled (brook) trout as a sport fish, and because of a lack of evidence that the Atlantic and Pacific salmons could be profitably cultured to marketable size. Thus government agencies were hesitant to change long-standing legislation prohibiting sales of prized sport fish to accommodate a new aspect of private enterprise. As a result, the development of a viable Canadian aquacultural industry based on food fish sales lags appreciably behind that of most countries which established fish farming industries. The transition of existing trout farming units, previously geared only to the production of trout for live sale, has been a rather slow process accomplished by slowly increasing the production to meet unknown but seemingly attractive market protentials. Despite expressed interest within the business community, aquacultural development in Canada has generally been left to the family unit level of operation. There are presently no large corporate aquacultural undertakings in Canada which have developed beyond the pilot experimental stage. The economic viability of large-scale production units remains to be proven. However, a number of previously established and new trout farms have become geared for the intensive culture of food fish. Their operators accumulated several years of invaluable experience in producing trout for human consumption.

In 1972 food sales of Canadian products totaled an estimated 1.1 million fish weighing 352,227 kg (775,000 lb), of which 99% was rainbow trout. The remainder was brook trout. Sales of speckled (brook) trout for human consumption is still prohibited by law in Quebec, although the species made up over 92% of the live sales in that province during the past year. Most fish marketed for human consumption in Canada were in the 170 to 280 g (6 to 10 oz) range, although one commercial unit in Quebec had begun to produce in volume rainbow trout of approximately 2.3 kg (5 lb)—principally for a specialized Montreal and New York market. The majority of rainbow trout produced for human consumption across Canada was sold by individual producers to local shops and restaurants, with some fish being distributed by chain store systems. Cooperative marketing of rainbow trout was limited to fish harvests from prairie potholes (sloughs) in western Canada, which were distributed on an experimental basis by the Freshwater Fish Marketing Board. At present the Ontario Fish Farmers' Association is advocating a cooperative approach towards the purchase of feed and other production items, and the collective marketing of their products.

Despite restrictive measures adopted by federal and provincial governments on the importation and transfer of eggs and fish, a reported 11 million rainbow trout and 2.5 million brook trout were imported into Canada from the United States in 1972. These eggs and fingerlings came from Washington, Idaho, Montana and the New England States. Reported transfers of live fish across federal and provincial boundaries totaled approximately 14.6 million eggs and 4.8 million fry and fingerlings.

In Canada there are, as yet, no specialized brood hatcheries geared to meet the egg requirements of the trout farming industry. However, most of the well-established trout farmers maintain their own brood stocks with a considerable "in-trade" market for eggs and young fish. Because of concern over the transmission of infectious fish diseases, there has been a general tightening of regulations by both federal and provincial governments pertaining to the importation and transfer of fish. Acting under authority of the Fisheries Act of Canada, the Fisheries and Marine Service of Environment Canada implemented new constraints on transfers of eggs and fish under the Fish Health Protection Regulations. The regulations apply to live fish, eggs of fish and dead products of cultured fish destined to move into Canada or cross provincial boundaries within Canada. Beginning January 1, 1977, it became illegal to make such transfers without a permit provided by a local Fish Health Officer, following periodic inspections of hatchery facilities and failure to detect any of the following pathogens: the kidney disease bacterium; the redmouth bacterium, *Aeromonas salmonicida;* the protozoans *Myxosoma cerebralis* and *Ceratomyxa shasta;* and the viruses causing viral hemorrhagic specticemia, infectious hematopietic necrosis, infectious pancreatic necrosis; myxobacteria; the motile aeromonads, and the vibrios. Further clarification of these regulations can be obtained from the National Registry of Fish Diseases, Environment Canada, Ottawa. Various provincial agencies have their own regulations as well, pertaining to fish propagation and transfers. At present the Great Lakes Fisheries Commission has proposed stringent regulations pertinent to that region of North America.

IMPORTS

Imports of cultured food fish into Canada during 1972 totaled 986 MT (2,169,406 lb) (Tables 2.1 and 2.2). Included were: 6234 kg (13,715 lb) of carp from Hong Kong, Taiwan, and the USA; 909 kg (2000 lb) of milkfish from the Philippines; and 978,950 kg (2,153,691 lb) of trout from Denmark, Great Britain, the USA, Japan and Portugal. Most of the trout were frozen, but shipments included about 3636 kg (8000 lb) of canned and smoked fish, principally from Denmark and Portugal. Imports of rainbow trout to Canada have more than doubled during the past five years despite the embargo imposed on Danish fish (Table 2.1).

TABLE 2.1

CANADIAN IMPORTS OF RAINBOW TROUT FOR HUMAN CONSUMPTION [1]

Country of Origin	1967 (lb)	1968 (lb)	1969 (lb)	1970 (lb)	1971 (lb)	1972 (lb)	1972 (kg)
Denmark	574,000	546,000	412,000	132,000	45,000	5,900	2,682
Great Britain				27,000	1,500	14,900	6,773
Japan	371,000	550,000	399,000	275,000	1,616,000	2,003,200	910,545
Portugal						1,200	545
USA	55,000	95,000	114,000	207,000	124,000	128,500	58,409
Total	1,000,000	1,191,000	924,000	641,000	1,786,500	2,153,691	978,954

[1] To determine kg imported during 1967-1971, multiply pounds by 2.20.

TABLE 2.2

IMPORTS OF DOMESTIC FRESHWATER FOOD FISHES INTO CANADA DURING 1972

Fish	(kg)	(lb)	Country of Origin
Trout			
Fresh	6,773	14,900	Great Britain
Fresh and frozen	910,545	2,003,193	Japan
Fresh and frozen	58,409	128,528	USA
Canned and smoked	2,682	5,898	Denmark
Canned	545	1,172	Portugal
Carp			
Fresh and frozen	1,045	2,300	Taiwan
Fresh and frozen	36	80	Hong Kong
Fresh and frozen	4,948	10,885	USA
Salted roe	205	450	USA
Milkfish			
Fresh or frozen	909	2,000	Philippines

MARKETING

Canadian aquaculturists have, with the exception of the prairie pothole farming operations, received no more than token government assistance in advertising or marketing the hatchery product, in contrast to the strong support given historically to agriculturalists and other aspects of the fisheries industry. Permission to sell "game" fish in Canada is, in itself, of comparatively recent origin. The developing fish farming industry has had to contend with various regulatory restrictions which, in combination with negligible government support until the present, have often been cited by fish farmers as being suppressive to growth and viability. Only in 1972 were prairie trout farmers permitted to sell directly to retailers, being previously restricted to direct sales to consumers or the Freshwater Fish Marketing Board. The Province of Quebec still prohibits the sale of native speckled (brook) trout for human food, whether of wild or hatchery origin, although there is extensive commercial propagation of the species for live release. Within the Canadian provinces there has been a gradual liberalization in attitude towards the domestic production and sale of salmonoid fishes. Consequently in most instances it is the ability of the fish farmer to profitably grow char, salmon and trout to marketable size that governs the production of Canadian aquaculture. The production and marketing of salmonoid fishes in Canada, therefore, continues to depend almost solely on the expertise and ingenuity of individual producers, although there has been very limited government support of private aquacultural systems by development grants and the like. Sales promotion among private fish farmers for the marketing of their products continues to be minimal, with a dependence on verbal communication with new and established clientele.

Sales of live trout for release into fishing waters are usually made on a local or regional basis, although not infrequently interprovincial and transcontinental transfers of eggs and fish (using oxygen-filled plastic bags) have been made by air freight. Sales of hatchery-produced salmonoids, principally rainbow trout, for human consumption are made in both fresh and frozen form. Until recent government authorization of the legal sale of these species on the food market, Canadian consumers were dependent on frozen packaged trout imported principally from Japan, Denmark and Britain. Frozen trout of foreign origin, usually dressing out at 170 to 280 g (6 to 10 oz) per fish, are usually retailed at 2 to a package for the home consumer or bulk packaged for the restaurant trade. Thus the Canadian production of domestic trout provided, for the first time, fresh and unfrozen trout for the local consumer. Trout farmers have found a ready market for this product in direct sales to consumers and to the restaurant and retail food trade. Only surplus trout grown in Canadian hatcheries are usually frozen. Hence, fresh locally-produced trout have a distinct market advantage over processed imports. Markets for Canadian domestic trout are best developed in Quebec, Ontario and British Columbia, although prairie trout farmers have found suitable markets for their harvests of rainbow trout from pothole lakes, including a smoked product. Quebec producers are, at the moment, developing a unique market for large smoked rainbow trout in eastern American cities. At existing levels of trout production across Canada there is a healthy market for domestic trout despite negligible to modest market promotion and product development.

It is estimated that about 1.3 kg (2.9 million lb) of domestic salmonoids were marketed in Canada during 1972 for human consumption. Of this quantity 979,000 kg (2,150,000 lb) were imported, with
93% of this amount coming from Japan and 6% from the USA. Thus
only 26.5% of the domestic trout marketed in Canada was produced
by the Canadian fish farming industry. Sales of hatchery-grown
freshwater fish in Canada in 1972 were solely rainbow trout, except
for a few speckled (brook) trout (0.3%) produced and sold in the
provinces of Nova Scotia and Ontario. While no precise figures
are available on the present production and sales of hatchery-produced
domestic salmonoids in Ontario, the trout farming industry has undergone substantial development since 1972. Efforts have been restricted
largely to the increased production of rainbow trout for the food
market. However, there is growing interest in the cage culture of
Atlantic and Pacific salmons in coastal waters, and in the potentials
of the Arctic char for rearing in cold waters typical of most Canadian
hatcheries on a seasonal if not continual basis.

The sale of salmonoid eggs is essentially an in-trade operation by
Canadian trout producers, although numerous stocks of ova and fry
(most notably of rainbow trout and salmon) have been made available by governments to private operators from time to time for
purposes of experimentation and commercial production. The trend
of ova importations from the United States has been one of decline
as the capability of Canadian trout farmers to meet their own production requirements has increased almost to the point of self-
sufficiency. However, exchanges of eggs, fry and fingerlings among
Canadian hatcheries, both government and commercial, is a continuing
activity and one which has come under much closer government
scrutiny with the new Federal Fish and Health Regulations and
various provincial enactments. The now considerable market for live
trout for stocking private and commercial "pay-fishing" ponds is bound
to expand as demands for recreational fishing near larger centers
of population increase. A number of trout farmers also profitably
operate "pay-fishing" ponds in conjunction with their production
hatcheries.

Price Structure

Live Sales.—Sales of fish eggs by private hatcheries in Canada are
limited to those species produced, namely the rainbow trout and
brook (speckled) trout. There is presently little price differential
between the two species. Eggs sold in 1972 for between $3.00 and $6.50

per thousand depending on locality, quality, quantity purchased and availability.

The price of fish, again restricted to brook and principally rainbow trout, is based on size, demand, numbers purchased and availability (Table 2.3). Live trout purchased by the trout farming industry are typically fry or fingerlings averaging 10 cm (2 to 3 in.) in length.

TABLE 2.3

PRICE STRUCTURE FOR LIVE SALE OF TROUT IN CANADA DURING 1972

Size (in.)	(cm)	Price per Thousand ($)
Eggs		3.00 – 6.50
1 – 2	less than 1	20.00 – 35.00
2 – 3	1	40.00 – 60.00
3 – 4	1.5	75.00 – 150.00
4 – 6	2	150.00 – 200.00
6 – 8	2.5–3	400.00 – 450.00
8 – 10	3–4	600.00 – 750.00
10 – 12	4–5	800.00 – 1000.00
Adults		3.00 each and up

The market for larger and older trout is restricted largely to sales of fish to owners of ponds or other fishing waters used for personal angling, or for the sale of recreation. The Ontario Fish Farmers' Association looks forward to the day when members may have the opportunity to bid on government contracts for fish to be released in public waters, thus welcoming the opportunity to compete with the government hatchery system in terms of cost and quality of product. In 1972 "keeper-size" trout sold across Canada for prices ranging from $600 per thousand for 15 cm (6 in.) fish to $1000 per thousand for 30 cm (12 in.) fish. Mature adults for brood stock sold at upwards of $3.00 per individual fish. Brook (speckled) trout were priced slightly lower than rainbow trout despite a generally slower growth rate.

The sale of recreational angling for hatchery-reared trout held in confinement is increasing in popularity across Canada. The operation takes several forms. Traditional fishing clubs, particularly in Ontario and Quebec where some clubs have been in existence for a century or more, charge substantial membership fees and may place a daily or seasonal creel limit on its members. Newer clubs have tended to expand their facilities to include other forms of family outdoor recreation with annual fees on a single or family basis. As membership in private clubs utilizing private ponds, streams, or small lakes is too costly, or perhaps not appealing to many Canadians, the "pay-fish" or "U-Catch-em" operations are providing anglers with an alternative to private fishing clubs or public waters. The charge normally consists of a "gate" or "fishing pole" fee, plus a charge for all fish caught. The usual cost of the trout lies between $0.25 and $0.38 per cm ($0.10 to $0.15 per in.). Thus, the price of a "single portion" sized trout averages between $1.00 and $1.50 each, or about $4.00 to $8.00 per kg ($2.00 to $3.00 per lb), plus admission or fishing fee. A comparison of sale prices for live hatchery-reared fish over the past decade has indicated surprisingly little change in value to compensate for increasing costs of production.

Food Sales for Human Consumption.—Canadian aquaculturists must compete with imported trout in terms of price and quality. Bulk prices of frozen, dressed Japanese rainbow trout to west coast packers averaged $1.25 per kg ($0.57 lb) in 1972. Wholesale prices for the frozen Japanese product averaged between $1.80 and $2.00 per kg ($0.80 to $0.90 per lb), while similar USA fish sold at $2.75 to $3.30 per kg ($1.20 to $1.50 per lb). Retail value of the Japanese product (2 fish per package) was $2.10 to $2.45 per kg ($0.95 to $1.10 per lb), whereas the U.S. product retailed at $3.00 to $4.20 per kg ($1.35 to $1.90 per lb) in west coast cities.

Prairie trout farmers received $1.10 to $1.35 per kg ($0.50 to $0.60 per lb) for dressed trout shipped to the Freshwater Fish Marketing Corporation. Wholesale prices for fresh fish processed by the Corporation averaged $1.90 per kg ($0.85 per lb) with the fish advertised on the retail market at $3.35 to $4.00 per kg ($1.50 to $1.80 per lb). However, those prairie farmers selling a fresh product directly to local markets realized a substantial financial advantage, receiving better than $2.00 per kg ($0.90 per lb) for dressed fish.

Hatchery-produced rainbow trout sold directly to local stores, restaurants and individual consumers in British Columbia, Ontario and Quebec brought prices of up to $4.25 per kg ($1.90 per lb) for gutted iced fish. The production and sale of trout for human consumption would seem to be the most advanced in the Province of Quebec, especially in the production of large fresh and smoked rainbow trout, exported to New York at prices similar to those received for prime Atlantic salmon at $6.50 to $7.50 per kg ($3.00 to $4.50 per lb).

The value of freshwater aquacultural fish products marketed in Canada during 1972, exclusive of aquarium-reared ornamental and tropical fishes, was estimated at $3,141,000. Live sales of fry and fingerlings (72%) and yearlings and older salmonoids (28%) were valued at $2,154,000, representing 69% of the total value of all fish produced. Sales for human consumption made up the remaining $987,000. Salmonoids propagated in hatchery facilities including rearing ponds and cages made up 86%, and prairie pothole production comprised 14% of the value of these sales.

The marketability of trout produced in Canadian aquacultural systems, according to the 1972 survey, was excellent. Only among trout harvested experimentally from prairie pothole lakes was there any evidence of problems of product quality; these stemming from difficulties in autumn harvesting procedures and an unusual taste imparted to the flesh from some waters. Many prairie trout, however, were of top quality. Like trout produced elsewhere in Canada for human consumption, they found ready markets in local and urban centers in competition with foreign imports.

NUTRITION

Rations purchased by government and licensed private hatcheries for feeding those fish produced in 1972 exceeded 1136 MT (2.5 million lb) with a cost of nearly $460,000. With the exception of a few pounds of minnows purchased to feed esocids, the feed was utilized solely for the production of salmonoids.

An overall (average) food conversion rate of 1.4 was realized for salmonoids produced in Canadian aquacultural systems based on reported food purchases yield data. This resulted in an average feeding cost of about $0.53 per kg ($0.23 per lb) of trout produced for food or live release. Despite limitations on the use of the data because of differences in hatchery objectives, and especially in the size and age of the final products, an analysis indicated that these feed conversions and costs per pound of fish were realized by the commercial fish farmers. Of the total weight of food fed to salmonoids, some 84% was formulated dry rations (avg $0.159 per lb, or $0.35 per kg), 10% fresh or frozen liver and spleen (avg $0.278 per lb, or $0.61 per kg), and 6% marine shrimp. Because of rapidly escalating food costs on Canadian and world markets, the cost of salmonoid production has undoubtedly increased substantially since 1973.

Dry rations were purchased for Canadian government and private hatcheries from 8 feed manufacturers in 1972, but considerably less than 10% of the marketed product was manufactured in Canada. The present situation is relatively unchanged with a dominance of foreign-manufactured salmon and trout feeds on the Canadian market although, particularly in the Province of Ontario, there has been a rather modest increase in the local manufacture of formulated trout diets. Canadian feed manufacturers continue to be hampered in their operations by the need to import some of the ingredients necessary to meet prescribed ration formulae. Significant research on salmonoid nutrition and on the formation of diets utilizing principally locally-grown agricultural products is presently under way at several federal and university laboratories across Canada.

WATER SUPPLIES AND USE

Canada is blessed with an abundance of ground and surface waters, yet only insignificant quantities are presently being used for fresh-water aquaculture. It is estimated that government and commercial fish hatcheries in Canada are currently using on a continuous basis about 13,000 liters per sec (205,000 gal. per min) of clean ground-waters provided by wells, springs and headwater streams. During the era when government jar hatcheries in the Great Lakes were popular for coregonid and percoid culture, it is probable that pumped lake water increased the total continuous consumption of water flowing through Canadian hatcheries to about 14,000 liters per sec (225,000 gal. per min).

In view of the emphasis presently placed on the production of salmonoid fishes in Canadian aquaculture, present hatcheries and rearing stations are situated principally on cold water sources. Headwaters (especially flowing springs, artesian wells and feeder creeks) continue to be the favored water sources for intensive salmonoid propagation in eastern Canada and, to a lesser extent, in western Canada. Water supplies to Canadian hatcheries have a reported hardness ranging from 5 to 1570 ppm (expressed as $CaCO_3$), while pH values of 6.5 to 8.0 are typical. Two characteristic ecological features of Canadian fresh waters used commonly for salmonoid production are low temperature and low nutrient levels. Aquifer (and hence well and spring water) temperatures across Canada are locally constant, ranging from as low as 4.4° C (40° F) in parts of Quebec to as high as 13.9° C (57° F) in parts of British Columbia. However, seasonal variability in surface waters and hatchery facilities typically ranges downward towards the freezing point. In some instances, these waters approach the lethal temperatures of salmon and trout during the summer months, causing significant seasonal difference in fish growth rates within and among government and commercial hatcheries.

Despite the abundance of fresh water in Canada, both government and private aquaculturists have experienced difficulty in locating new potential hatchery sites with adequate groundwater discharge to meet requirements for the volume production of salmonoid fishes. Further, fish production in a number of government and private establishments across Canada is limited by water shortages on a permanent or seasonal basis. Consequently, more novel aquacultural procedures utilizing prairie pothole lakes on a live trout release and annual recovery basis, and cage culture in freshwater lakes and coastal saline waters, offer seemingly attractive alternatives to the traditional dependence on surface and pumped waters. Although cage culture has been used for the commercial production of salmonoids and other food fish species in other countries for some time, this method is still at the experimental stage in Canada, with the exception of a proven operation in the Province of Quebec now serving a Montreal and New York market.

As the use of recirculated water would seem to solve simultaneously the problems of inadequate water supply and cold water temperatures characteristic of north temperate climates, the partial recirculating of water is now a reality in several government and private hatcheries. About 7.5% of the production hatcheries in Canada re-use some warmed water, although usually only for the incubation of eggs and the early rearing of fry. Major recirculation systems using heated waters have been initiated in a few government hatcheries. However, at present the necessary testing has not been done which will indicate the economic justification for major investment by commercial operators. While there is considerable government and private interest in the use of heated wastewaters from power generating plants for fish production, feasibility studies undertaken to date are less optimistic than might be expected, despite the attractiveness of such potential sources of low cost energy.

RESEARCH AND EDUCATION

Canada has an outstanding record in terms of marine and freshwater fisheries research dating back to the early formation of the Biological Board of Canada and, until recently, the Fisheries Research Board of Canada with its directors drawn from representatives of governments, universities and the fishing industry. At present, fisheries research at the federal level is being undertaken by the various laboratories of the Fisheries and Marine Service of Environment Canada at St. John's, Newfoundland; Halifax, Nova Scotia; St. Andrew's, New Brunswick; the Canada Centre for Inland Waters, Burlington, Ontario; the Freshwater Institute at Winnipeg, Manitoba; and at Nanaimo and Vancouver in British Columbia. Most provincial governments also support fisheries research laboratories with studies related to the management of inland fisheries resources.

Graduate studies in one or more of the disciplines embraced by fisheries science are offered by most Canadian universities. However, integrated fisheries programs are given at comparatively few. The following seven institutions of higher learning in Canada are presently receiving major federal financial support for aquatic studies: Memorial University (Newfoundland), Dalhousie University (Nova Scotia), McGill University (Quebec), University of Toronto (Ontario), University of Guelph (Ontario), Univertsity of Victoria (British Columbia), and the University of British Columbia. Also, the Freshwater Institute of Environment Canada is located on the University of Manitoba campus. Considerable provincial support is given to a number of universities for fisheries research of particular benefit to the management of local inland waters. In addition, the Huntsman Marine Laboratory on the east coast (Bay of Fundy) and the Bamfield Marine Laboratory on the west coast (Vancouver Island) provide marine training and research facilities for Canadian universities on a consortium basis.

Despite the very significant Canadian investment in fisheries research at the federal, provincial and university levels, studies directed towards the solving of biological problems inherent in the development of commercial aquaculture in Canada have, until comparatively recently, been a token consideration. Thus it seems fair to state that private aquaculture in Canada has evolved to its present status largely as a result of private enterprise working within regulatory constraints imposed by governments, particularly for purposes of enforcement, statistical record and disease control. During the past decade there has been increasing federal and provincial support for environmental, fish disease and nutrition studies. Progressively better provisions have been made for diagonostic services to assist both government and private aquaculturists. However, it is only very recently that the Fisheries and Marine Service of Environment Canada made the decision to redirect substantial resources to aquaculture, thereby giving essential impetus to the solving of development, operational and production problems in the growing fish farming industry.

Representative aquacultural research and development programs presently activated across Canada by federal and, to a lesser extent, by provincial funding include: studies on the nutrition, biochemistry, physiology and behavior of oysters, lobsters, mussels and salmonoids; harvesting and processing methods for aquatic vegetation, most notably marine kelp and Irish moss and freshwater macrophytes in eutrophic lakes; hatchery facility design and technology for salmonoid fishes and marine invertebrates; recirculation hatchery systems; experimental cage culture of Atlantic salmon, rainbow trout, brook (speckled) trout, Arctic char, splake and the Pacific salmons; prairie pothole trout farming; eel farming; modular aquacultural systems;

genetics and fish breeding; fish processing and marketing; fish pathology; trout nutrition and diet synthesis with Canadian-grown ingredients; feasibility studies on use of thermal energy from power generating plants; and a wide variety of fundamental biological studies being undertaken within government and university laboratories. The North American Salmon Research Center (a combined project of the International Atlantic Salmon Foundation, the Huntsman Marine Laboratory, and the Fisheries and Marine Service of Environment Canada) is a new facility with its principal activity in the area of genetics and selective breeding of salmonoids, especially the Atlantic salmon. Among the several Canadian universities actively engaged in fisheries research, a comprehensive aquaculture program has been best developed at the University of Guelph where, as a component of undergraduate and graduate teaching programs in the aquatic sciences, there is long-standing research and practical experience in the environmental, physio-behavioral, nutritional and pathological aspects of fish propagation.

SUMMARY

This chapter has attempted to present the reader with an overview of the growth and present status of government and private aquaculture in Canada; and, in particular, to indicate the exciting developments which are just beginning to take form with increased government commitment to scientific aquaculture. Despite over a century of successful and often innovative culture of aquatic life in Canada, aquaculture as a resource industry is just in its infancy. Endowed with an abundance of water and the dedicated support of the scientific community, government agencies, and practicing aquaculturalists, the principal limiting factor to the development of aquaculture as a significant protein-producing industry in Canada would seem to be its as yet unproven economic viability in other than small-scale operations. Finally, it should be noted that in addition to its support and participation in fisheries research and management programs in Canada, the Canadian government maintains a strong international commitment towards aquacultural research and development in Third World countries.

REFERENCES

MACCRIMMON, H.R., STEWART, J.E., and BRETT, R. JR. 1975. Aquaculture in Canada. Research Board of Canada (1973 and 1975), Bull. *188.*

Chapter 3

Norway

In 1975 Norway had about 300 fish farmers. About 250 were engaged in rainbow trout production, but 100 of these, particularly the older ones, were more or less out of production. The remaining 50 were culturing Atlantic salmon (Fig. 3.1). These were the only two cultured fish species grown in Norway. Hence this chapter is divided into two parts, one for Atlantic salmon and one for rainbow trout. Most of the cultured fish industry is concentrated in the area north and south of Bergen (Fig. 3.2). Restricted production in northern Norway is due to difficulties in raising fingerlings in the colder waters.

There are four types of fish-culturing establishments in Norway: (1) freshwater farms on land; (2) seawater farms on land; (3) enclosures in seawater; and (4) coves, sounds or fiords closed off by concrete, net or wire.

Most freshwater farms on land are sited so that they are fed by gravity flow. In only a few cases is pumping of groundwater done. The ponds or raceways are of concrete, earthen or lined with wooden materials.

Seawater farms on land usually use brackish water. Sea water is pumped and mixed with fresh water to increase water availability and water temperatures. The ponds, or raceways, are constructed of concrete, earthen or lined with wooden materials. Both freshwater farms and seawater farms on land produce eggs, fry and fingerlings. From these two types of establishments the fingerlings are transferred to enclosures in seawater and coves, sounds or fiords for growing out as food fish. There are about 70 freshwater farms on land and seawater-on-land farms; their production, however, is small.

Type (3), enclosures in seawater, is further divided into two subtypes. One is a floating enclosure of nylon netting in the sea. This

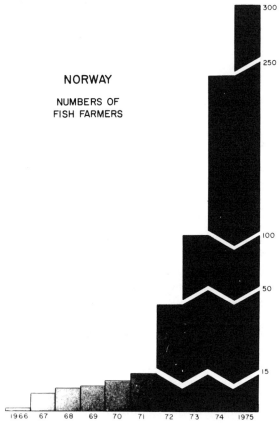

NORWAY

NUMBERS OF
FISH FARMERS

FIG. 3.1.   CHANGES IN THE NUMBER OF FISH FARMERS, 1966-1975

may be circular, rectangular or square, and of varying depths. Floating enclosures are the most common type of fish farm, accounting for
about 200 of the 300 different fish farms in Norway. The second sub-
type involves nets which go out from the shoreline into the sea or
brackish water area. The shore constitutes one side and netting suspended from driven poles constitutes the other three sides.

Coves, sounds or fiords rely on tides and currents to exchange water.
However, sometimes seawater is pumped through the enclosure if
the natural flow of water is not sufficient to maintain a given density
of fish. There are 10 to 15 of this kind of farm.

ATLANTIC SALMON *(Salmo salar)*

Atlantic salmon resemble sea trout in color and can grow to more
than 1 m (3 ft) in length. It is an anadromous fish, migrating from

fresh to salt water until it reaches sexual maturity, and then migrating back to fresh water to spawn. The sea stage can last from 1 to 3 years, after which the salmon migrate to fresh water, swimming upstream to spawn in the waters from which they originated. Young salmon may remain in the fresh water for 1 to 2 years before the return to the sea.

In Europe, Atlantic salmon are found in waters from the Arctic Ocean, the North Sea and Baltic Sea, the waters around Great Britain and Ireland, and on south to Portugal. Only in very recent years have they been cultured by man as a food fish.

In culturing, spawning usually occurs in November and December with eggs hatching the following April or May. During the first summer of growth, the fry should be fed every hour; continuous feeding is best. These fry eat very little when the water temperature is below 10° C (50° F) and growth ceases below the 5° to 7° C (41° to 45° F) range. During the second summer of growth the fry, which are now considered fingerlings, can be fed every other hour. The fingerlings are able to tolerate lower temperatures than the first summer fry. However, growth still ceases at about 5° C (41° F). During the winter seasons,

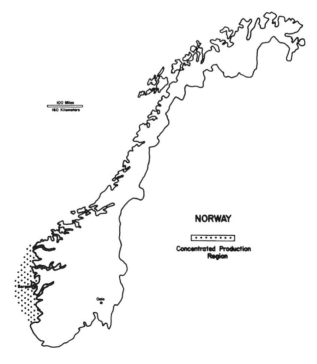

FIG. 3.2.   CONCENTRATED PRODUCTION AREA FOR CULTURED FISH, 1975

when water temperatures are below 5° C (41° F), the fish are given only one small feeding every second day. By about May 1, 1 or 2 years after hatching, the fingerlings turn into so-called smolts, averaging 13 to 20 cm (5 to 8 in.). If heated water is used smolts can be produced in one year. Using water with natural temperatures generally requires two years to develop smolts. The smolts are stocked in the early spring into net enclosures (fiords closed off by nets or concrete impoundments). After a minimum period of one year in the sea, some of the salmon have grown to a size suitable for sale in the spring. However, most of the food fish are sold after 18 to 24 months of feeding (i.e., for Christmas or Easter). Average weight at this time is about 5 kg (11 lb). The record salmon produced in 4 years after hatching was 16 kg (35 lb).

Optimum water temperatures for feeding salmon are unknown, but good growth is reported up to about 15° C (69° F). Because of the colder fresh waters in northern Norway there are cultural problems with fry and fingerlings. Hence, salmon fish farmers in the north import their fingerlings by wellboat from the southern areas. This has not proven to be entirely satisfactory due to high mortality en route.

Salmon prices are highest during the Christmas and Easter seasons. Hence, production is keyed to marketing at these two peak seasons. In 1974 the lowest price received by salmon culturists was $3.00 to $4.00 per kg ($1.36 to $1.82 per lb). In August 1975 the price was $4.50 per kg ($2.05 per lb). In Norway, prices for cultured salmon are determined by world market prices for wild fish.

About 1972, in an effort to spread the benefits of fish farming to as many people as possible and to avoid concentration among a few individuals or firms, the authorities passed a law stating that the maximum size of any production facility would be 8000 m³. Translated into practical terms this would mean about 60 to 80 MT of salmon production per facility. At present very few fish farmers produce over 50 MT of salmon, while the average is less than 25 MT. Investigations are still under way to determine what a practical-sized economical production unit should be. The salmon farms that did contain over 8000 m³ of water prior to passage of the law were permitted to keep operating. The largest farm is reported to produce more than 500 MT annually. While some farms are presently producing 50 MT per man-year of labor, some Norwegians think that the optimum size for economical production might be 250 MT, employing 3 to 4 persons.

A typical management may consist of a fiord enclosure, or floating pens, on the southwestern coastline. All production is in salt or brackish water. The first 1 to 2 years (depending on water temperature) will bring the fish up to 1 to 2 kg (2.2 to 4.4 lb). During the final year of feeding, from about May 1, the fish will reach marketable size by September-November (3.5 to 5.0 kg, or 7.7 to 11 lb). Harvesting and marketing starts then and continues until Easter (March or April) of the following year. Small frozen trash fish bought in blocks of several hundred pounds each are thawed and fed to the fish. The frozen trash fish are brought by boat or truck to the production area from nearby fishing villages.

FIG. 3.3.   SALMON FARMING IN NORWEGIAN FIORD, FEEDING RING
SHOWN IN CENTER

A fiord enclosure usually consists of blocking off a fiord that may be 70 to 80 m wide (200 to 250 ft) by using nets on both the seaward and the land side. These two nets may be up to 300 m (more than 900 ft) apart. Pumps are sometimes used to keep water flowing through the enclosures instead of relying solely on the moderate tidal action.

Growth of the salmon culturing industry has been rapid. In 1973 there were an estimated 275 MT produced. By 1974 production increased to 800 MT with a further increase to 950 MT in 1975. The production in 1976 is estimated to be more than 1500 MT. This is a significant amount when compared to the 1600 to 1800 MT of wild salmon caught in Norwegian waters.

An estimated 80% of the cultured salmon is exported to other countries. The major importing countries are France, the Federal Republic

FIG. 3.4.   UNLOADING FROZEN TRASH FISH FOR SALMON FEEDING

of Germany and Sweden. The salmon are sold fresh, iced, to Sweden and France, while about one-half of the German imports are frozen. Some of the fresh salmon sold to Germany is then smoked.

Typically harvested salmon are killed and dressed at the point of production and then go by truck or boat to a central distributor The distributor then sells domestic fish to fishmongers or retail markets. The exported salmon go to importing firms in the recipient countries. There are large efforts by producers to organize a cooperative to serve the first distributor functions.

RAINBOW TROUT *(Salmo gairdneri)*

Spawning time for rainbow trout in Norway is usually from February to April. Because of the cold water, hatching is usually in May. After hatching, the fry are raised in rearing tanks made of fiberglass using fresh water. If they are grown in warmer waters they may grow to 70 to 100 g (up to 4 oz) by October and to about 330 g by the next May. In colder waters they may only reach 10 g (0.4 oz) by October. In October some fingerlings, about 10 cm (4 in.) in length, are stocked in saltwater enclosures. The usual practice is to stock in salt water in May of the year following hatching. At this time the fingerlings are about 15 cm (6 in.) and weigh about 45 g (1.6 oz) each. The fish, whether stocked in October or May, are often harvested the following fall in October or November. Harvest weights will average between 1.0 to 2.0 kg (2.2 to 4.4 lb) after only one grow-out season. Many fish farmers produce larger trout up to 3.5 kg (7.7 lb) after 18 months in the sea. If kept longer most of the trout reach sexual maturity, which spoils both quality and growth for a considerable time.

Trout farms are generally located in salt or brackish waters along the coast. Very few of the 250 trout farms are located in fresh water, partly because of the stoppage of growth caused by lower water temperatures in winter and partly because of more fish diseases, compared with saltwater farming. Salt and brackish waters are part of the public domain and there are no fees charged by the government for use of such waters as in some other countries. According to Norwegian authorities, about 30 hatcheries produce eyed eggs or fingerlings for their own use or for sale to other fish farmers.

Of the remaining 220 fish farms, about 175 are small-scale production units producing less than 20 MT (22 ST) annually. Only about 30 farms produce over 50 MT of food fish annually.

Until the fingerlings are stocked into grow-out facilities the fish are fed pelleted food. When ready for stocking in October or May they are usually transferred to float net type enclosures (either round, square or rectangular) up to 10 m (40 ft) around or per each side. They are then continued on 40% protein pellets for one month and shifted to eating wet food. The wet food is chopped or diced trash fish with premixed vitamins added and made into a paste. Because of the adding of shrimp waste or other color components to the wet food (minced trash fish) the trouts' flesh color is red and texture is a little softer and

FIG. 3.5.   FLOATING RAINBOW TROUT PENS IN NORWEGIAN FIORD

more delicate than that of salmon. Red-colored flesh is preferred in Norway and white-fleshed trout produced in other countries on pelleted feed is not in demand. The trout production in Denmark is based solely on wet food, but with no shrimp added they get only white-fleshed trout.

There are two main reasons why wet food (small, waste or trash fish) is used in Norway: (1) the demand for red-colored flesh, which can be obtained with the proper mixture[1] of wet food but not with artificial, pelletized foods and (2) the price relationship between feeding cost of wet food versus artificial. The price of wet food generally varies between $0.055 and $0.074 per kg ($0.025 to $0.034 per lb) for waste or trash fish. With a feed conversion of 7:1, the feed cost of producing 1 kg of food fish is $0.385 to $0.518 per kg ($0.175 to $0.235 per lb). Dry feed usually costs between $0.276 and $0.368 per kg ($0.125 and $0.167 per lb). With the feed conversion of 1.8:1, feed production cost is $0.497 to $0.662 per kg ($0.225 to $0.362 per lb).

Ample natural food is available for further expansion of the cultured fish industry. At present the industry is using less than 25,000 MT of trash fish annually. It is estimated that there are between 250,000 and 300,000 MT (276,000 to 331,000 ST) of trash fish produced yearly. Hence, the cultured fish industry has enough cheap feed to expand to about 12 times its present size.

Floating net enclosures (pens) may be as small as 3 m (9 to 10 ft) square and 3 m deep. Usually there is no pumping of water; tides bring fresh water into the pens. Production per cubic meter depends on water flow rates, the area of the country and water temperature. Average production per cubic meter (34 ft^3) is 20 to 30 kg, or 1.3 to 1.9 lb per ft^3.

Placement of floating net enclosures is important. If the pens are inside a fiord where water is only slightly brackish, the fiord may freeze over in winter, thus reducing the oxygen supply. These farms harvest their fish in October and November, resulting in temporary low prices and the need to freeze some fish to avoid breaking the market. About one-third to one-half of the cultured trout is frozen for

---

[1] Red-colored flesh is obtained only by adding color components in sufficient quantities and time to the food (for instance, 1/7 shrimp waste, 6/7 trash fish). For wet food Norwegians use shrimp waste, red capelin oil (of *Mallotus villosus*) containing 50 mg astaxanthin per kg. However, artificial red pigment powder containing 10% canthaxanthin may also be added to the wet food. For dry food capelin oil and canthaxanthin is added and red-colored flesh is thereby obtained. This color or shade is particularly suitable for smoking or marinating processes. However, if insufficient color components are added the red color of the flesh can be weakened and will turn into pink when cooked. This occurs also when only canthaxanthin is added to the dry food. However, a correct balance of capelin oil and canthaxanthin secures a better and more stable color of the flesh.

later use. In addition, many trout become sexually mature at this time, with growth nearly stopping. Hence these fish are also harvested.

In 1974 market prices to trout culturists were depreed as a result of rapid industry expansion. Prices ranged from $1.50 to $2.00 per kg ($0.68 to $0.91 per lb). In 1975 trout fish prices delivered from the farm were averaging about $2.20 per kg ($1.00 per lb) in the round. Fresh fish at retail averaged about $4.96 per kg ($2.25 per lb) for gutted fish with heads on. Frozen trout at retail averaged $4.04 per kg ($1.84 per lb). A limited quantity of trout was smoked and prices averaged $14.00 to $15.00 per kg ($6.59 per lb). However, about 50% of the weight is lost during the smoking process. Unlike salmon, which is duty free in the Common Market, trout prices are mainly determined by domestic supply and demand. The present duty on trout is 12%, which helps restrict the sale of the large rainbow trout in export channels. In 1974 and 1975 the domestic market became oversupplied and prices became erratic. The situation was not aided by the existing marketing system, wherein some producers market trout directly to consumers and retail outlets while some producers market through numerous fish buyers. However, in 1975-76 there was a considerable increase in the prices of salmon, which improved the sales and exports of large rainbow trout.

In 1962 production was an estimated 200 MT (220 ST). In 1972 and 1973 the number of fish farmers increased dramatically (Fig. 3.1) and production soared to 1300 MT (1433 ST) by 1973 and increased to about 2200 MT in 1974 (2400 ST). Only about 300 MT were exported. In 1975 exports increased to 450 MT (496 ST). These were fresh or frozen and were sold to Sweden, West Germany and France. A limited amount was also sold to the USA. In 1975 production declined and was estimated at about 1800 MT total, largely because of lower stocking in October, 1974 and May, 1975 as a result of low producer prices encountered in 1974. In 1976 an increase of trout production to 2000 MT (2204 ST) was expected.

SPECIAL ACKNOWLEDGEMENTS

MR. LARS BULL-BERG, Norske Fiskeoppdretteres Forening, Rosenkrantzgt 8, Oslo 1

PROFESSOR G.M. GERHARDSEN, Institute of Fisheries Economics, The Norwegian School of Economics and Business Administration, 500 Bergen

MR. TORFINN GRAV, Institute of Marine Research, 5001 Bergen

MR. PER L. MIETLE, Director, Fisheries Economic Division, Directorate of Fisheries, Box 185-186, 5001 Bergen

DR. DAG MOLLER, Institute of Marine Research, 5001 Bergen

Chapter 4

Sweden

RAINBOW TROUT *(Salmo gairdneri)*
and ATLANTIC SALMON *(Salmo salar)*

In Sweden only salmonidaes are cultured. These include rainbow trout *(Salmo gairdneri)*, Atlantic salmon *(Salmo salar)*, sea trout (not identified), brown trout *(Salmo trutta)*, brook trout *(Salvelinus fontinalis)*, lake trout *(Salmo trutta lacustris)*, and Arctic char *Salvelinus alpinus)*. However, rainbow trout and Atlantic salmon are the only two species cultured for food consumption.

There are about 100 private fish farms raising fish for the domestic food market. These are generally small-scale enterprises varying from 0.5 to 50 MT annually and averaging about 4.0 MT of production (4.4 ST). There is a total domestic production of about 400 MT (440 ST) of cultured fish destined for immediate consumption. Of this amount, 300 MT or 75% of the total are rainbow trout and 100 MT are Atlantic salmon. Rainbow trout culture is essentially done in fresh water with some experimental work being conducted in brackish and salt water. Production is in earthen ponds, concrete raceways, tanks, nets and cages. Salmon culture is done in salt water using net enclosures.

In addition to the 400 MT of domestically cultured fish, Sweden in 1974 imported about 550 MT (606 ST) of rainbow trout with nearly all of it coming from Denmark as live and frozen shipments. Less than 100 MT (110 ST) of rainbow trout were imported from Norway. In addition, between 100 and 150 MT (110 to 165 ST) of fresh iced salmon are imported from Norway. There are no exports of any consequence; hence the total supply of cultured fish for consumption is slightly less than 1000 MT (1102 ST).

The domestically produced fish are generally sold locally by producers as live fish. Trout and salmon from Norway are iced. About one-half of the Danish rainbow trout are iced and the remainder frozen.

Rainbow trout spawn in April and May, and between 15 and 25 months are required to raise the fish to the 250 to 300 g (9 to 11 oz) fish sold by domestic producers. Rainbow trout imported from Denmark and Norway are about the same size as domestically produced trout. Imported salmon from Norway average about 5 kg (11 lb). Domestically produced Atlantic salmon take about 2 years of culturing to reach market sizes of 5 kg (see Chap. 3).

Nearly all cultured fish production comes from southern Sweden where the water is warmest and most conducive for farming activities (Fig. 4.1). Production of both trout and salmon is expected to increase as research findings on culturing these two species in brackish and salt water are adopted.

FIG. 4.1.   PRODUCTION OF RAINBOW TROUT BY REGIONS, 1972

In addition to production of cultured fish for food consumption, there are intensive efforts by the government in restocking. Major emphasis is on producing Atlantic salmon and sea trout. About 1.8 million smolts are released annually in an effort to restock the Baltic Sea. Additionally, the government restocks brown trout, brook trout, lake trout and Artic char in fresh waters for public fishing.

SPECIAL ACKNOWLEDGEMENTS

DR. OLLE LJUNGBERG, B.C., Statens Veterinäemedicinska Anstalt, National Veterinary Institute, S-104-05 Stockholm 50, Sweden

Chapter 5

Denmark

RAINBOW TROUT (*Salmo gairdneri*)

The first Danish trout farms were established about 1890 in Jutland, the western portion of Denmark, which is attached to the European continent (Bregnballe 1963). In the beginning brown trout (*Salmo trutta*) was the dominant species cultured. Within a short period of time rainbow trout were introduced and soon became predominant. More recently rainbow trout has become the exclusive food fish cultured.

The Danish trout industry was the first in Europe to become established and grow to maturity. Other countries are still rapidly increasing the number of fish farms and volume of production. However, by 1961 there were already 500 to 525 trout farms in operation in Denmark; only 6 of them outside Jutland. In 1975 there were an estimated 530 trout farms, indicating very little change in numbers over a 15 year period. The output of rainbow trout has increased, but at a much slower rate than for other European countries such as Norway, France, Italy, Spain, Switzerland and West Germany. In 1962 export of rainbow trout from Denmark was 7781 MT (8575 ST). By 1972 it reached 14,600 MT (16,100 ST) and then began to decline. In 1974 output was below 13,000 MT (14,265 ST). In 1975, 14,763 MT (16,269 ST) were exported.

Of the 1975 total of 530 rainbow trout producers, about 150 produced eggs, fry and fingerlings only. In general these producers are located nearest the headwaters of small streams and rivers west of the Jutland ridge area. Production is mainly concentrated in the western and central part of Jutland (Fig. 5.1).

Of the 530 producers all but 3 produce trout in fresh water. Three producers with a total production of about 200 MT (220 ST) are loca-

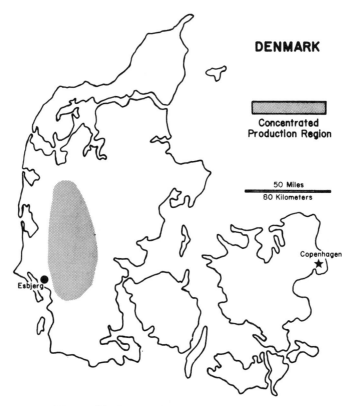

FIG. 5.1.   CONCENTRATED FISH CULTURING AREA, 1975

ted in Ringkobing Fiord North of Esbjerg. This production is in brackish water with a salt content of about 1 ppm.

A typical Danish trout farm is constructed in the following way:
the water in a river or brook is dammed and led through two inlet
channels to two rows of parallel rectangular ponds; from the ponds
the water can be released into the outlet channel and flow back into
the river. The outlet channel is provided with a fish screen and is
also used for trout production. The water is used first in the ponds
and then in the outlet channel. Though each pond receives only a
relatively small water supply, the channel receives water from all
ponds (Fig. 5.2 and 5.3). All ponds are earthen and the channels are
also excavated from earth. Inlet and outlet pipes are made of wood
or P.V.C. A middle-sized Danish trout farm will have 35 to 60 ponds.

There is no net or cage culture in Denmark. All culturing of
grow-out fish is in earthen ponds which are usually about 300 m²
(2916 ft²). The typical pond measures 30 m (97 ft) long by 10 m (39
ft) wide and is from 0.5 to 1 m (18 to 39 in.) deep. The water exchange

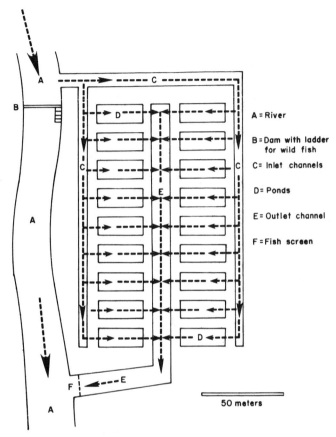

FIG. 5.2.   TYPICAL TROUT FARM, 1975

A = River

B = Dam with ladder
for wild fish

C = Inlet channels

D = Ponds

E = Outlet channel

F = Fish screen

50 meters

rate is three times daily. Each pond produces about 2.5 MT of fish yearly (1 lb per ft^3). This is far less than could be produced in a raceway system. However, water temperatures in the winter approach $0°C$ ($32°F$) and at times small streams are partially frozen and water flow may stop for several hours. With less fish per unit of water flow and water volume in ponds, there is still sufficient "oxygen reserve" to keep the fish alive until the water flow is reestablished. This would not be true if raceways and denser stocking were followed.

Even with this less intense stocking rate in earthen ponds, many fish farmers mechanically aerate their ponds in the summer and recirculate part of the water back through the ponds.

Concrete tanks are used in Denmark, but only for special purposes. Whirling disease (caused by the sporozoan *Lentospora cerebralis*) was a real problem during the 1950s. This problem was

FIG. 5.3.   TYPICAL TROUT FARM

FIG. 5.4.   FLOATING AERATOR IN TROUT POND

FIG. 5.5.   DRYING AND CLEANING EARTHEN TROUT POND
FOR DISEASE CONTROL

solved by keeping the fry in concrete tanks until they reached a length of 5 cm (2 in.). These fingerlings could then be kept in earthen ponds infected with *Lentospora* without being damaged by the parasite.

As stated earlier, the Danish rainbow trout industry has probably reached its peak in production (Table 5.1). Less than 10% is consumed domestically. The remaining 90% of production has been exported.

Though slightly less than 10% of domestic production was consumed internally during the 1971-74 period, this was still a considerable increase over 1962 when less than 2% of production was consumed internally (Fig. 5.6). Denmark is essentially a nation of saltwater fish consumers. This custom is probably perpetuated by the fact that the price of cultured freshwater trout is higher than that of wild, saltwater fish.

The leading and increasingly important trout export market is Germany. In 1973 and 1975 it purchased between 56 and 65% of all exports (Table 5.1). Other importing countries in order of importance are: (1) Belgium-Luxembourg, (2) Great Britain, (3) Switzerland, (4) Sweden, (5) France, (6) Austria and (7) Holland. Minor volumes are also exported to Italy, Finland, Norway and other countries (Fig. 5.7).

Comparison of 1962 data for exports and total production with 1975 data is very revealing. (1) There were considerable increases in exports to West Germany and Belgium-Luxembourg. (2) Exports to Italy declined from over 1800 MT to almost nothing. (3) Exports to the USA and Canada completely vanished. These shifts are explained by (1) Italy increasing rainbow trout production from that of a deficit nation to a position of exportation, and (2) restrictions on importation of trout into the USA and Canada to prevent the spread of European fish diseases to North America — especially Whirling disease (*Myxosoma* [*Lentospora*] *cerebralis*). Losing these markets, the Danes shifted to the nearer markets of West Germany and Belgium.

Since 1971 exports to Great Britain have been slowly declining as production has increased in that country. Switzerland has been shifting to imports from Italy, and Sweden to imports from Norway.

Of the maximum 10% of domestically consumed trout about 30% is smoked. The remainder is sold to local restaurants where the fish are generally boiled or broiled, or directly to the consumer for boiling, broiling or frying in the home.

Market sizes of trout range from 180 to 350 g (6.5 to 12 oz). The sizes from 180 to 250 g are sold as portion fish; those from 250 to 350 g are often smoked.

In 1975, 43% of the export trout were sold live, 37% were frozen and 21% were sold iced. A normal shipment of live trout is about 2400 kg

TABLE 5.1

PRODUCTION AND EXPORTS OF RAINBOW TROUT, DENMARK

Country	(MT)					
	1962	1971	1972	1973	1974	1975
West Germany	789	5113	6617	7092	6985	9582
Belgium-Luxembourg	948	2096	2492	2157	1754	2231
Great Britain	1077	1459	1485	1164	991	876
Switzerland	508	1001	1160	985	838	778
Sweden	846	771	726	573	426	424
France	127	200	236	250	279	448
Austria	42	151	151	ˑ121	129	128
Holland	86	106	96	136	148	164
Italy	1804	43	79	43	60	58
Finland	68	87	114	60	17	1
Norway	78	25	14	0	30	14
USA	967	—	—	—	—	—
Canada	168	—	—	—	—	—
Others	149	300	150	99	111	60
Total exports	7657	11,352	13,320	12,680	11,768	14,764
Total production	7781	12,600	14,600	13,950	12,945	?

Source: Anon. (1971-1974); Anon. (1975); Bregnballe (1963).

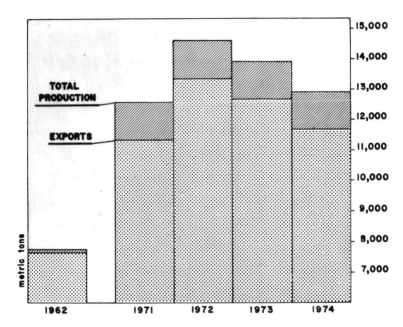

FIG. 5.6. TOTAL PRODUCTION, EXPORTS AND DOMESTIC CONSUMPTION
OF RAINBOW TROUT IN DENMARK, 1962, 1971-74

(2.6 ST) per truck. When a trailer is attached a load can contain as much as 4800 kg (5.3 ST) of live fish.

The usual spawning season is January-March, but there are certain strains that produce eggs in the November-December period. Hatching time varies depending on water temperatures. This is usually calculated as 300 to 320 day degrees. Each 1°C of temperature is calculated as 1 degree day. Hence, if hatching water is a constant 10°C (50°F) each day counts as 10 degree days and it would take 30 to 32 days for hatching. After hatching it takes 1 to 1.5 years to grow a 200 g trout, depending on water temperatures. Most fry are hatched in April and after consumption of the yolk sac are ready to feed in May. Fry are then reared in concrete tanks until June-July and then, at about 6 cm (2 to 2.5 in.), are transferred to earthen ponds. They can be sold the following April-July period or later.

Very few fish farmers produce eggs only. The usual custom is to produce eggs and raise them through fingerling or food sizes. However, there are a number of producers who purchase fingerlings for grow-out operations only. Egg and fingerling producers are usually the farms nearest the headwaters of streams or small rivers where water quality is best. They, in turn, sell fingerlings to grow-out operations located further downstream on the same river.

DANISH EXPORTS
OF RAINBOW TROUT
1974

(in metric tons)

FIG. 5.7.   EXPORT SALES OF DANISH RAINBOW TROUT TO
EUROPEAN COUNTRIES, 1974

Nearly all fry after hatching are grown in concrete tanks. Dry
pelletized food of 43% protein is fed until the fry reach 6 cm (2 to 2.5
in.). They are then transferred to earthen ponds and gradually shifted to a wet diet, which is composed of fresh minced trash fish obtained from the marine fishing fleet. Most marine fishing is near the
coast and the fleet returns daily, six days per week. From the ports
all trout ponds are within 2 to 4 hr driving time. The minced fresh
fish, essentially herring, sand eels, whiting and sperling, are made
into a "slurry" with alginate added as a binder. Trash fish in 1975
averaged about $0.075 per kg ($0.035 per lb), not including transportation costs, binder or mixing costs. With a feed conversion of 5:1,
the cost of adding 1 kg of weight to cultured fish was about $0.375
($0.175 per lb). In Denmark this is a cheaper method than feeding
commercial fish pellets. Denmark is the only country in western
Europe (possibly the only one in the world) where fresh wet food is
fed. This is possible only because of the fishing fleet returning to port

daily and the short haul from the ports to cultured fish production areas. Norway feeds a wet food but, because the fishing fleet ranges further offshore, there is not a daily supply. Hence, the trash fish are frozen into blocks for purchase by fish culturists (see Chap. 3). As a result of the diet, Denmark's rainbow trout are considered as "fat" fish. This is believed to give them a superior flavor. The trout's flesh is white instead of pink or red.

FIG. 5.8.   TANK FOR FEEDING MINCED TRASH FISH TO RAINBOW TROUT

Average price received at the farm by fish farmers in 1975 was $1.64 per kg ($0.75 per lb). Very efficient producers are reported to have a total production cost as low as $1.05 per kg ($0.48 per lb), while the average producer was reported to have production costs of $1.35 per kg ($0.61 per lb).

There are about 12 fish buyers in the concentrated area of production. The largest of these, Dantrout, a farmer's cooperative, buys about one-third of total production. The remainder is purchased by other fish dealers, some of whom are large fish producers themselves. About one-half (five) of these other fish buyers also process and freeze trout. These 12 large buyers export about 95% of Danish trout. However, in 1974 there were 60 other buyers who exported the re-

FIG. 5.9.   EARTHEN TROUT PONDS SHOWING LIMITED WATER USE

FIG. 5.10.   MECHANICAL FEEDER ON TRACKS FOR
FEEDING MINCED TRASH FISH TO TROUT

FIG. 5.11.   EQUIPMENT FOR MINCING TRASH FISH FOR
TROUT FEEDING

FIG. 5.12.   UNLOADING SIZED FISH INTO TROUT POND

maining 5%. Many of these were red meat sellers who included a small amount of fish in their total shipments.

One of the major costs of marketing Danish trout is transportation. With 42% of all trout sold being delivered live, delivery cost of live fish is very important. The average size truck hauls 2.4 MT of live fish. With a trailer attached, 4.8 MT can be hauled. For a 500 km (310 mi.) one-way trip the cost is $725. This amounts to $0.15 per kg ($0.068 per lb). Fish shipped on ice is more economical. About 15 MT of fish can be transported per load and transportation cost drops to $0.048 per kg ($0.022 per lb). Frozen fish is still cheaper to transport since there is no ice or water to haul.

FIG. 5.13.   PREDATOR BIRDS CAUGHT IN PROTECTIVE NETTING OVER TROUT PONDS

Reported prices for 1975 include eyed eggs at $3.03 per 1000. Fry sold for $13.47 per 1000. These were 6 cm (2.5 in.) in length. Fingerlings 12 to 15 cm (5 to 6 in.) sold for $0.051 each.

There is no record of the number of rainbow trout eggs sold through export channels. The numbers are certainly in excess of 100 million eggs, however. Nearly every country in western Europe imports some eggs. In addition, limited quantities are sold to countries in eastern Europe.

No rainbow trout are stocked in public waters since they are considered a foreign species. Some, however, are stocked in old

FIG. 5.14.   TANK TRUCK FOR HAULING LIVE TROUT TO PROCESSING PLANT

sand and gravel pits for use as fee fish-out ponds. The number of these is small and they are operated almost solely for use by German tourists during the summer.

Outlook

Forecasts of future Danish rainbow trout production by various knowledgeable individuals range from no increase beyond the present 15,000 MT of production, to a possible increase of 12% to the 16,800 MT level. The latter is optimistically forecast, assuming effective disease control measures will reduce losses, and pollution control and water use standards do not become more restricting. Such restrictive measures would result in higher capital costs for cleanup techniques and equipment to treat efflux from present fish facilities, and would reduce production by an amount equal to gains expected from disease control measures now being instituted.

BROWN TROUT (*Salmo trutta*)

Brown trout was the original species of trout living in Danish waters. It was also the first trout cultured in Denmark. In recent years no brown trout have been raised as food fish. However, brown trout is still cultured for restocking domestic waters and for export. Most restocking of domestic waters is with 12 to 15 cm (5 to 6 in.) fingerlings. Restocking is carried out by fishery clubs and by those producers who are required to furnish fingerlings to the river as payment for use of river water.

About 15 million brown trout fry are sold to Germany for restocking. Estimates of total brown trout fingerling production for domestic use and export sale is 50 MT annually. This means that about 1.1 million brown trout fingerlings are produced annually. In addition, 50 to 100 million eggs are exported to various European countries and probably 100 MT (110 ST) of trout from 100 to 200 g (4 to 8 oz).

EUROPEAN EELS (Anguilla anguilla)

No eels are cultured in Denmark. The water is too cold for rapid growth and the estimated cost of production is higher than anticipated returns. However, some elvers are captured for restocking of lakes to increase fishing catches. The elvers are normally caught in April and May on the North Sea coast of Jutland. In addition there

FIG. 5.15.   MECHANICAL TROUT SIZING IN PROCESSING PLANT

is commercial fishing for wild silver eel. A comparison of the catch of silver eel over time is:

1948	4242 MT
1958	3287 MT
1969	3624 MT
1970	3309 MT

These figures suggest that the eel catch may be declining. This could be due to less fishing effort as well as pollution effects and natural conditions.

Silver eels are normally caught in the August 1-November 1 period. They are captured as they pass through the narrow straits of the Baltic Sea going into the North Sea. Nearly all of the silver eels are consumed domestically, smoked or fried. The limited volume exported is shipped live to Holland, Belgium and Germany.

SPECIAL ACKNOWLEDGEMENTS

DR. F. BREGNBALLE, Director, Danish Trout Research Station, Brøns, Denmark

DR. N.O. CHRISTENSEN, Den KgL. Veteriner - Og Landbøbojskole, Ambulatorisk Klinik, Bulowsvej 13, 1870 Købehaun V, Copenhagen, Denmark

MR. ERIK HANSEN, Dantrout, Brande, Denmark

DR. N.P. KEHLET, Veteranian Officer, 7361 Ejstruphohm, Denmark

REFERENCES

ANON. 1975. Bulletin for members of the freshwater fishing association. The Journal of the Freshwater Fishing Industry. Danish Association, Viborg, Denmark (Danish)

ANON. 1976. European Federation of Salmonoids — Breeders (FES), Treviso, Italy, 1971 - 1974.

BREGNBALLE, F. 1963. Trout culture in Denmark. Progressive Fish-Culturist 25, No. 3, 115-120.

Chapter 6

Netherlands (Holland)

In Holland several species of fish are cultured. These are: (1) roach (*Rutilus rutilus*), (2) pike-perch (*Lucioperca lucioperca*), (3) carp (*Cyprinus carpio*), (4) pike (*Esox lucius*), and (5) rainbow trout (*Salmo gairdneri*). All, with the exception of 50 to 60 MT of rainbow trout, are cultured for restocking or stocking of ponds, lakes, streams and canals.

Four different cultural systems are used. These are: (1) pond culture, (2) raceways for trout, (3) controlled conditions in glass houses where water temperature and oxygen levels are maintained artificially, and (4) cage culture in the warm water discharge of a power plant. In the cage culture system, cages are 6.5 m³ (229 ft³) in size, with each one producing 1625 kg of fish (3575 lb or 15.6 lb per ft³). In this operation there are 48 cages.

The fish are released into public waters at various sizes: roach at a 15 cm minimum size (6 in.); pike-perch at 10 cm in length (4 in.); carp as either 2-year-old fish of 25 cm length (10 in.) or as 3-year-old fish of 40 to 45 cm (16 to 18 in.); pike as either 5 or 10 cm fish (2 to 4 in.); and rainbow trout at either 25 cm (10 in.), the legal size for anglers, or at 200 g (7 to 8 oz).

Between 20,000 and 50,000 roach are stocked annually. Stocking of pike-perch varies between 70,000 and 1.5 million; carp between 150,000 and 200,000; pike about 1.5 million and trout about 150,000 annually. Total production of these stocking fish approximates 200 MT annually for angling by about 1.25 million sports fishermen. All these fish are raised and released under government auspices. As mentioned earlier, between 50 and 60 MT (55 to 66 ST) of rainbow trout are raised as food fish by one commercial fish grower.

In addition, the commercial catch of eel (*Anguilla anguilla*) var-

FIG. 6.1.   USING WARM WATER DISCHARGE FROM AN
ELECTRICAL POWER PLANT FOR REARING RESTOCKING FISH

ies between 823 and 875 MT (1971-73 data) (907 to 964 ST). In 1973 there were 3500 MT of eel imported. These came from Denmark, Turkey, the USA and other countries. These were received alive, dressed and smoked. About 800 MT (882 ST) of smoked eel were then exported to various western European countries.

Imports of trout and salmon were estimated at 1000 MT in 1973 (1102 ST). These were either fresh or smoked. About 40% of these fish were, in turn, exported to other European countries, generally as a smoked product.

Available data concerning rainbow trout imports and exports suggest that Holland is similar to Denmark in that consumption of rainbow trout is very low. In 1973 only 137 MT (151 ST) were imported—mainly from Denmark with a few tons coming from Germany. In 1974 only 148 MT (163 ST) were imported with all of it coming from Denmark. Negligible quantities of 5 to 10 MT (5.5 to 11 ST) annually are transshipped to Germany and France.

SPECIAL ACKNOWLEDGEMENTS

DR. H.O. HEERMA VAN, University of Utrecht, Transitorium 2, Utrecht, Netherlands

PROF. DR. E.A. HUISMAN, O.V.B. Viskwerery, Karperweg 8-10, Lelystrad, Netherlands

MR. and MRS. JACK VAN DEN BRINK, Grevenhofsweg 27, Harderwijk, Netherlands

MR. D.E. VAN DRIMMELEN, Chr., Organization for Improvement of Inland Fisheries, Stadhoudenslaan 53, Utrecht, Netherlands

Chapter 7

Belgium and Luxembourg

RAINBOW TROUT (*Salmo gairdneri*)
and BROWN TROUT (*Salmo trutta*)

In 1975 there were 18 rainbow and brown trout fish farms in Belgium and none in Luxembourg. Total production of food fish was estimated to be only 300 MT (331 ST) of rainbow trout annually.

Rainbow and brown trout are produced in both pond and raceway production units. Most rainbow and brown trout eggs are imported from Italy, Denmark and France. Some fingerlings are also imported. All of the brown trout are destined for restocking for angling in private and public waters, as either fingerlings or small adult fish. Some catchable-size rainbow trout are also stocked in public waters for angling. However, most of the rainbows are destined for the table as food fish.

Available data indicates that Belgium and Luxembourg are large and growing importers of rainbow trout. In 1968 an estimated 2350 MT (2590 ST) were imported. By 1974 imports reached 3619 MT (3988 ST). The major sources of supply are Denmark, France and Italy (Table 7.1 and Fig. 7.1). Minor amounts, less than 100 MT (110 ST) annually, are also imported from Germany. Amounts varying from 65 to 231 MT (71 to 255 ST) are imported from Japan.

Belgium-Luxembourg is thought by many individuals concerned with European cultured fish to be a large transshipper of trout. In other words, food fish are thought to be imported into the two countries for exportation outside the area. However, available data for 1970, 1973 and 1974 indicate that less than 13% of imported and domestically raised trout is exported. In 1970 only 255 MT were exported. The volumes for 1973 and 1974 were 170 and 521 MT, respectively. Export sales were essentially to France and Germany with small

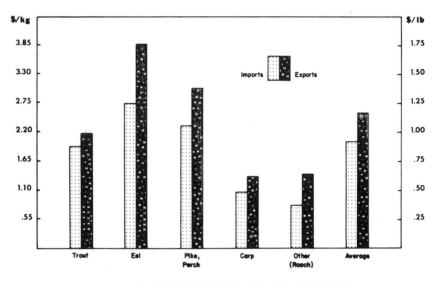

FIG. 7.1.   VALUES OF FRESHWATER FISH—IMPORTS AND EXPORTS

amounts sold to Holland.

In 1972, of the total imports of 3175 MT (3499 ST), 1642 MT (51.3%) of living rainbow trout were transported in water, 930 MT or 29.0% were ice packed fresh fish, and 628 MT or 19.7% were deep frozen fish from both Denmark and Japan. These data indicate the market demand for nonfrozen fish (more than 80%). Data were not available to determine if makeup of the 1972 imports is the general rule or whether it pertains only to 1972. Available data indicate that frozen trout

TABLE 7.1

IMPORTS OF RAINBOW TROUT INTO BELGIUM-LUXEMBOURG, 1968-74 [1]

				Metric Tons			
Origin	1968	1969	1970	1971	1972	1973	1974
Denmark	[1]	[1]	[1]	2087	2433	2141	1754
France	[1]	[1]	[1]	294	311	433	1161
Germany	[1]	[1]	[1]	53	75	56	[1]
Japan[2]	186	231	112	65	143	222	153
Italy	[1]	[1]	[1]	7	213	252	704
Totals	2350	2350	2234	2502	3175	2882	3619

Source: Anon. (1971-74); Anon. (1968-74).
[1] Data obtained from European Trout Federation Office, Treviso, Italy.
[2] Japanese data secured from Japan Fisheries Agency, Tokyo, Japan.

importing prices and retail prices are 7 to 8% higher than for live or ice packed fish.

In 1974 a total of 6371 MT (7021 ST) of freshwater fish were imported into Belgium-Luxembourg. These fish had total value of over $12.5 million (Table 7.2). Major imports, in order of tonnage and value, were: (1) trout, both brown and rainbow, (2) eel and (3) other, mostly roach.

The 3686 MT of trout imported were divided into about 3619 MT (4062 ST) of rainbow trout and about 67 MT (74 ST) of brown trout. Imports were valued at $6.9 million and exports at $1.1 million.

CARP (Cyprinus carpio)

In 1975 there were 19 culturalists of cyprinids such as carp (Cyprinus carpio), tench (Tinca tinca) and roach (Rutilus rutilus), with small quantities of pike (Esox lucius) produced by some.

All carp were produced in earthen ponds. Production was essentially for sport fishing in private angling club ponds or streams. However, a few tons were sold immediately for food.

The supply of domestically raised carp was estimated at less than 100 MT annually. In addition, 213 MT were imported. Most of these imports were from France. Thus, total supply of carp was about 313 MT (345 ST). Forty-seven percent of the total supply or 146.5 MT were exported (Table 7.2). These were nearly all destined for West Germany.

TENCH (Tinca tinca) and
ROACH (Rutilus rutilus)

Tench and roach are cultured extensively in earthen ponds by some of the 19 culturalists of cyprinids. Little or no feeding of the fish is done, and often the ponds are not even fertilized. Production is between 200 and 300 kg per ha (180 to 270 lb per acre). Total supply of domestically raised tench and roach is probably less than 200 MT. To this supply must be added the 710 MT of imported fish (Table 7.2). Hence, the total supply is about 800 to 900 MT (880 to 990 ST) annually.

Only a few tons of this amount are destined for immediate food consumption. Essentially these fish are destined for restocking. Since there is no state restocking, all restocking of angling waters is under the auspices of angling and local fishing clubs. Fish are restocked at catchable sizes of 250 to 300 g (9 to 11 oz).

TABLE 7.2

IMPORTS, EXPORTS AND VALUES OF FRESHWATER FISH FOR BELGIUM-LUXEMBOURG, 1974

Species	Imports (MT)	Exports (MT)	Value in U.S. Dollars for Imports			Value in U.S. Dollars for Exports		
			(total)	(per kg)	(per lb)	(total)	(per kg)	(per lb)
Trout, brown and rainbow	3,685.8	520.5	6,916,383	1.87	0.85	1,121,481	2.15	0.98
Eel	1,711.6	414.3	4,708,572	2.75	1.25	1,609,815	3.89	1.77
Pike and perch	50.5	10.6	116,662	2.31	1.05	32,057	3.02	1.37
Carp	212.8	146.5	222,800	1.05	0.48	189,424	1.29	0.59
Other (mainly roach)	710.4	116.1	573,928	0.81	0.37	162,516	1.40	0.64
Totals	6,371.1	1,208.0	12,538,345	—	—	3,115,793	—	—

PIKE (Esox lucius) and
PIKE PERCH (Lucioperca lucioperca)

There is no separate culture of pike or pike-perch in Belgium-Luxembourg. However, most carp producers stock 10 to 20 fingerlings of each specie per ha (4 to 8 per acre) in their carp ponds. At the end of the first or second summer of growth for the carp, pike and pike-perch are also harvested. Total annual production of pike and pike-perch is probably not over 20 MT. To this amount one can add the 50.5 MT imported (Table 7.2). Total supply is thus around 70 MT. About 10 MT are exported so total available supply is about 60 MT (66 ST).

These fish are stocked at sizes ranging from fingerlings to food sizes. Pike in particular are stocked in waters overpopulated or threatened to be overpopulated by native fish. The pike, a voracious fish, keeps these waters in balance by consuming most of the fry and fingerlings of these fish. Only small quantities, perhaps less than 2 to 3 MT, are consumed as a food fish.

EUROPEAN EEL (Anguilla anguilla)

There is no eel culturing in Belgium-Luxembourg. However, between 300 and 400 kg (660 to 880 lb) of elvers and young black eel are captured in the rivers for restocking in more fertile inland waters. An estimated 100,000 such fish are restocked annually.

About 50 MT of eels are captured annually in native waters. In 1974 an additional 1712 MT were imported. Most of these were live eels from France. Hence, total supply was about 1762 MT. Of this

TABLE 7.3

CATCHES OF WILD FISH BY ANGLERS IN PUBLIC WATERS, BELGIUM
AND LUXEMBOURG, 1972-74

Species	Metric Tons 1972	1973	1974
Brown trout (Salmo trutta)	60.1	64.2	63.4
Roach (Rutilus rutilus)	135.6	152.6	142.3
Eel (Anguilla anguilla)	49.4	50.0	50.5
Carp (Cyprinus carpio)	28.8	29.0	27.9
Pike (Esox lucius)	22.3	21.7	21.9
Tench (Tinca tinca)	9.4	8.6	10.2
Pike perch (Lucioperca lucioperca)	1.0	1.1	1.5
Others[1]	138.9	138.3	129.6
Totals	445.5	465.5	447.3

[1] About one-third of all others was common bream (Abramis brama).

amount, 414 MT were re-exported as live or smoked eels to Holland and West Germany. Hence, the total domestic consumption was about 1348 MT (1485 ST).

The total value of eel imported in 1974 was $4.7 million, while exports amounted to $1.6 million. Value of imported eel was $2.75 per kg, while value of exports was $3.89 per kg (Table 7.2). The difference in values is accounted for by the fact that a higher proportion of exported eel are smoked and have a higher value per unit of sale (Fig. 7.1).

RECREATION FISHING

As previously mentioned, there is no state restocking of public waters. Restocking fish come almost solely from fish raised by private fish farmers or imports. Sales are made to local angling clubs. The most important wild fish caught by anglers in public waters is roach, followed by brown trout, eel, carp, pike, tench and pike-perch (Table 7.3). The total wild catch annually is about 450 MT (496 ST).

TOTAL SUPPLY

The total supply of freshwater fish annually in Belgium-Luxembourg is about 5307 MT (5848 ST). This is composed of 300 MT of rainbow trout produced domestically, plus 3094 MT of net imports, a few tons of brown trout, perhaps 100 MT of carp, a few tons of tench and roach, a few tons of pike and pike-perch and 1348 MT of eel. To this volume of 4857 MT must be added the wild freshwater catch of 450 MT.

OUTLOOK

Future production can be expected to be relatively static. No large increases in production can be foreseen. This is largely due to inadequate volumes of suitable water for fish culture.

SPECIAL ACKNOWLEDGEMENTS

PROF. MARCEL HUET, Station de Recherches des Eaux et Forets, Groenendaal, 1990 Hoeilaad, Belgium

MR. J.A. TIMMERMANS, Station de Recherches des Eaux et Forets, Groenendaal, 1990 Hoeilaad, Belgium

Chapter 8

Federal Republic of Germany (West Germany)

In Germany, six species of fish are cultured for food. These are: (1) rainbow trout *(Salmo gairdneri)*, (2) brown trout *(Salmo trutta)*, (3) carp *(Cyprinus carpio)*, (4) tench *(Tinca tinca)*, (5) eel *(Anguilla anguilla)* and (6) pike *(Esox lucius)*.

RAINBOW TROUT *(Salmo gairdneri)* and BROWN TROUT *(Salmo trutta)*

About 5300 MT (5841 ST) of trout are produced annually in West Germany by 1400 to 1500 enterprises. The major trout areas are: (1) Baden-Württemberg, with 31% of total production, and (2) Bavaria, with 27% of total production (Fig. 8.1). Nearly all of these enterprises produce trout in earthen ponds with running water. Only a few producers use concrete raceways. In 1975 there were between 30 and 50 cage producers, who produced between 10 and 50 MT each. There were numerous other producers having only one or two cages, who produced 1 and 2 MT annually as a hobby. Cage culture is usually in old gravel pits containing 10 to 40 ha (25 to 100 acres) of surface water and being from 8 to 15 m deep (25 to 50 ft). Cages range from 50 to 70 m³ (1776 to 2472 ft³) and produce about 1 MT annually. The number of cage producers is increasing.

Cage culture is year round. In summer the water is cold enough for feeding and growth. Swimming activity of the fish keeps the water in and around the cages from freezing in winter when the pond or lake freezes over.

In earthen ponds the usual calculation is 1 liter of water flow per second for 100 kg of annual production. This means that 14 lb of fish are produced per gallon of water flow per minute.

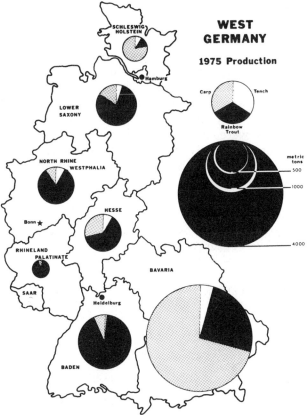

FIG. 8.1.   PRODUCTION OF CULTURED FISH, BY SPECIES, BY REGIONS,
WEST GERMANY, 1975

It is estimated that less than one-half of the rainbow trout eggs are produced domestically. Most remaining eggs are imported from Denmark. Domestic eggs are produced in the September-May period.

German trout are sold in portion sizes of 250 to 300 g (9 to 11 oz) for cooking. About 70% of all trout is sold for this purpose. About 30% of all trout by weight is sold as 300 to 500 g (11 to 18 oz) for smoking. Producers try to sell their production locally within a 50 to 80 km (30 to 50 mi.) radius. Sales are to restaurants or individual families. The 1975 price was about $3.00 per kg ($1.36 per lb). Domestic trout which cannot be sold through this channel must be sold at wholesale. The wholesale price is determined by live or frozen imports and the 1975 price was $1.91 per kg ($0.87 per lb), which was only about 64% of the live domestic price. Some of the live imports are fish that weigh less than the 250 to 300 g sizes in demand, and which are stocked by producers and fed-out until they reach market size and then resold.

Increasing amounts of the larger (300 to 500 g) fish are being produced for smoking. Retail prices of smoked fish averaged about $7.80 per kg ($3.54 per lb) in 1975.

One unique German cooking feature is that nearly all fresh or frozen fish are boiled or steamed. This is one of the few countries in western Europe that does not fry or grill trout.

In 1972 a study was made of the trout industry for West Germany (Anon. 1972). The study indicated a total of 1465 salmonoid producers (Table 8.1). These were classified by size of pond area. Pond area was construed to mean not only the water surface area but also the surrounding land which the farmer considered to be an integral part of the pond area.

TABLE 8.1

NUMBER OF SALMONOID ENTERPRISES, NUMBER OF PONDS, AND TOTAL
WATER AREA BY SIZES OF ENTERPRISES; WEST GERMANY, JUNE, 1972

Pond Area (ha)	Enterprises		Ponds		Total (ha)	Water Area	
	(no.)	(%)	(no.)	(avg)		Avg Each Pond (ha)	(acres)
Under 1	911	62.2	3,781	4.2	130	0.03	0.08
1-2	259	17.7	2,439	9.4	139	0.06	0.14
2-5	189	12.9	2,567	13.6	203	0.08	0.20
5-10	55	3.7	849	15.4	101	0.12	0.29
10-20	26	1.8	479	18.4	70	0.15	0.36
20-50	15	1.0	199	13.3	47	0.24	0.58
50 and over	10	0.7	134	13.4	24	0.18[1]	0.44
Totals and Averages	1465	100.0	10,448	7.1	713	0.07	0.17

[1] Pond size was smaller than for farms having a pond area of 20 to 50 ha (49 to 124 acres). Evidently these farms had a larger land area and a smaller water area.

There was a wide array of salmonoid producers by size of enterprise as measured by "pond area." Over 62% of the enterprises had less than 1 ha (2.47 acres) of "pond area." This group averaged 4.2 ponds each, with an average size per pond of 0.03 ha (0.07 acre). The total water area for each of these farmers was 0.13 ha (0.29 acre). At the other extreme were 15 farmers with 18.4 ponds each and with each pond containing 0.24 ha (0.59 acre). These larger producers had 4.42 ha (10.9 acres) of water area. The average salmonoid farmer had 7.1 ponds with each pond covering 0.07 ha (0.16 acre), for an average total water area of 0.50 ha (1.24 acres).

These 1465 enterprises contained 10,448 separate ponds, having a total water area of 731.4 ha (1807 acres). A complete estimate of total annual production of food-size fish was not obtained for all 1456 enterprises. Production data were obtained for 1386 enterprises with total annual output of 3446 MT (3797 ST). If it can be assumed that the missing enterprises had the same average output as those reporting, then annual production would have been 3642 MT (4013 ST). In addition, several hundred MT are produced in concrete raceways, some 1200 MT in cages and several hundred more MT by hobbyists. When added together, allowing for some growth in the industry between 1972 and 1975, we arrive at the 1975 level of 5300 MT (5841 ST).

For 1386 salmonoid enterprises of a total 1465 there were 139,630,000 eggs produced, 92,383,000 fry, and 32,093,000 fingerlings (Table 8.2). With an average market weight of 275 g for the two-summer fish, there was a total of 12,406,000. Thus, there were 7.4 fry and 2.6 fingerlings for each fish marketed at the end of the second summer of growth.

TABLE 8.2

NUMBER OF SALMONOID ENTERPRISES BY SIZE AND TYPE OF PRODUCTION; WEST GERMANY, JUNE, 1972

Pond Area (ha)	Enterprises (no.)	Eggs (000's)	Fry (000's)	One-summer Fingerlings (000's)	Two-summer Food Fish (MT)
Under 1	843	11,586	11,039	5,904	731
1-2	248	29,887	18,013	7,026	881
2-5	185	47,461	36,007	9,270	1,013
5-10	57	15,501	6,704	3,841	433
10-20	25	15,850	11,050	3,666	277
20-50	17	7,055	961	1,330	67
50 and over	11	12,290	8,608	1,056	84
Totals	1,386	139,630	92,382	32,093	3,446

On 1465 salmonoid enterprises in West Germany in 1972, rainbow trout were dominant, with nearly 75% of the total fry, nearly 83% of the total one-summer fingerlings, and nearly 95% of the two-summer or food-size fish (Table 8.3). The remainder was brown trout.

Rainbow trout are raised almost exclusively for food fish or for stocking fee fish-out ponds. The fish used for stocking fee fish-out ponds are food-size (two summers). Hence, the data shown in Table 8.4 are really an indication of survival. This table shows that out of every 100 fry there are only 43.1 fish at the end of the first year of growth, and only 19.3 fish reach market size of approximately 275 g (10 oz). Brown trout and other salmonoids are used almost exclusively for restocking angling waters. Most brown trout are restocked at the end of the first summer of growth, as shown by the decline from 37.8 finger-

TABLE 8.3

NUMBER AND DISTRIBUTION OF SALMONOIDS BY STAGE OF GROWTH;
WEST GERMANY, JUNE, 1972

Stage of Growth	Species of Trout							
	Rainbow		Brown		Other[1]		Totals	
	(000's)	(%)	(000's)	(%)	(000's)	(%)	(000's)	(%)
Fry	61,678	74.9	11,012	13.4	9,692	11.7	82,382	100.0
One-summer (Fingerlings)	26,564	82.8	4,166	13.0	1,362	4.2	32,092	100.0
Two-summer[2] (Food size)	11,889	94.8	465	3.7	184	1.5	12,538	100.0

[1] Brook, lake, etc.
[2] Assumed to be 275 g fish (10 oz).

ling fish to 4.2 food-size fish. Some two-summer fish are also stocked. Other salmonoids such as brook or lake trout *(Salvelinus fontinalis* and *Salmo trutta lacustris)* are commonly stocked in angling waters as fry, or at the end of the first summer of growth (Table 8.4).

TABLE 8.4

NUMBER OF ONE-SUMMER AND TWO-SUMMER SALMONOIDS PER 100 FRY BY
SPECIES OF SALMONOID: WEST GERMANY, JUNE, 1972

Stage of Growth	Species of Trout		
	Rainbow	Brown	Other[1]
Fry	100.0	100.0	100.0
One-summer (Fingerlings)	43.1	37.8	14.1
Two-summer[2] (Food size)	19.3	4.2	1.9

[1] Brook, lake, etc.
[2] Assumed to be 275 g fish (10 oz).

The stocked waters are usually leased by angling clubs. About 80% of the 600,000 licensed anglers fish in leased waters or at daily fee fish-out ponds. About one-fourth of total fish caught on rod and reel equipment is at daily fee fish-out ponds. Hence this outlet is an important market for fish culturalists.

Density of Production

Table 8.5 indicates the density of salmonoid production by stages of

TABLE 8.5

DENSITY OF SALMONOID PRODUCTION BY TYPE OF PRODUCTION AND SIZE OF ENTERPRISES, WEST GERMANY, JUNE, 1972

Pond Area Enterprises (ha)	Enterprises (No.)	Avg No. Ponds	Avg Water Area Per Pond (ha)	Total Water Area Per Enterprise (ha)	Avg No. Eggs		Avg No. Fry		One-summer Fingerlings		Avg Kg Two-summer or Food Fish	
					Per Enterprise	Per ha of Water Area	Per Enterprise	Per ha of Water Area	Per Enterprise	Per ha of Water Area	Per Enterprise	Per ha of Water Area
(1)	(2)	(3)	(4)	(5)	(6)	(7)	(8)	(9)	(10)	(11)	(12)	(13)
Under 1	843	4.2	0.03	0.13	13,744	105,723	13,095	100,731	7,004	53,877	876	6,669
1-2	248	9.4	0.06	0.56	120,512	215,200	72,633	129,702	28,331	50,591	3,391	6 055
2-5	185	13.6	0.08	1.09	256,546	235,363	194,632	178,561	50,108	45,971	5,476	5,024
5-10	57	15.4	0.12	1.85	271,947	146,998	117,614	63,675	67,386	36,425	7,596	4,106
10-20	25	18.4	0.15	2.76	634,000	229,710	442,000	160,145	146,648	28,346	11,080	4,014
20-50	17	13.3	0.24	3.19	415,000	130,094	56,529	17,721	78,235	24,525	3,941	1,235
50 and over	11	13.4	0.18	2.41	1,117,276	463,598	782,545	324,707	96,000	39,834	7,636	3,168
Totals and Averages	1,386	7.1	0.07	0.50	100,743	201,486	66,654	133,308	23,155	46,310	2,486	4,972

growth listed by size of enterprise. With the two-summer or food fish there is a perfect correlation between output per hectare of water uses and size. The larger the farm, the smaller the output of food fish per hectare of water. This was also true for one-summer fingerlings. These data suggest that larger enterprises are more likely to produce eggs and fry than the smaller producer, while the smaller producer is more likely to raise fingerlings up to market-size fish.

Imports and Exports

Imports of rainbow trout into West Germany have been increasing rapidly. In 1970 total imports were 4374 MT (4820 ST); by 1975 imports had reached 10,996 MT (12,118 ST). This was an increase of 151% in 5 years, or an average annual increase of 30% (Fig. 8.2 and Table 8.6). Denmark is the most important country from which trout are imported. In 1975 Denmark accounted for 87%. Imports from Italy are increasing rapidly, but they are small compared to those from Denmark.

FIG. 8.2.   IMPORTS OF RAINBOW TROUT INTO WEST GERMANY BY COUNTRIES, 1975

TABLE 8.6

IMPORTS OF RAINBOW TROUT INTO WEST GERMANY

Country	1970	1971	1972	1973	1974	1975
			Metric Tons			
Denmark	3314	5021	6253	7188	7324	9580
Italy	195	338	410	549	1014	755
Belgium	85	129	103	162	387	385
Japan	773	80	143	160	22	109
France	5	21	3	10	46	68
Norway	1	2	4	68	15	66
Others[1]	1	11	7	21	8	33
Total (MT)	4374	5602	6323	8158	8816	10,996
Total (ST)	4820	6173	6968	8990	9715	12,118

Source: Professor Pachmann, customs data, Federal Research Institute for Fisheries, Hamburg, Germany.
[1] During various years this included Holland, Poland, USSR, Yugoslavia and Canada.

West Germany imports rainbow trout from about ten countries. Information obtained by the author indicates about 59% of the fish arrive live by truck and the remaining 41% arrive as iced or frozen fish.

In 1975 the average import price was $1.91 per kg ($0.98 per lb). Prices paid by countries varied considerably. For example, trout from Yugoslavia were worth $1.53 per kg ($0.70 per lb) while those from Norway were $3.10 per kg ($1.41 per lb). This difference was probably due to some Norwegian trout being larger and arriving as smoked fish, while fish from Yugoslavia are portion sized and unprocessed.

Only minor amounts of trout are exported from West Germany. In 1975 total exports were 85 MT (94 ST). Total value of these exports was $251,751. Average price received was $2.96 per kg ($1.35 per lb).

Total Supply

Total supply of rainbow trout for West Germany in 1975 was about 16,211 MT (17,865 ST). Of this amount, 5300 MT were produced domestically, 10,886 MT were imported and 85 MT were exported.

CARP (Cyprinus carpio)

In general, production of desired sizes of carp in West Germany takes 3 years compared to only 2 years in central and southern Europe where the waters are warmer.

In the first year the eggs are hatched and fry are raised in spawning and first-rearing ponds. In the second year carp of two summers are grown to larger sizes. In the third year (or third summer) the carp reach market sizes. In general, carp ponds have a minimal flow of water coming into the pond. Usually only enough new water is added to replace evaporation losses.

Distribution of a farmer's water area between different sizes of fish is dependent on that farmer's individual arrangement. Because of difficulty in raising the exact number of fry needed as two- or three-summer fish, some farmers raise only fry, and other farms buy one- or two-summer fish for rearing or growing-out. For a complete operation roughly 12.5% of the water area will be devoted to first-summer fish, 25.0% to second-summer fish and the remainder, or 62.5%, to third-summer fish.

Carp is a warmwater fish. Spawning is delayed until water temperature reaches 18° to 20°C (65° to 68°F). These first-summer fish will weigh 35 to 50 g (1 to 2 oz) in the autumn, with a length of 9 to 12 cm (3.5 to 5 in.). At the end of the second summer of growth they average 350 g (12 oz) and vary from 250 to 500 g (8 to 17 oz). After the third summer of growth they weigh an average of 1.25 kg (2.75 lb), varying between 900 and 1600 g (2 to more than 3 lb). Nearly all carp are harvested at a weight of 1 to 2 kg (2.2 to 4.4 lb). These are intended to be family-size fish as compared to individual portions for trout.

Most carp farms are very extensive. This means that there is little fertilization carried out and little or no artificial food fed. A study conducted (Anon. 1972) indicated that only 13% of the enterprises were feeding any artificial foods. Amount fed based on total production was 1 kg (2.2 lb) for every 4 kg (8.8 lb) of fish produced. In actual practice the amount fed per unit of production is lower than these figures indicate, since a considerable amount of feed goes to newly-hatched and first-summer fish.

In 1972 a study was made of the West German carp industry. The basic month, or benchmark month, was June. This study indicated a total of 4295 separate carp enterprises as producers (Table 8.7). Eighty-one percent of total production was in Bavaria (Fig. 8.1). These enterprises were classified by size of pond area, which was construed to mean not only water surface area, but also the surrounding land which the farmer considered an integral part of the pond area.

There was a wide array of carp farms by size of enterprise measured by "pond area." Over 48% of the enterprises had less than 1 ha of "pond area." This group averaged 1.7 ponds each with an average size per pond of 0.17 ha (0.4 acre) each, for a total water area of about 0.25 ha (less than 0.75 acre). At the other extreme were 63 farm enterprises with 30.6 ponds each and with each pond having 2.7 ha of water surface. These larger producers had an average of 83.2 ha of water area (206

TABLE 8.7

NUMBER OF CARP ENTERPRISES, NUMBER OF PONDS, AND TOTAL WATER
AREA BY SIZES OF ENTERPRISES; WEST GERMANY, JUNE, 1972

Pond Area	Enterprises		Ponds		Water Area	
					Total	Avg Each
(ha)	(no.)	(%)	(no.)	(%)	(ha)	Pond (ha)
Under 1	2065	48.1	3585	1.7	625	0.17
1-2	870	20.2	2477	2.8	861	0.35
2-5	786	18.3	3549	4.5	1749	0.49
5-10	281	6.5	2003	7.1	1450	0.72
10-20	140	3.3	1406	10.0	1475	1.05
20-50	90	2.1	1415	15.7	2272	1.61
50 and over	63	1.5	1929	30.6	5252	2.72
Totals and Averages	4295	100.0	16,364	3.8	13,684	0.84

acres). The average carp farmer had 3.8 ponds, each covering 0.84 ha (2.1 acres), for an average total water area of 3.2 ha (nearly 8 acres).

These 4295 enterprises contained 16,364 separate ponds, covering a total water area of 13,684 ha (33,800 acres). A complete estimate of total annual production of food-size fish was not obtained for all 4295 enterprises. Production data was secured for 4015 enterprises having a total annual output of 3368 MT (3705 ST) of food fish. If it can be assumed that the missing enterprises had the same average output as was found for those reporting, then annual production would have been 3602 MT (3962 ST).

With approximately 62.5% of all carp waters devoted to third-summer fish for food, the 3602 MT were produced from 8553 ha (21,126 acres), for an annual average yield of 421 kg per ha (375 lb per acre).

For the 4015 separate enterprises out of the total of 4295, there were 19,461,000 one-summer fingerlings and 6,300,000 second-summer fish (Table 8.8). With an average harvest weight of 1 kg (2.2 lb) for the three-summer fish, there was a total of 3,368,000 of these fish. Thus, there were 5.8 first-summer fish and 1.9 second-summer fish for each third-summer fish harvested. These data indicate that large sales of first- and second-summer fish are made to angling clubs and fee fish-out ponds, and for restocking of public waters. While mortality does account for part of the decline in numbers, it is also known that sale of first- and second-summer fish is high.

Intensity of Production

Table 8.9 clearly shows the makeup of different sizes of fish produced, by year of production. For example, as size of enterprise

TABLE 8.8

NUMBER OF CARP ENTERPRISES FOR SIZE AND TYPE OF CARP
PRODUCTION; WEST GERMANY, JUNE, 1972

Pond Area (ha)	Enterprises (no.)	One-summer Fingerlings (000's)	Two-summer Fingerlings (000's)	Three-summer Food Fish (MT)
Under 1	1893	291	179	245
1-2	812	592	387	274
2-5	750	1284	614	542
5-10	266	1374	565	425
10-20	137	2555	757	311
20-50	89	4770	1365	469
50 and over	68	8598	2431	1102
Totals and Averages	4015	19,461	6300	3368

increases (shown by total water area per enterprise), the percentage of total fish made up of one-summer fish increases from 40.7 to 72.2%. Also, as size of enterprise increases, the proportion of food-size fish (three summers) declines. These data indicate that larger enterprises concentrate more heavily on one-summer fish and smaller enterprises on producing food-size or three-summer fish. The average carp farmer with more than 10 ha of water surface in ponds may have 75% more concentration of one-summer fish than the small enterprise, while the small enterprise may be nearly 5 times as concentrated on food-size or three-summer fish (compare 7.1% to 34.1%).

Density of Production

Table 8.10 indicates density of carp production by years of growth, by size of enterprise. With food fish (three summers) there is a perfect correlation between size of enterprise and output per hectare of water surface. Production was 445 kg per ha (396 lb per acre) of water surface for the smallest enterprise and continually decreased to 195 kg per ha (174 lb per acre) for the largest enterprise. Also, there was almost a perfect correlation between size of enterprise and density of one-summer fish. Fish numbers increased from 531 fish per ha (125 fish per acre) of water for the smallest enterprise to 2120 per ha (858 per acre) for the next largest enterprise. These data indicate that larger enterprises are more likely to produce one-summer and two-summer fish (fingerlings), while smaller enterprises are more likely to produce food or three-summer fish.

TABLE 8.9

INTENSITY (PERCENTAGE) OF CARP PRODUCTION BY YEARS OF GROWTH AND BY SIZES OF ENTERPRISES, WEST GERMANY, JUNE, 1972

Pond Area (ha)	Enterprises (No.)	Total Water Area Per Enterprise (ha)	Avg No. Per ha of Water				Proportion (%) Fish By Years			
			One-summer Fingerlings	Two-summer Fingerlings	Three-summer Food Fish[1]	Total No. Fish	One-summer	Two-summer	Three-summer	Total (%)
			(1)	(2)	(3)	(4)	(5)	(6)	(7)	(8)
Under 1	1893	0.29	531	328	445	1304	40.7	25.2	34.1	100.0
1-2	812	0.98	744	487	344	1575	47.2	30.9	21.9	100.0
2-5	750	2.21	775	371	327	1473	52.6	25.2	22.2	100.0
5-10	266	5.11	1011	416	313	1740	58.1	23.9	18.0	100.0
10-20	137	10.50	1776	526	216	2518	70.5	20.9	8.6	100.0
20-50	89	25.28	2120	607	208	2935	72.2	20.7	7.1	100.0
50 and over	68	83.23	1514	454	195	2163	70.0	21.0	9.0	100.0
Totals and Averages	4015	3.19	1519	492	263	2274	66.8	21.6	11.6	100.0

[1] Assumes an average harvest weight of one kg per fish.

Note: Columns (1) through (3) come from Tables 8.7 and 8.8; Column (4) is summation of Columns (1), (2) and (3); Column (5) is proportion that Column (1) is of Column (4), Column (6) is the proportion that Column (2) is of Column (4); and Column (7) is the proportion that Column (3) is of Column (4).

TABLE 8.10

DENSITY OF CARP PRODUCTION BY YEARS OF GROWTH AND BY SIZES OF ENTERPRISES, WEST GERMANY, JUNE, 1972

					Avg No. One-summer Fingerlings		Avg No. Two-summer Fingerlings		Avg Kg Production of Three-summer Fish	
Pond Area (ha)	Enterprises (No.)	Avg No. Ponds	Avg Water Area Per Pond (ha)	Total Water Area Per Enterprise (ha)	Per Enterprise	Per ha of Water Area	Per Enterprise	Per ha of Water Area	Per Enterprise	Per ha of Enterprise Water Area
(1)	(2)	(3)	(4)	(5)	(6)	(7)	(8)	(9)	(10)	(11)
Under 1	1,893	1.7	0.17	0.29	154	531	95	328	129	445
1-2	812	2.8	0.35	0.98	729	744	477	487	337	344
2-5	750	4.5	0.49	2.21	1,712	775	819	371	723	327
5-10	266	7.1	0.72	5.11	5,165	1,011	2,124	416	1,598	313
10-20	137	10.0	1.05	10.50	18,650	1,776	5,526	526	2,270	216
20-50	89	15.7	1.61	25.28	53,596	2,120	15,377	607	5,270	208
50 and over	68	30.6	2.72	83.23	126,000	1,514	35,750	454	16,206	195
Totals and Averages	4,015	3.8	0.84	3.19	4,847	1,519	1,569	492	839	263

Note: Column (2) from Table 8.9; Columns (3) and (4) from Table 8.7; Column (5) determined by multiplying Column (3) by Column (4); Columns (6), (8), and (10) determined by dividing appropriate data in Table 8.8 by Column (2); Columns (7), (9) and (11) determined by dividing Columns (6), (8), and (10) by Column (5).

The fact that density of fish numbers for one- and two-summer fish, as well as kilograms of three-summer fish, declined for the enterprise group having largest water area (83.23 ha or 206 acres per enterprise) suggests one of several observations: (1) these producers are more extensive and do not feel the need to intensify production; (2) these farmers have a lesser degree of pond and fish management skills; or (3) the waters in these enterprises are less productive.

Commercial Wild Catch

In 1971, the last year for which data are available, total commercial wild catch of carp and cyprinids was 715 MT (788 ST) (Table 8.11). This catch was 53% from rivers and 47% from lakes. Thirty-six percent of the wild catch came from the state of Bavaria, and 28% from Schleswig-Holstein and Hamburg (Table 8.12). The remaining 36% came from the other six German states, including West Berlin.

TABLE 8.11

CATCH OF WILD FOOD FISH BY SPECIES, WEST GERMANY, 1971

Species	River Fisheries (kg)	Lake Fisheries (kg)	Total (kg)
Eel	121,367	124,374	245,741
Pike	34,539	56,073	90,612
Perch	8,178	44,453	52,631
Pike-perch	19,744	27,266	47,010
Carp	18,672	43,280	61,952
Cyprinids	362,279	291,072	653,351
Coregonids	1,682	375,596	377,278
Others	266,011	21,247	287,258
Total food fish	832,472	983,361	1,815,833

Imports and Exports

In 1975 a total of 3763 MT (4147 ST) of carp were imported into West Germany. Imports have increased continually since 1970. Total increase in imports since 1970 is 50% (Table 8.13 and Fig. 8.3). Imports are essentially from eastern Europe where carp are cultured extensively. Major imports each year come from Yugoslavia, Hungary, Poland and the USSR, while lesser amounts are imported from France, Belgium, Holland and Switzerland.

Only minor amounts of carp are exported from West Germany. In

TABLE 8.12

CATCH OF WILD FOOD FISH BY SPECIES BY STATES, WEST GERMANY, 1971

River Fisheries By States (kg)

Species	Schleswig-Holstein & Hamburg	Lower Saxony	North Rhineland-Westphalia	Hesse	Rhineland-Palatinate	Baden-Württemberg	Bavaria	West Berlin	Total
Eel	9,472	51,615	8,712	8,116	6,621	13,081	20,095	3,655	121,367
Pike	1,138	3,177	833	3,738	3,140	7,298	13,446	1,769	34,539
Perch	666	911	103	200	610	3,534	1,638	516	8,178
Pike-perch	2,702	1,048	126	495	2,348	1,449	5,543	6,033	19,744
Carp	420	1,146	489	965	2,241	1,442	11,856	113	18,672
Cyprinids	2,682	28,008	5,878	44,588	66,122	60,094	149,491	5,416	362,279
Coregonids	—	156	—	3	1,342	181	—	—	1,682
Others	228,713	5,484	1,310	1,442	2,496	11,649	14,161	756	266,011
Total food fish	245,793	91,545	17,451	59,547	84,920	98,728	216,230	18,258	832,472

Lake Fisheries By States (kg)

Species	Schleswig-Holstein & Hamburg	Lower Saxony	North Rhineland-Westphalia	Hesse	Rhineland-Palatinate	Baden-Württemberg	Bavaria	West Berlin	Total
Eel	90,774	16,790	1,052	4,214	—	937	10,607	—	124,374
Pike	36,868	3,918	750	55	—	2,039	12,443	—	56,073
Perch	38,848	1,692	1,855	46	—	433	1,579	—	44,453
Pike-perch	6,308	8,993	802	2,725	—	1,520	6,918	—	27,266
Carp	19,239	590	1,400	1,190	—	1,810	19,051	—	43,280
Cyprinids	176,253	18,572	3,260	3,331	—	12,155	77,501	—	291,072
Coregonids	44,296	—	22,000	—	—	—	309,300	—	375,596
Others	10,151	2,608	—	300	—	901	7,287	—	21,247
Total food fish	422,737	53,163	31,119	11,861	—	19,795	444,686	—	983,361

Total Lake and River Fisheries By States (kg)

Species	Schleswig-Holstein & Hamburg	Lower Saxony	North Rhineland-Westphalia	Hesse	Rhineland-Palatinate	Baden-Württemberg	Bavaria	West Berlin	Total
Eel	100,246	68,405	9,764	12,330	6,621	14,018	30,702	3,655	245,741
Pike	38,006	7,095	1,583	3,793	3,140	9,337	25,889	1,769	90,612
Perch	39,514	2,603	1,958	246	610	3,967	3,217	516	52,631
Pike-perch	9,010	10,041	928	3,220	2,348	2,969	12,461	6,033	47,010
Carp	19,659	1,736	1,889	2,155	2,241	3,252	30,907	113	61,952
Cyprinids	178,935	46,580	9,138	47,919	66,122	72,249	226,992	5,416	653,351
Coregonids	44,296	156	22,000	3	1,342	181	309,300	—	377,278
Others	238,864	8,092	1,310	1,742	2,496	12,550	21,448	756	287,258
Total food fish	668,530	144,708	48,570	71,408	84,920	118,523	660,916	18,258	1,815,833
% of Total	36.8	8.0	2.7	3.9	4.7	6.5	36.4	1.0	100.0

TABLE 8.13

IMPORTS OF CARP INTO WEST GERMANY, 1970-75

Country	Metric Tons					
	1970	1971	1972	1973	1974	1975
Yugoslavia	635	592	751	689	1190	1225
Hungary	459	566	790	723	1430	916
Poland	386	469	561	425	444	708
USSR	192	192	378	340	408	514
France	792	773	735	636	426	300
Belgium	34	92	115	195	175	99
Others[1]	3	4	2	8	5	1
Total (MT)	2501	2688	3332	3016	4078	3763
Total (ST)	2756	2962	3672	3324	4494	4147

Source: Professor Pachman, customs data, Federal Research Institute for Fisheries, Hamburg, Germany.

[1] Includes various years, Holland and Switzerland.

1975 total exports were 51 MT (56 ST). Total value of these exports was $73,541. Average price received was $1.44 per kg ($0.65 per lb).

Total Supply and Prices

Total supply of carp in Germany in 1975 was 8029 MT (8848 ST). This was composed of an estimated 3602 MT of cultured fish, 715 MT of commercial wild catch, and 3763 MT of imports minus 51 MT of exports. Hence, total supply was 45% cultured fish, 9% wild catch and 46% net imports.

Since imports play such a large role, domestic producers' prices are nearly the same as import prices. In 1975 farmer prices averaged $1.04 per kg ($0.47 per lb). Wholesaler prices ranged between $1.05 and $1.36 per kg ($0.48 to $0.62 per lb). Retail prices were between $1.95 and $2.33 per kg ($0.89 to $1.06 per lb).

EUROPEAN EELS (Anguilla anguilla)

In general, water temperatures in West Germany are too cold for good growth of eels. Under natural conditions it takes 7 to 15 years for eels to mature in natural waters. Since the Germans prefer the mature silver eel rather than the immature yellow eel, this means that culturing is at a distinct economic disadvantage. This disadvantage is due to high costs of producing a fish which has such a long growth period. Hence, the only German eel culture is found in the warm water discharge of electrical generating plants. Since availability of such water

FIG. 8.3.   IMPORTS OF CARP INTO WEST GERMANY BY COUNTRIES, 1975

is limited, cultured eel production is also limited. In 1975 total cultured production was about 10 MT (11 ST).

Because of the German fondness for eel, particularly smoked, there is considerable commercial effort to catch wild eels in rivers and lakes. In 1971, the latest year for which the author could secure wild catch data, the wild catch totaled 246 MT (271 ST) (Table 8.11). About one-half the catch came from rivers and one-half from lake fishing. More than two-thirds came from the two states of (1) Schleswig-Holstein and Hamburg and (2) Lower Saxony (Table 8.12). However, eels are caught in every state, including West Berlin.

Imports of Eel

West Germany imports significant amounts of eels. Between 1970 and 1976 a total of 31,489 MT (34,701 ST) were imported. This averaged 5248 MT (5783 ST) annually (Table 8.14 and Fig. 8.6). Eels are imported

Courtesy of Mr. Harold Koops

FIG. 8.4.   FEEDING EELS IN WEST GERMANY

FIG. 8.5.   SIMPLE DEVICE FOR CAPTURING ELVERS

TABLE 8.14

IMPORTS OF EEL INTO WEST GERMANY, 1970-75

Country	1970	1971	1972	1973	1974	1975
			Metric Tons			
Denmark	2056	1949	2186	2224	2057	2023
New Zealand	406	588	682	441	388	576
Holland	715	646	507	396	569	547
Canada	171	455	593	419	475	421
Sweden	176	234	256	285	288	380
Poland	256	298	250	390	333	348
USA	319	442	323	310	282	286
Australia	44	77	92	133	191	212
Belgium	116	107	97	125	219	179
France	90	188	167	253	164	110
Italy	2	7	15	17	46	88
Turkey	68	142	73	67	32	32
Great Britain	23	11	101	72	96	21
Red China	45	47	36	—	30	27
Greece	22	—	—	43	—	26
Ireland	56	60	37	54	70	25
Others[1]	93	72	111	97	64	41
Total (MT)	4658	5323	5526	5326	5304	5352
Total (ST)	5133	5866	6090	5869	5845	5898

Source: Professor Pachmann, customs data, Federal Research Institute for Fisheries, Hamburg, Germany.

[1] Includes Norway, Switzerland, USSR, Hungary, Austria, Spain and Taiwan.

from practically every country in the world where they are captured or cultured. Major imports are made annually from Denmark, which accounts for nearly 38% of Germany's total.

In 1975 eel imports accounted for $21,494,000. Average price paid for imported eel was $4.02 per kg ($1.87 per lb). Price paid per unit of import varied from country to country, depending on the degree of processing undergone by the imported eels. For example, the lowest price was for eels imported from New Zealand. These eels—worth $1.60 per kg ($0.73 per lb)—are all unprocessed and are received frozen. At the other extreme were eels imported from Sweden, which were worth $5.11 per kg ($2.32 per lb). Many of these were imported as a smoked product. The remainder was imported as live fish. The eels from Denmark, which in 1975 accounted for 38% of all imports, amounted to $4.86 per kg ($2.21 per lb). This price difference was probably due to a smaller percentage of Danish eels being processed than Swedish eels.

While West Germany imports significant volumes of eels, it also exports smaller amounts. In 1975 exports were made to nine European

FIG. 8.6.   IMPORTS OF EEL INTO WEST GERMANY BY COUNTRIES, 1975

countries. A total of only 208 MT (229 ST) were exported; chiefly to Holland and Sweden, with minor amounts to Finland, Denmark, France, Poland, Belgium, Switzerland and Austria. Total eel exports amounted to $591,000. Average selling price for exports was $2.84 per kg ($1.29 per lb), indicating that most exports were unprocessed.

As a result of exports and imports, West Germany's net supply of eels increased by 5144 MT (5669 ST) in 1975. The net amount paid out was $20,903,000.

Net Supply

Net eel supply in West Germany in 1975 was an estimated 5400 MT (5951 ST). This was derived from 10 MT of cultured fish, 246 MT of wild catch and net imports of 5144 MT.

Prices

As indicated previously, prices for imported eels varied widely, depending on the proportion of fish received in some processed form. The New Zealand price of $1.60 per kg ($0.73 per lb) was for unprocessed frozen fish. The domestic price for wild and cultured eels would be somewhat higher because of the higher quality of live fish, which are superior for smoking. Fishermen's prices would, of course, be lower than retail prices. In 1975 retail prices for fresh live eels ranged between $4.67 and $5.45 per kg ($2.12 to $2.48 per lb). The smoked fish, however, retailed for $11.67 to $13.62 per kg ($5.30 to $6.19 per lb).

TENCH (Tinca tinca)

Tench is raised as a food fish in West Germany. It is nearly always grown in conjunction with carp production. However, the growth rate is slower than that of carp.

Very little information is available about West Germany's tench production, which is about 300 MT per year (330 ST). The major production area is Bavaria, with about 45% of total production (Fig. 8.1).

While carp are harvested for food fish at the end of the third summer of growth, averaging 1.5 to 2 kg (3.3 to 4.4 lb), the slower-growing tench may weigh only 250 g (9 oz), and are rarely grown past 700 g (1.5 lb).

Tench prices are not affected by imports, so in essence there is no wholesale price level. Fish are sold directly to consumers or to retail outlets. Average prices for these sales varied between $2.33 and $2.72 per kg ($1.06 to $1.24 per lb) in 1975.

COMMON OR NORTHERN PIKE (Esox lucius)

There is no separate culture of pike in West Germany. However, many carp producers stock 10 to 20 pike fingerlings per ha (4 to 8 per acre) in their carp ponds. Pike normally attain 15 to 20 cm (6 to 8 in.) during the first summer they are stocked with carp. In the fall carp ponds are drained or seined for harvest. At this time many of the pike fingerlings are sold to sportsmen's clubs for stocking angling waters. Some of the pike fingerlings are returned to the carp ponds for a second summer of growth. At the end of the second summer these pike usually measure over 40 cm (16 in.) and average about 1 kg (2.2 lb) in weight. They are then harvested and sold as either fish for angling waters or for immediate consumption as food. When pike are grown with carp, the usual production rate is between 5 and 15 kg per ha (4 to 13 lb per acre). If it can be assumed that an average of 10 kg per ha (9 lb per acre) is produced by one-fourth of the carp producers, who have a total water area of 13,700 ha (33,800 acres), then total annual production would be about 350 MT (385 ST). Even by the most liberal interpretation, annual production must be considerably less than 1000 MT (1100 ST).

Prices of pike vary widely. In 1975 an average of $2.33 per kg ($1.06 per lb) was received by fish farmers, while pike sold in southern West Germany may have been worth 50% more than in the northern part of the country because of differences in local and regional demand. Only a few tons of pike are imported annually. These come in from Poland and Denmark.

RECREATION FISHING

Angling clubs lease fishing waters, and 80% of fishing by some 600,-000 anglers is in club waters.

About 25% of all fish caught on rod and reel are from daily fee fish-out ponds or enclosures. These types of fishing establishments are in great demand in West Germany. In fact, this is one of the few countries in western Europe that has a large pay-and-fish system. Unfortunately the number of such establishments or total volume of the catch is not known.

OUTLOOK

Trout production is unlikely to increase significantly because of (1) limited water area and (2) the availability of relatively cheap imports from Denmark and Italy where production costs are less.

Production of carp is also unlikely to expand significantly for the same reasons as with trout. Presently domestic producer prices are set by import prices. Since carp can be imported at relatively low prices from Yugoslavia, Hungary, Poland and the USSR, there is little economic incentive to increase domestic production. However, domestic output is unlikely to decline since there is no other economical use for carp waters.

"Other" cultured fish production is related to carp production. Since the area devoted to carp production is unlikely to expand, these other species will always be found in limited supply.

Germany's waters are too cold for eel production. Hence, in spite of high prices, production will remain small.

SPECIAL ACKNOWLEDGEMENTS

MR. HAROLD KOOPS, Institut fur Kusten und Binnen fischerei, Hamburg, Federal Republic of Germany

DR. H. ZOLMER KUHLMANN, Institut fur Kusten und Binnen fischerei, Hamburg, Federal Republic of Germany

PROFESSOR HANS MANN, Institut fur Kusten und Binnen fischerei, Hamburg, Federal Republic of Germany

PROFESSOR PACHMANN, Budesamt fur Ernahrung und Forstwirtschaft, Hamburg, Federal Republic of Germany

DR. GERT RAUCK, Institut fur Kusten und Binnen fischerei, Hamburg, Federal Republic of Germany

PROFESSOR UDO RIEMANN, Institut fur Landwirtschaftliche Verfahronstechnick der Christian-Albrechts Universitat, Kiel, Federal Republic of Germany

REFERENCES

ANON. 1970-75. Customs data. Federal Research Institute for Fisheries, Hamburg, Germany.

ANON. 1971-72. Results of Freshwater Fish Production, According to the Census of Agriculture.

ANON. 1972. A Study of the Trout Industry. Federal Research for Fisheries, Hamburg, Germany.

Chapter 9

Switzerland

RAINBOW TROUT (*Salmo gairdneri*)
and BROWN TROUT (*Salmo trutta*)

There are two species of salmonoids cultured for food fish production in Switzerland — rainbow trout and brown trout. Total production of both species in 1974 was about 300 MT (331 ST). Reliable estimates of proportions of each species produced are difficult to obtain. However, a generally accepted estimate is about 270 MT of rainbow and about 30 MT of brown trout. There are no exports. However, Switzerland does import rainbow trout from Denmark, Italy and France. In 1974 a total of 1032 MT (1137 ST) was imported (Table 9.1). Denmark accounted for more than 80% of all imports, followed by Italy with 15%, and with minor quantities from France. Total supply, including domestic production and imports, was an estimated 1332 MT (1468 ST). Domestic production accounted for only 23% of total supply, with imports accounting for the bulk of total supply.

For domestically produced trout, rainbow trout spawning season is in February and March, while the brown trout spawning season is in November, December and January. After spawning, fry are raised in concrete rearing tanks until they are about 6 cm (2.5 to 3 in.) and are then placed in grow-out facilities. Grow-out facilities are primarily earthen ponds with some concrete raceways used. Commercial pelletized trout food is fed throughout the production cycle.

It takes between 12 and 24 months to raise rainbow and brown trout to market sizes. Market demand is for fish between 180 and 200 g (6.5 to 7.5 oz).

An output of 300 MT of trout is achieved by 30 fish farmers. Hence, the average farm produces only 10 MT. This indicates that trout enterprises are essentially supplementary farm or nonfarm enterprises.

Nearly all sales are direct from farmer to housewife. The housewives

TABLE 9.1

PRODUCTION AND IMPORTS OF RAINBOW TROUT FOR SWITZERLAND

| | Metric Tons | | | |
	1971	1972	1973	1974
Production[1]	200	240	275	300
Imports[2]				
Denmark	1001	1160	985	838
Italy	NA	NA	NA	159
France	NA	NA	20	35
Total Supply[2]	1201	1400	1280	1332

[1] Estimated.
[2] Source: European Federation of Salmonoids — Breeders (FES).

then prepare the trout in a variety of ways. Cooking methods include fried, boiled and grilled. All domestic production is sold live and about 80% of imported trout are also sold live. Only about 20% of imported trout are frozen. Imports are generally sold in fish markets and super-markets.

CARP (Cyprinus carpio)

There is limited production of carp for food use in Switzerland. This is because of less than optimum water temperatures, which result in slow growth and limited demand. In 1975 there were 5 carp farmers averaging only 0.4 MT each, with a total output of 2 MT annually *in toto.*

The usual spawning period is in April and May. Fry are then placed in small earthen ponds, with little if any artificial foods fed. Total pro-duction time to market sizes of 1.5 kg (3.3 lb) varies between 3 and 4 years. All of the limited production is found in the French (western) part of Switzerland. The fish are sold live directly from farm to house-wife. She then either bakes or boils the fish for her family.

OTHER CULTURED FISH

In addition to commercial raising of rainbow trout and carp as food fish by private growers, the Swiss government has an intensive re-stocking program under way to build up or maintain fishing in public waters. During 1974 production of governmental fish hatcheries was 25,216,258 fish (Table 9.2). This included young fry, fingerlings and year-old fish. The species of fish released in greatest numbers were: (1) northern pike (*Esox lucius*), accounting for 37.8% of all released fish;

TABLE 9.2

PRODUCTION OF THE SWISS FISH HATCHERIES DURING THE HATCHING PERIOD, 1974

Species	Number of Fish Placed in Open Water Under Official Control Over a Multi-monthly Period							
	Pre-summerlings		Summerlings		Yearlings		Total	
	Head	(%)	Head	(%)	Head	(%)	Head	(%)
Lake trout (*Salmo trutta lacustus*)	815,905	4.8	335,652	4.6	6,314	0.8	1,157,871	4.6
Brown trout (*Salmo trutta*)	1,829,455	10.7	4,795,850	66.0	559,719	69.8	7,185,024	28.6
Steelhead trout (*Salmo irideus*)	280,100	1.6	1,184,966	16.3	156,801	19.5	1,621,867	6.4
Artic char (*Salvelinus alpinus*)	252,900	1.5	169,500	2.3	10,162	1.3	432,562	1.7
Grayling (*Thymallus thymallus*)	531,912	3.1	135,387	1.9	22,279	2.9	689,578	2.7
Brook trout (*Salvelinus fontinalis*)	1,100	0	13,700	0.2	2,300	0.3	17,100	0
Lake whitefish (*Coregonus lavaretus*)	4,392,777	25.6	10,000	0.1	—	—	4,402,777	17.5
Northern pike (*Esox lucius*)	9,040,678	52.7	486,541	6.7	5,935	0.7	9,533,154	37.8
Pike-perch (*Lucioperca lucioperca*)	—	—	48,900	0.7	—	—	48,900	0.2
Perch (*Perca fluviatilis*)	—	—	—	—	700	0	700	0
Danubian wels (*Silurus glanis*)	—	—	—	—	4,400	0.5	4,400	0
Carp (*Carassius carassius*)	—	—	—	—	3,310	0.4	3,310	0
Eel (*Anguilla anguilla*)	—	—	15,000	0.2	—	—	15,000	0
Other	—	—	73,570	1.0	30,445	3.8	104,015	0.5
Total	17,144,827	100.0	7,269,066	100.0	802,365	100.0	25,216,258	100.0
% Production	68.0		28.8		3.2		100.0	

Source: Federal Office for Environmental Protection, Bern, Switzerland.

TABLE 9.3

CATCH OF WILD FISH BY LAKES AND TRIBUTARIES, SWITZERLAND

Lake Waters and Tributaries[1]	Area (ha)	1967	1968	Annual Production (kg) 1969	1970	1971	Production by Hectare (kg per year)
Genfersee	34,745	215,294	669,369	367,628	317,236	854,619	13.95
Neuenburgersee	21,581	318,018	197,157	319,812	502,852	213,228	14.37
Bodensee mit Untersee	17,144	527,951	442,950	521,093	700,232	716,575	33.93
Vierwaldstättersee	11,380	167,322	167,646	214,345	200,598	266,428	17.86
Zürichsee mit Obersee	8,852	143,872	155,265	196,506	183,241	171,730	19.22
Thunersee	4,780	56,852	48,996	91,155	121,221	172,565	20.16
Bielersee	3,920	125,745	81,693	105,302	114,320	134,183	28.63
Zugersee	3,824	209,481	195,970	97,678	79,934	91,676	35.28
Brienzersee	2,918	23,961	13,168	16,421	23,500	18,101	6.52
Walensee	2,423	34,385	26,504	36,469	31,765	42,174	14.14
Murtensee	2,282	22,755	21,236	17,118	37,091	17,418	10.13
Hallwilersee	1,030	8,148	6,317	4,112	5,223	6,514	5.88
Lac de Joux	953	8,331	18,962	20,825	16,569	11,449	15.97
Greifensee	856	15,447	9,977	8,848	8,794	7,169	11.74
Sarnersee	773	14,875	12,778	18,270	11,596	13,683	18.42
Aegerisee	724	not given	9,067	12,199	9,906	11,520	14.74
Pfäffikersee	334	3,062	5,798	2,697	2,244	2,475	9.74
Lungernsee	201	3,377	3,899	4,286	3,931	4,227	19.62
Total	118,720	1,898,876	2,086,752	2,054,764	2,361,344	2,755,734	18.79

Source: Federal Office for Environmental Protection, Bern, Switzerland.
[1] Excluding the portion from foreign countries.

TABLE 9.4

CATCH OF FISH IN SWITZERLAND¹, BY SPECIES

Species	1967 (kg)	(%)	1968 (kg)	(%)	1969 (kg)	(%)	1970 (kg)	(%)	1971 (kg)	(%)	Total kg 1967-1971
Perch (Perca fluviatilis)	744,984	39.2	987,662	47.3	871,601	42.4	1,022,565	43.3	1,335,038	48.5	4,961,850
Lake whitefish (Coregonus lavaretus)	584,058	30.8	521,794	25.0	682,795	33.2	672,969	28.5	757,072	27.5	3,218,688
Assorted rough fish	426,085	22.4	406,752	19.5	343,847	16.7	500,843	21.2	501,196	18.2	2,178,723
Common bream (Abramis brama)	27,739	1.5	50,731	2.4	46,350	2.3	63,611	2.7	47,727	1.7	236,158
Northern pike (Esox lucius)	52,760	2.8	53,377	2.6	47,231	2.3	40,638	1.7	41,995	1.5	236,C01
Lake trout (Salmo trutta lacustus)	32,358	1.7	32,583	1.6	31,208	1.5	27,324	1.2	31,284	1.1	154,757
Artic char (Salvelinus fontinalis)	8,602	0.5	8,182	0.4	12,507	0.6	9,145	0.4	10,120	0.4	48,556
Other	22,290	1.1	25,671	1.2	19,223	1.0	24,249	1.0	31,302	1.1	122,735
Total	1,898,876	100.0	2,086,752	100.0	2,054,764	100.0	2,361,344	100.0	2,755,734	100.0	11,157,470

Source: Federal Office for Environmental Protection, Bern, Switzerland.
¹Water area involved, in hectares — 118,720.

(2) brown trout (*Salmo trutta*), accounting for 28.6%; (3) lake whitefish (*Coregonus lavaretus*), accounting for 17.5%; (4) steelhead trout (*Salmo irideus*), accounting for 6.4%; (5) lake trout (*Salmo trutta lacustris*), accounting for 4.6%; and (6) grayling (*Thymallus thymallus*), accounting for 2.7%. Other fish released included artic char (*Salvelinus alpinus*), brook trout (*Salvelinus fontinalis*), pike-perch (*Lucioperca lucioperca*), perch (*Perca fluviatilis*), Danubian wels (*Silurus glanis*), carp (*Carassius carpio*) and European eel (*Anguilla anguilla*).

Success of the restocking program has been significant. The catch in the 118,720 ha (293,238 acres) of public waters in 1967 was 1,898,876 kg (4,178,000 lb). By 1971 the catch had increased to 2,755,734 kg (6,063,000 lb) (Table 9.3). This table also shows the catch per hectare of water per year. It ranged from a low of 6 kg to a high of 35 kg per ha (from 5 to 31 lb per acre). This variation was caused by natural fertility of the water and restocking intensity.

The commonly caught wild fish, many stocked as fry or fingerlings, are shown in Table 9.4. The three most commonly caught fish were perch, lake whitefish, and assorted rough fish. The wild catch increased from 1899 MT in 1967 to 2756 MT in 1971. A breakdown of the wild fish catch by species is shown in Table 9.4.

OUTLOOK

The future appears bright for Switzerland's cultured fish industry. Production of rainbow trout is predicted to keep increasing at the rate of 10% annually. However, because of lack of suitable terrain for production sites, cost of land, high production costs, and cheaper imports, production may never exceed 800 to 1000 MT. This means that, even if demand does not increase, Switzerland will never become self-sufficient in trout production unless imports are restricted by tariffs or quotas.

Production of carp is very limited and, because of lack of good growing conditions, will probably stay small.

SPECIAL ACKNOWLEDGEMENTS

DR. WILLY MEIER, Universitat Bern, Eidg. Untersuchungsstelle, Fur Fischkrankeiter, Langgesstrasse 122, CH.-3012 Bern, Switzerland.

REFERENCES

ANON. 1967-71. Federal Office for Environmental Protection, Bern, Switzerland.

ANON. 1971-74. European Federation of Salmonoids—Breeders (FES), Treviso, Italy.

ANON. 1974. Federal Office for Environmental Protection, Bern, Switzerland.

Chapter 10

Austria

There are several species of fish cultured in Austria. Among these are: (1) rainbow trout (*Salmo gairdneri*), (2) carp (*Cyprinus carpio*), (3) tench (*Tinca tinca*), (4) grass carp (*Ctenopharyngodon idella*) and (5) lake whitefish (*Coregonus albula*). Since Austria is a landlocked country all fish raised are freshwater species.

RAINBOW TROUT (*Salmo gairdneri*)

The major fish cultured in Austria, by tonnage produced, is rainbow trout. In 1975 an estimated 1000 MT (1100 ST) were produced. Major trout producing areas are Upper Austria, Tirol, Vorarlberg, Styria, Carinthin and Salzberg (Fig. 10.1). Trout are produced in concrete and earthen raceways with flowing water and in ponds and reservoirs, such as old gravel pits, using both net enclosures and cages.

Since Austria is a major tourist country, it is deficient in domestic production of cultured fish. Less than 20 MT of trout are exported yearly while 230 to 250 MT are imported. Imports are essentially from Denmark and Italy. In 1974, 129 MT (142 ST) were imported from Denmark while 101 MT (111 ST) were imported from Italy. A few metric tons are also imported from West Germany.

Market size of trout is between 250 and 300 g (9 to 11 oz). The usual spawning months are February and March. After spawning, between 10 and 20 months are required to grow the fish up to market sizes. The exact time for growth depends upon water temperatures, which vary from very cold Alpine waters to warmer waters in eastern Austria. The most usual grow-out time varies between 16 and 18 months.

Only very small quantities of eyed eggs are exported. They are sold to Switzerland, Turkey and Persia. Sales to Turkey and Persia are in small quantities for breeding experiments.

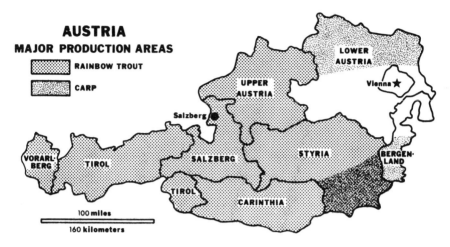

AUSTRIA
MAJOR PRODUCTION AREAS

RAINBOW TROUT

CARP

LOWER AUSTRIA

UPPER AUSTRIA

Vienna ★

Salzberg ●

VORARL-BERG

TIROL

SALZBERG

STYRIA

BERGEN-LAND

TIROL

CARINTHIA

100 miles

160 kilometers

FIG. 10.1.   RAINBOW TROUT AND CARP PRODUCTION AREAS,
AUSTRIA, 1975

Limited water resources do not permit high production by individual growers. A large trout farm produces only 20 to 50 MT (22 to 55 ST). In 1975 there were 72 members belonging to the Association of Austrian Trout Producers. These producers might be considered as commercial operators. In addition, there are an estimated 800 to 825 other producers who culture trout for their household consumption or as a hobby.

Nearly all trout are sold live. Almost every large city has one or more fish buyers. There are an estimated 10 to 12 buyers in the country. The trend, however, is toward direct sales of live fish to hotels, restaurants and housewives, the major outlets being hotels and restaurants. A minor outlet for trout is as a smoked product. At present only about 2% of the cultured trout is sold smoked. However, the market is increasing rapidly.

Because of the unique geographic position of Austria and its German-related culture, some trout are boiled as in West Germany. However, trout are also broiled, grilled and pan-fried, as is customary in most other European countries.

CARP (Cyprinus carpio)

The second most important fish species cultured is carp. About 600 MT (660 ST) are raised annually. Major production areas are northern Lower Austria (Waldviertel), Burgenland and southeastern Styria (Fig. 10.1). All carp are produced in earthen ponds, usually adding water only to keep up with evaporation.

As in the case of rainbow trout, Austria is also a deficit carp-producing country. Only some 20 to 30 MT (22 to 33 ST) are exported yearly. More than offsetting these exports is the import of 200 MT (220 ST) of carp from East European countries and West Germany.

Market size of carp is from 1000 to 2000 g (2.2 to 4.4 lb). Usual spawning months are May and June. After spawning, between 2 and 3 years are required for the fish to reach market size. For further information about production practices, the reader is referred to the carp section in Chapter 8, West Germany.

The third summer of production averages 300 kg of food fish per ha (267 lb per acre). After making allowances for maintenance of brood fish, fry and fingerlings, about 2000 ha (4940 acres) are devoted to production of third-summer fish and 3200 ha (7900 acres) devoted to the entire production cycle.

There are two carp-producing associations with total membership of about 40 producers. In addition, there are an estimated 160 small enterprises and hobbyists who are not commercial producers. The average commercial producer has about 80 ha of water area or less and produces about 15 MT annually (about 17 ST from 200 acres of water area or less).

Nearly all carp are sold live. Marketing channels differ considerably from those for trout in that most sales are made directly to fish buyers instead of hotels and restaurants.

As is found in West Germany and other carp-producing and carp-consuming countries, most carp are cooked slowly by boiling or baking. This permits the flesh to peel away from the bones during eating.

OTHER CULTURED FISH

Other species cultured include: (1) tench (*Tinca tinca*), (2) grass carp (*Ctenopharyngodon idella*) and (3) lake whitefish (*Coregonus albula*). About 30 MT (33 ST) of tench, 20 MT (22 ST) of grass carp and 5 MT (5.5 ST) of lake whitefish are produced annually.

Most of these fish are produced in conjunction with carp, hence there are no meaningful figures available on production per hectare or per unit of water flow.

Spawning times for tench are May and June, for grass carp the month of June, and for whitefish, April. After spawning, it takes between 2 and 3 years to raise the tench to market weights of 200 to 300 g (0.5 lb) and between 3 and 4 years to raise the grass carp to market weights of 3000 to 4000 g (6.6 to 8.8 lb). Market weights and the production cycle of the lake whitefish could not be obtained by the author.

SPORTS FISHING

Because of intense interest in sports fishing and inability of the government to keep public waters well stocked, fishing clubs and fee fishing are important. Nearly all running water streams and lakes are in the public domain. The provinces issue yearly fishing licenses for public fishing. However, fishing clubs also lease fishing rights from private owners and stock these areas. The individual club members' licenses may vary from one day to one year and are expensive. About 5% or 50 MT (55 ST) of cultured trout and about the same volume of cultured carp are stocked annually in private waters for special licensed fee fishing.

OUTLOOK

Because of the strong demand from Austrians, as well as from tourists, trout production is increasing. It is estimated that production will increase 10% during the next few years. The limiting factor is water supply, rather than a market with profitable producer prices.

Unlike trout, producer prices for carp are such that the profitability of carp farming is low. For this reason, production is stagnating and is unlikely to increase under present cost-benefit relationships. However, production is not likely to decrease because there is little alternative use for the carp ponds.

Production of tench, grass carp and lake whitefish is usually in conjunction with carp. Unless the water area for carp expands, production of these other species is unlikely to increase significantly.

SPECIAL ACKNOWLEDGEMENTS

DR. JENS F. HEMSEN, Director, Austrian Federal Institute for Water Research and Fishery, 5310 Scharfling-am-Mondsee, Austria
DR. KURT IGLER, 8775 Kalwang, Austria

Chapter 11

Italy

RAINBOW TROUT *(Salmo gairdneri)*

Trout culture in Italy was originated in the mountains, in a region called Trentino, roughly at the beginning of the century. Development in this region was related to the common belief that, since natural trout live in mountain brooks, trout farms should be built in the mountains. These early trout farmers erroneously believed that altitude was an important factor. They did not know that the limiting factor was high quantity of water with constant flow and constant temperature. Consequently, for many years trout farming was done in mountain areas where water was scarce, winters were long and land was restricted. In this way trout farming was born, but its development in the mountains stopped shortly because of restricted land areas and water supplies. Marketable trout were produced in mountain trout farms in over 24 months, and the amount produced was not enough to supply 10% of the national demand. Italy, until about 1965, was the best market for the trout production of other countries, such as Denmark.

Beginning about 1960, several Italian innovators found that trout could be produced in the plains south of the Alps. Parts of this area contained large marshes fed by springs. Water temperatures at the springs were about 12° to 13°C (54° to 55°F). Individual springs had flows of 2000 to 3000 liters per sec (528 to 792 gal. per sec). The marshes were drained and the spring flows channeled. By 1964-65, modern scientific rainbow trout fish farming was in full swing and production increased by leaps and bounds. The major production areas are shown in Fig. 11.1.

By 1965 or 1966, Italy had changed from a trout-importing nation to a trout-exporting nation. During the 1970s Italy became the largest producer of rainbow trout in western Europe, surpassing France and

156

Denmark. Production increased from 12,000 MT (13,224 ST) in 1968[1] to 16,430 MT (18,106 ST) in 1974, a dramatic increase of 37% in 7 years (Table 11.1). Trout exports increased even more dramatically, increasing from 1000 MT (1102 ST), or 8.3% of production in 1968, to 3819 MT (4209 ST), or 23.3% of production in 1974. Total increase from 1000 MT to 3819 MT was 282%, which was an outstanding feat.

FIG. 11.1.   MAJOR TROUT PRODUCTION AREAS, ITALY, 1975

Data from Table 11.1 indicate that domestic consumption of trout in Italy stayed near 12,000 MT during the 7-year period, with production beyond this figure going into export channels. Export data for the years 1971 through 1974 indicate that major markets for Italian trout exports were: France, about 60%; Germany, about 25%; and other countries about 15% in 1971 and 1972. In 1973 Belgium became a significant

[1] 1968 was the first year for which the author was able to gain reliable data on production and exports.

TABLE 11.1

PRODUCTION AND EXPORTS OF RAINBOW TROUT, ITALY, 1968-74

Year	Production (MT)	Exports[1] (MT)
1968	12,000	1000
1969	13,000	1500
1970	13,500	1500
1971	14,000	1800
1972	15,500	2200
1973	15,000	2043
1974	16,430	3819

[1] Imports were only a few MT annually.

importer. More recent data for 1974 indicate that exports from Italy were distributed as follows:

	(MT)
France	1744
Germany	1060
Belgium	704
Switzerland	159
Austria	101
Denmark	42
Others	109
Total	3818

Disposition of domestic trout (about 12,000 MT annually) was about evenly divided between sales to wholesalers and sales to other outlets. About 40% of the trade of "other outlets" was in fee fish-out lakes where customers pay so much per kilogram for the fish caught. Remaining fish sales are divided nearly equally into direct sales to consumers, retailers and restaurants. Before the oil crisis of late 1973 and early 1974 a larger proportion went to fee fish-out lakes. With higher gasoline prices, trips of fishermen to these outlets has declined.

Data for 1974 indicate that average production cost of rainbow trout in Italy was $1.21 per kg ($0.55 per lb). Prices received by farmers, at the farm, were $1.43 to $1.47 per kg ($0.65 to $0.67 per lb) for live fish. Ice-packed fish in the round (whole) sold for $1.50 per kg ($0.68 per lb), and deep frozen fish for $2.31 per kg ($1.05 per lb). A small speciality market for smoked trout existed; average producer price was $5.26 per kg ($2.39 per lb).

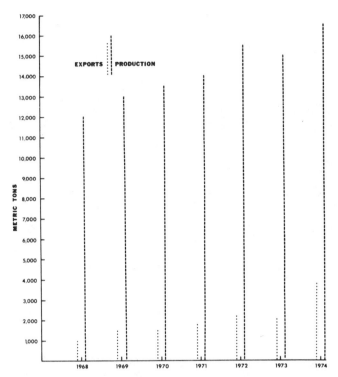

FIG. 11.2.   PRODUCTION AND EXPORTS OF RAINBOW TROUT
FROM ITALY, 1968-1974

Farm Examples

Case 1.—One of the largest fish farming complexes in the world
(if not the largest) is the Salvador farms near Treviso, Italy. These are
a total of eight different farms under one private ownership, most
of which are specialized. Total production is several thousand tons
of rainbow trout annually. Several farms raise only spawning fish
and produce eyed eggs. One raises fish up to about 50 g (2 oz) and 3
farms then finish them from 50 g size up to market weights of 180 to
300 g (6 to 11 oz), depending on which country buys the fish.

Typically each farm uses the stream flow from one or more large
springs for raising fry, fingerlings and food fish. Well water of 12.6°C
(55°F) is used for hatching purposes. About 20 million eggs are produced annually and eyed eggs and fry are sold to other growers.

Water used for food fish production is from streams flowing from
nearby springs. Because of the distance from the springs, water temperature may drop to as low as 9°C (48°F) in winter and rise to as high as

FIG. 11.3.   LARGE TROUT FARM OF 1000 MT, ITALY

FIG. 11.4.   LOADING WEIGHING BUCKETS WITH TROUT HARVESTED
FROM RACEWAY

16°C (61°F) in summer. Each concrete raceway system may be up to 500 m long (1640 ft). These systems are divided into several different segments with the water being aerated throughout the system. The raceways are 10 m (37 ft) wide and 1 to 2 m deep (3 to 6 ft).

At one of the Salvador farms the operation is completely integrated: spawning fish are raised, eggs hatched and fish grown out to market sizes in the same production facility. At this farm, one spring with a constant year-round temperature of 12°C (53°F) is used for hatching eggs. Spring flow is 500 liters per sec (132 gal. per sec). Over nine million eggs are hatched annually. A second spring with an additional 500 liters per sec is used for growing the young fry. At about 4 to 5 cm (2 to 2.5 in.) the fry are transferred by machine to the growing-out segments of various raceways. Here, water is fed by a large stream of 4000 liters per sec (1056 gal. per sec), with temperatures varying from 10.5°C (51°F) in winter to 13°C (55°F) in summer. During the growout period of 4.5 to 5 months, fish are graded and sorted by automatic equipment and transferred to different raceway segments by pumped water pipes. Fish are always fed by tractor-driven feeders. Resulting labor efficiency is 500 MT of fish annually or over 80 MT of fish for each of 6 workers.

FIG. 11.5.   WATER TRANSPORT PIPES LEADING FROM MECHANICAL
TROUT SIZER TO DIFFERENT RACEWAYS

FIG. 11.6.   MECHANICAL FEEDING OF RAINBOW TROUT

Case 2.—One large rainbow trout producer near Milano produces between 500 and 1000 MT of food fish annually. The farm is supplied by numerous springs, the largest of which has a flow of 4000 liters per

FIG. 11.7.   RAINBOW TROUT HATCHERY

sec (1056 gal. per sec). Total spring flow is 7000 liters per sec (1846 gal. per sec). The temperature of the water at the springs is 12° to 13°C (54° to 55°F). The water effluent (waste water) temperature is between 17° and 18°C (63° to 64°F).

FIG. 11.8.   SEINING EARTHEN TROUT RACEWAY, TROUT FEEDING SILOS IN BACKGROUND

The farm has numerous earthen and concrete raceways. They measure up to 400 m (1313 ft) long by 9 m (28 ft) wide. Depth of the water is 1.3 m (4 ft). This typical production unit brings in eyed eggs from the Alps, hatches the eggs and grows the fish through fry and fingerling stages to food sizes. Sales are made in Italy, France and Germany. Sizes of fish sold are 200 to 240 g (8 to 9 oz) in Italy, 180 to 200 g in France[2] (6 to 7 oz) and 250 to 300 g (9 to 11 oz) in Germany. About 50% of total production is exported. About 90% of the total fish are sold alive and 10% are sold as dressed, iced trout. The live fish are sold through brokers or to fish markets, and may be sold to the consumer in the round (whole) or dressed.

---

[2] Some of these small fish are not destined for immediate consumption bet for finishing out in France. See Chapter 12.

FIG. 11.9.   EGG TRAYS USED IN HATCHING RAINBOW TROUT

FIG. 11.10.   MECHANICAL SIZING OF RAINBOW TROUT

FIG. 11.11.   BAILING RAINBOW TROUT INTO MECHANICAL SIZING MACHINE

EUROPEAN EELS *(Anguilla anguilla)*

There are several eel farms in Italy. The first was established in 1969. Some are located in brackish water and others in fresh water. Total output is about 2200 MT annually. Production is expanding rapidly and could double or triple in a few years. Production of 3 kg (6.6 lb) per m^3 (10.7 ft^3) is possible under intensive culturing practices. All of the 2200 MT of cultured eel are produced in northern Italy while 1800 MT of wild eel are captured in southern Italy. Hence, total domestic production is about 4000 MT. In addition, about 4000 MT of live eel are imported from France, Greece, Algeria and other countries. Several hundred MT of deep frozen eels are also imported from the USA, Japan and New Zealand. Hence total supply is in the nature of 8200 MT (9036 ST) annually. However, because of inadequate distribution channels and techniques, about 3000 MT of the domestically produced eel are exported to Germany from northern Italy and the imports are brought into southern Italy. The main demand for eels is in southern Italy; there is only limited demand in the northern part of the country.

About 90% (2000 MT) of the cultured eel is produced in brackish water and only about 200 MT in fresh water. Present planning calls for an additional 1500 MT of brackish production within 15 years and 300 to 500 more MT of freshwater production within 5 years.

FIG. 11.12.   ELVER REARING TANKS FROM 0.1 TO 5 G SIZES

The elvers (larvae eels) are secured mainly from western coast areas, such as a 45 km (30 mi.) area north of Naples. Elvers are captured in November, March and April. In December, January and February the water is cold and they do not migrate. The elvers are trans-

FIG. 11.13.   PADDLE WHEEL AERATION OF SPRINGWATER PRIOR
TO USE IN EEL POND

FIG. 11.14.   EEL RECIRCULATION POND SHOWING CATCH BASIN
AND FEEDING STATION

ported from the catching area to fresh water and brackish water
(Valli - culture or lagoons) culturing areas in northern Italy. Average
size is 7 cm (3 in.) with a weight of about 0.16 to 0.20 g (5500 per kg,
or 2500 per lb). They are stocked directly in the extensive, brackish
water of the Valli-lagoons at the rate of between 2500 and 6500 per
ha (1000 to 2600 per acre), depending on water quality and degree of
intensification of production. No food is fed during the grow-out

FIG. 11.15.   EEL RECIRCULATION POND UNDER CONSTRUCTION

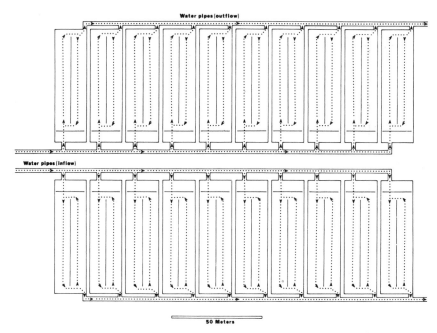

FIG. 11.16.   EXAMPLE OF INTENSIVE EEL CULTURING FARM

stages. Elvers for intensive freshwater culturing are placed in concrete tanks about 5 m in diameter (16 ft) and 0.5 m (19 in.) deep. Stocking is at the rate of 80,000 to 100,000 per tank. Water is constantly circulated through the tanks. The elvers are raised in these tanks until
they are about 2 g in weight. During the 0.2 to 2.0 g cycle they are fed
worms, finely ground fish meal made into a paste, and finely chopped
sea fish. At about 2.0 g weight, the black eels are transferred to rearing or finishing ponds. After stocking these ponds, in which water
is circulated and aerated, the fish are fed fish meal paste and fresh
raw fish.

The elvers are fed 6 to 10% of body weight daily. This declines to
about 1% at harvest weights.

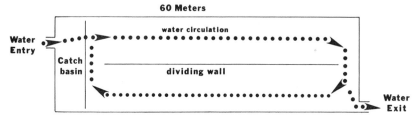

FIG. 11.17.   DETAIL OF ITALIAN EEL POND SEGMENT SHOWN IN FIG. 11.16

Mortality of the fish is about 50% during the 0.2 to 2.0 g cycle, and an additional 50% of the remainder die between the 2.0 g size to finish sizes of 200 g maximum for males and about 350 g for females. From elvers to maturity takes three growing seasons in northern Italy. Hence, to produce 100 MT requires about 400 to 500,000 eels. To secure this many live adults requires nearly 2 million elvers, or about 310 kg (682 lb) of larvae eels. To produce all of the 2200 MT of cultured eel in Italy requires about 7 MT of elvers.

Elver prices were about $18 per kg ($8.20 per lb) in 1975. This means that each elver costs about 0.3 cents. With a 25% survival rate, each fish at harvest would thus cost about 1.25 cents. Prices paid to producers for the 150 to 200 g males and 200 to 400 g females in 1975 ranged between $4.50 and $6.00 per kg ($2.05 to $2.73 per lb). This price was about three times higher than that paid for live trout. Retail prices for mature live eel ranged between $7.46 and $8.95 per kg ($3.39 to $4.07 per lb).

One typical intensive eel farm has nine different pond segments in production with more under construction. Each segment is 60 m long, 18 m wide and 1 m deep (194 ft long, 58 ft wide and 3 ft deep) (Fig. 11.17). Fresh water comes from a well at a constant temperature of 13°C (55°F). Production is 2 MT per segment or 18 MT total. This is at a production rate of about 2 kg per m³ (4.4 lb per yd³). With experience, intensity is expected to be doubled.

BLACK BULLHEAD *(Ictalurus melas)*

There is limited production of black bullheads in Italy. One or two farms established in 1971 produce about 100 MT annually. Small amounts are consumed locally, but the major market is reported to be to fee fish-out ponds or lakes. Prices received are about $2.10 per kg ($0.95 per lb), which is considerably higher than live trout prices.

Drs. P. Ghittino and E. Vigliani (1975) believe that one of the major factors restricting farming of the black bullhead *(Ictalurus melas)*, brown bullhead *(Ictalurus nebulosus)* and other catfish species is an existing Federal law (March 2, 1931, No. 442) which forbids any trade in live catfish. This act was passed when the species were regarded as extremely noxious to fish fauna living in Italian waters and without redeeming value. Until this act is eliminated, catfish farming will remain small.

CARP *(Cyprinus carpio)*

Carp is not a popular food fish in Italy. Therefore there is no com-

FIG. 11.18.   HARVESTING CARP FROM RICE FIELD

mercial production by specialized producers. Before the advent of specialized weedicide sprays, the rice farmers in the Po Valley of northern Italy commonly grew carp in conjunction with rice production. With the discovery of modern weedicides, rice farmers discovered that the value of the additional rice produced using sprays was much higher than having reduced rice yields and producing carp as part of the rotation. Hence, joint production of rice and fish has nearly stopped. A few rice producers still set aside a few hectares of rice for joint production with fish. This is solely to have some protein food for their workers. In 1975 fish yields of 30 to 35 kg per ha could still be found in isolated fields. Total volume of such production would certainly be less than 5 MT.

BRACKISH WATER CULTURE

It is reported that ancient Romans were able to maintain and raise some marine fish along the Italian coast. They probably learned methods of primitive fish farming from the Etruscans who in turn learned it from the Phoenicians. The practice of brackish water culture in fenced lagoons is quite ancient along the North Adriatic coast around Venice and the Po River delta (Ghittino 1975). About 40,000 ha (100,-000 acres) are utilized for fish culture. In addition, the federal govern-

FIG. 11.19.   BRACKISH WATER CULTURING PONDS NEAR VENICE

ment is constructing a 12,000 ha (30,000 acre) facility in Valli de Comácchio south of Venice. When completed, brackish water culture will be conducted on 52,000 ha (130,000 acres).

There are four main species of fish cultured. These are: (1) mullets *(Mugil cephalus, Mugil auratus, Mugil chelo* and *Mugil saliens)*, (2) sea bass *(Dicentrarchus labrax)*, (3) gilthead *(Sparus auratus)* and (4) European eel *(Anguilla anguilla)*. These fish generally belong to the euryhaline group, which can tolerate a high degree of salinity. These fish exhibit a seasonal or life cycle of migrating from sea to fresh water and *vice versa.* The inward or initial migration usually consists of young fish, and the second or outward migration of more or less adult fish returning to the sea to breed or to avoid colder, shallow waters in the lagoons.

Young fish either enter the fenced lagoons, which permit entry but not exit, or are caught in nets by specialized fishermen who, in turn, sell them to owners of lagoons. Descending fish are caught by special traps, which are made of concrete and aluminum screens. In addition, some lagoons can be almost emptied by low tide and then pumped nearly dry to expedite fish harvesting.

Water salinity in cultured lagoons ranges from 10 to 30 ppm. Water temperature ranges from a low near the freezing point to a high of more than 30°C (86°F) in summer. In winter it is often necessary to pump additional seawater into the lagoons to keep temperatures

FIG. 11.20.   MECHANICAL RACEWAY CLEANER

above the critical point for some of the young fish. Average depth ranges from 0.7 to 1 m (2 to 3 ft).

Young estuarine fish are bought to stock cultivated lagoons. The last estimate of the amount spent for stocking fish was in 1954 at $250,000 (De Angelis 1954).

Production of 150 kg of fish per ha (134 lb per acre) is considered satisfactory. The Italian fish market requires mullets and sea bass of 300 to 1000 g (11 to 35 oz), which requires 3 to 5 years; gilthead of 150 to 400 g (5 to 14 oz), requiring 1 to 2 years; and adult silver eel averaging 400 g (14 oz) and requiring 7 to 9 years.

Italian brackish water fish culture is extensive. With a water area of 40,000 ha and an average yield of 150 kg per ha, the total brackish water production is 6000 MT (6612 ST). This volume should expand by over 30% when the new federal project is completed. Composition of these 6000 MT by species is not known. Estimates indicate that it is about equally distributed among the four species.

At present, little or no artificial food is fed in brackish water culture. If artificial foods can be successful, fish production could reach 2.5 MT per ha for an increase of 1567%. This means total production might reach 90,000 MT.

Scientific advancements under way indicate production will be intensified in the near future. For example, fingerlings are currently be-

FIG. 11.21.   GILTHEAD HATCHING TANKS

ing prophylactically treated with formalin baths for external parasites prior to stocking the lagoons. Survival rates have increased to 80% in treated fish, compared to 20 to 30% in untreated fish. In addition, sea bass and giltheads have been successfully hatched and grown under artificial conditions. The incentive for artificial reproduction is the high value of the fry. In 1975, 2 to 3 cm (1 in.) sea bass fry were selling for $0.15 each. Similar sized gilthead fry were selling for $0.45 each. One breeding and spawning plant is now in production to produce fry (Anon. 1975). The goal is to produce 8 million fry annually. The plant is jointly sponsored by federal government and private capital. The difficulty in raising fry has been the need to feed the larva fish phytoplankton and zooplankton. Hence, it was necessary to develop an economical way of culturing large quantities of phytoplankton and zooplankton before fry could be raised economically.

OUTLOOK

Italy still offers opportunities for increased cultured fish production of rainbow trout, eel, catfish (bullheads) and brackish water species. Fish farmers and professional fisheries experts predict that rainbow trout production could, with favorable prices and control of viral hemorrhagic septicemia (VHS), increase by 15 to 20% above the 1974

FIG. 11.22.   MECHANICAL NOISEMAKER TO FRIGHTEN AWAY
PREDATOR BIRDS

level of 16,430 MT (18,106 ST). This increase would result from fuller use of existing freshwater flows and could amount to 3300 MT (3637 ST). The author also predicts, by intensifying production per unit of water flow, total production could increase by 50% or an additional 9715 MT (10,706 ST). This would result in production per unit of water flow increasing from an average of no more than 143 kg per liter of flow per sec to at least 210 kg (increasing from about 600 to 1800 lb per gal. per sec). Additional increases might result from better disease controls. Hence it is conceivable that trout production could increase from 16,430 MT (18,106 ST) to 29,000 to 30,000 MT (31,958 to 33,060 ST).

Increases in eel culture are indicated by the fact that freshwater eel culture began only five years ago. The present cultured volume of 2200 MT might, with favorable price relationships, increase tremendously. A projected forecast of 8000 to 10,000 MT appears a possible achievement.[3]

Catfish (bullhead) production is, at present, almost totally prohibited by the 1931 law. Removal of this restrictive law could well result

---

[3] Extensive eel culture, such as is found in brackish water, has in recent months been practically destroyed by a parasitic disease, resulting in 90% losses. In intensive freshwater culture it is easier to control this disease.

in a new major fish species for culturing. Total production would be limited only by market acceptance and price relationships.

Brackish water production of the mullets, sea bass and giltheads is increasing under existing conditions. A 30% increase over the present level of 6000 MT is expected by an ongoing Federal project. In addition, if artificial foods can be successfully fed and production per given volume of water intensified, production could increase more than 15-fold, or to over 90,000 MT (nearly 100,000 ST).

SPECIAL ACKNOWLEDGEMENTS

DR. S. CANCELLIERI, Direttore, Associazione Piscicoltori Italiani, Via Indipendenza 5, 31100 Treviso, Italy

DOTT. FRANSESCO GHION, Centro Ittiologico Valli Venete, Valpisani Ca'Venier 45010, Italy

PROF. DR. PIETRO GHITTINO, Fish Disease Laboratory, Instituto Zooprofillattico Sperimentale del Piemonte e delle Liguria, Via Bologna 148, 10154 Torino, Italy

DOTT. IGINO MANDELLI, 27023 Cassolnovo (Pavia), Italy.

DR. GINO RAVAGNAN, Societá Industriale Riproduzione Artificiale Pesce, Via Euganea 17, 35100 Padova, Italy

MR. LUIGINO SALVADOR, Azienda Agricola di Bortolo Salvador & Canizzano, 31100 Treviso, Italy

REFERENCES

ANON. 1975 Industrial Society for the Reproduction of Artificial Fish (S.I. R.A.P.), (Italian).

DE ANGELIS, R. 1954. The capture and stabling of young marine fish for the seeding of inland waters. Fish Bull. Vol. 5-6. (Italian).

GHITTINO, P. 1975. Estuarine fish culture in Europe. Italian Fish Cult. and Fish Pathol. 2, 37-41. (Italian).

GHITTINO, P., and VIGLIANI, E. 1975. Difficult take-off of catfish farming in Italy. Italian Fish Cult. and Fish Pathol., Vol. 2. (Italian).

France

RAINBOW TROUT (Salmo gairdneri)

The rainbow trout industry of France, not yet having reached maturity, is still undergoing rapid expansion. For example, in 1965 total production was less than 3000 MT. In 1974 production reached 15,000 MT. The index of production changed from 100 in 1965 to 501 in 1974 (Table 12.1), increasing production 401% in 10 years.

In 1975 there were 700 trout farmers in France. However, about 200 of these buy fish weighing about 160 g each (6 oz) and, in a few weeks, increase their weight to market size of 180 to 200 g (about 7 oz). While technically these finishers may be considered fish farmers, it may be more realistic to consider them as fish "jobbers." Their aim is to increase weights by 20 to 40 g (1 to 1.5 oz) and resell the fish to hotels, restaurants and retailers at a much better price. They buy in large quantities and sell in small quantities.

Of the remaining 500 *bona fide* trout farmers, most are small producers averaging between 10 and 20 MT per year. It is estimated that 50 large-scale farmers account for one-half of total production.

However, of these 500 farmers, about 150 farmers raise only part of their stock and buy the remainder for growing-out. These 150 farmers act a dual role of fish farmer and jobber. Hence, only about 350 of the 700 enterprises do not perform some marketing role.

Production Regions

Trout culturing is unequally distributed throughout France. The French divide the country into eight regions. These are: (1) North, (2) East, (3) Parisian Basin, (4) Normandy, (5) Brittany, (6) Southwest Central, (7) Southeast, and (8) Pyrenees (Fig. 12.2).

TABLE 12.1

CHANGES IN PRODUCTION OF RAINBOW TROUT FOR FOOD FISH IN FRANCE, 1965-74

Year	Index	Production (MT)	Production (ST)
1965	100	2992	3298
1966	148	4418	4870
1967	201	6006	6620
1968	238	7108	7835
1969	262	7827	8628
1970	308	9206	10,148
1971	352	10,516	11,592
1972	388[1]	11,595[2]	12,781
1973	451	13,500[3]	14,877
1974	501	15,000[3]	16,530

[1] Indexes for 1965-72 secured from Desplanques *et al.* (1972).
[2] Production data for 1972 in metric tons secured from Desplanques *et al.* (1972), Indexes for 1965-72 were used to compute MT of production for 1965-71.
[3] Metric tons for 1973 and 1974 secured from Tessier (1975).

In 1972 Brittany was the largest trout producer of any region in France. Its production was 3068 of a total 11,595 MT. This was 26.4% of total production (Table 12.2). Other regions with high production were North (16.7%), Normandy (14.7%), Southeast (12.4%) and Pyrenees (11.1%).

The northern half of France, where trout production began, represents approximately 70% of total production while the southern half accounts for about 30%. The southern half is an autonomous region and

FIG. 12.1.   LARGE TROUT FARM (1000 MT) FOR RAINBOW TROUT

FIG. 12.2.   PRODUCTION OF RAINBOW TROUT BY AREAS IN METRIC TONS,
FRANCE, 1972

exchanges little of its production with other regions. Nearly every producer has a distinct local market, or markets, for his production. In
the northern half of France, including Brittany, Normandy, North,
East and Parisian Basin regions, production is exchanged between regions. Their excess production during different times of year is sold at
the National Market at Rungis in Paris. Their contributions to this
National Market are growing. In 1965 only 688 MT were committed,
while in 1971, 1655 MT or 21% of their production was committed. The
size of enterprises also varies between northern and southern France.
One study concluded that northern France had about two-thirds of
production by only about one-third of the producers, while southern
France had nearly two-thirds of the producers and produced only
slightly more than one-third of total output. Fish producers with production greater than 10 MT per year represent more than half the producers in Brittany, North and Normandy. The average volume of production per producer in Brittany was 66 MT, North 43 MT, and Normandy 32 MT. The average for all other regions is about 15 MT.

Because of the distinct nature of the two areas, northern and southern France, producer prices in southern France are semi-independent
of the National Market at Rungis in Paris. In general, their prices are

higher by at least the cost of transportation from the northern part of France, and usually bear some premium. This is particularly true when seasonal shortages occur in the North.

In the northern half of France, trout prices at both producer and consumer levels vary more widely. This is due to more concentrated production with more seasonality in production. The regions of northern France which affect production most and have the widest swings in seasonal production rates are Brittany and Normandy. These two regions accounted for 41.1% of all France's production in 1972. However, all of northern France, which also includes the North, East and Parisian Basin regions, accounted for 66% of production.

Low prices occur in most years at the National Market at Rungis in Paris during the April-May-June period, which corresponds to peak periods of marketing for both the Brittany and Normandy areas. Highest prices usually occur in the December-January-February period, which also corresponds to the low months of marketing for both the Brittany and Normandy regions.

Table 12.2 gives monthly marketing of trout in metric tons for each of the eight regions of France. For 1972, these two regions accounted for 41.1% of total French production. However, during the three months of April, May and June they reached peak levels, and during December, January and February they reached their lowest levels. During April they sold 58.5% of the French total, 56.7% in May, and 51.5% in June. By comparison, during December they accounted for only 35.8%, 27.7% in January, and 37.7% in February (Table 12.3).

The North region was somewhat of an equalizer by having different low and high marketing periods. This region's low period of marketing was in the April-May-June period, when Brittany and Normandy flooded the market. However, this was not sufficient to stem the tide, and prices fell on the central market. The effects of all three of these regions are shown in Fig. 12.3. An even clearer presentation is shown in Fig 12.4, which gives average percentages of production for Brittany, Normandy, North and France for the lowest and for the highest three months of marketing. The difference between lows and highs for Brittany was 365%; for Normandy, 91%; for North, 79%. For all of France, marketing during the peak months of June, July and August was exactly 100% greater than during the low months of December, January and February.

These data show clearly that Brittany was the arch villain in price swings, that Normandy was a culprit, and that North was a fledgling hero. While it is easy to point an accusing finger at both Brittany and Normandy for causing low prices, the reader should also realize the cause of these wide fluctuations in marketing. In both Brittany and Normandy a disproportionate share of annual production must be

TABLE 12.2

ESTIMATION OF ACTUAL TROUT MARKETS IN METRIC TONS BY REGIONS OF FRANCE, 1972

Region	Jan.	Feb.	Mar.	April	May	June	July	Aug.	Sept.	Oct.	Nov.	Dec.	Total 1972	% of Total
Brittany	50	110	277	449	521	483	358	198	205	119	144	154	3068	26.4
Normandy	86	118	162	237	193	177	130	118	116	122	122	131	1712	14.7
North	162	134	124	125	102	124	141	206	195	210	226	185	1934	16.7
Parisian Basin	14	14	36	49	65	52	22	45	38	29	25	34	423	3.6
SW Central	32	50	55	62	81	90	140	183	116	61	59	73	1002	8.6
East	36	48	60	72	76	80	80	84	60	52	44	76	768	6.6
Southeast	67	77	96	102	134	163	181	233	160	70	74	84	1441	12.4
Pyrenees	43	55	65	77	82	112	168	203	116	142	124	60	1247	11.1
France	490	606	875	1173	1254	1281	1220	1270	1006	805	818	797	11,595	100.0

Source: Desplanques *et al.* (1972).

TABLE 12.3

MONTHLY RELATIVE IMPORTANCE OF THE PRODUCTION OF EACH REGION IN PERCENT OF TOTAL
TROUT MARKETING, FRANCE, 1972

Regions	Jan.	Feb.	Mar.	April	May	June	July	Aug.	Sept.	Oct.	Nov.	Dec.	Total 1972
Brittany	10.2	18.2	31.7	38.3	41.3	37.7	29.3	15.6	20.2	14.8	17.6	19.3	26.4
Normandy	17.5	19.5	18.5	20.2	15.4	13.8	10.6	9.3	11.6	15.1	14.9	16.5	14.7
North	33.1	22.1	14.1	10.6	8.1	9.7	11.6	16.2	19.4	26.1	27.6	23.2	16.6
Parisian Basin	2.9	2.3	4.1	4.2	5.2	4.1	1.8	3.5	3.8	3.6	3.1	4.0	3.6
SW Central	6.5	8.2	6.3	5.3	6.5	7.0	11.5	14.4	11.6	7.6	7.2	9.2	8.6
East	7.3	7.9	6.9	6.1	6.1	6.3	6.6	6.6	6.0	6.5	5.4	9.6	6.6
Southeast	13.7	12.7	11.0	8.7	10.7	12.7	14.8	18.4	15.9	8.7	9.0	10.6	12.4
Pyrenees	8.8	9.1	7.4	6.6	6.5	8.7	13.8	16.0	11.5	17.6	15.2	7.6	11.1
France	100.0	100.0	100.0	100.0	100.0	100.0	100.0	100.0	100.0	100.0	100.0	100.0	100.0

Source: Computed from Table 12.2.

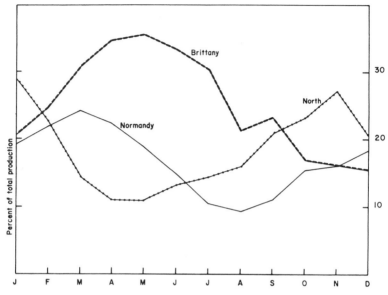

FIG. 12.3.   RELATIVE IMPORTANCE OF THE THREE MAJOR RAINBOW
TROUT PRODUCTION AREAS BY MONTHS, FRANCE, 1972

sold in April, May and June before water shortages occur in the summer and water temperatures increase. As temperature increases,
the oxygen-carrying capacity of the remaining water flow decreases,
and fish numbers must be reduced. During these months, fish prices
are depressed in northern Paris markets and fish from Brittany move
to other countries as well as to southern France. By August the temporary glut of food-size fish is over and a temporary shortage of the
right sizes occurs. During August and September imports are made,
chiefly from Italy and Denmark.

Imports and Exports

France has long been a deficit country in regard to rainbow trout
production. For example, in 1960, 1905 MT were imported and only
158 MT exported, resulting in a deficit of 1747 MT (Table 12.4).
Then from 1962 to 1968 imports declined. Beginning in 1968 imports
and exports increased, and a large annual deficit of trout occurred
each year. For the 5 years of 1970-74, the annual deficit was 1058 MT
annually. This amounted to an average deficit of 8.8%.

During the 6 year period 1968-73, a total of 8741 MT (9633 ST) of
rainbow trout was imported into France (Table 12.5). These imports
had a total value of $13,552,000. On a yearly basis, 1457 MT (1605 ST)
worth $2,258,666 were imported. Exports amounted to 282 MT (311

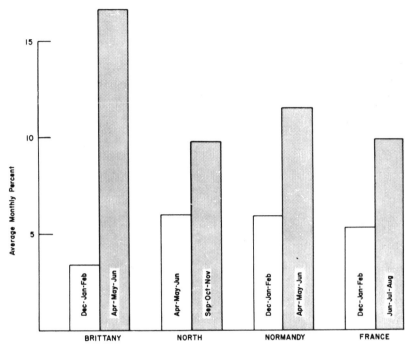

FIG. 12.4.   AVERAGE PERCENTAGES OF LOW AND HIGH THREE MONTHS TROUT
MARKETING PERIODS FOR THREE DISTRICTS AND FOR FRANCE, 1972

ST) annually, worth $464,000. These figures indicate that France had a deficit of 1175 MT (1294 ST) of rainbow trout annually. In spite of French exports selling for more than French imports on a per kilo-gram basis, the annual outflow of money in dollars was $1,794,666.

Observing France's trout export and import locations is interesting. In 1973 and 1974 the chief country for exports was Belgium (Table 12.6), which accounted for 89% of total exports. This is contrary to the belief of many French fish farmers that chief markets are West Germany and Switzerland. Seventy-five percent of all imports came from Italy, the primary importer, in 1973. In 1974 this increased to 80%. Denmark accounted for more than one-half of the remainder.

Trout Prices

The 1975 farm price of trout in southern France averaged about $2.20 per kg. Depending on location and local demand, this price was for live fish delivered within a radius of 100 km (62 mi.), or fish at the producer's pond, with an extra charge made for icing, dressing and packaging. However, some retailers furnish their own packaging

TABLE 12.4

PRODUCTION, IMPORTS, EXPORTS, DEFICIT AND TOTAL SUPPLY OF RAINBOW
TROUT IN FRANCE FOR THE 15 YEAR PERIOD 1960-74

Year	Production (MT)	Imports (MT)	Exports (MT)	Deficit (MT)	Total Supply (MT)
1960	NA[1]	1905	158	1747	NA
1961	NA	1780	187	1593	NA
1962	NA	36	2	34	NA
1963	NA	481	10	471	NA
1964	NA	468	12	456	NA
1965	2992	565	22	543	3535
1966	4418	433	65	368	4786
1967	6006	520	68	452	6458
1968	7108	1287	60	1227	8335
1969	7827	1425	43	1382	9209
1970	9206	1000	380	620	9826
1971	10,516	1037	333	704	11,220
1972	11,595	1983	406	1577	13,172
1973	13,500	1979	466	1513	15,013
1974	15,000	2167	1290	877	15,877

Source: 1961-74 production data secured from Table 12.1. 1960-69 export and import data from Desplanques *et al.* (1972). Data for exports and imports for 1970-74 from Anon. (1974).

[1] NA—Not Available.

materials. If the producer furnished packaging and boxes, about $0.20 per kg ($0.09 per lb) was added on. In southern France, only the very large producers sell outside the local areas. Large producers may sell directly to dealers in Belgium and other countries.

Producer prices in southern France bear little relationship to Paris prices. Southern prices are strongly affected by tourist influx between April 1 and September 1. Some producers sell more than 80% of their annual output during these five months. Southern prices are reasonably stable throughout the year.

Producer prices in northern France are generally 10 to 15% below those found in southern France. The average producer price in northern France in 1975 was about $2.00 per kg ($0.90 per lb). The price in northern France is usually determined by the National Market at Rungis in Paris.

In 1971 total production of rainbow trout in northern France from the regions of Brittany, North, Normandy, Parisian Basin and East was 7329 MT. The National Market at Rungis, Paris handled 1679 MT of this amount, or 22.9%. In 1972 production from the five regions was 7905 MT. Estimated sales at the National Market were 2004 MT or 25.4%. These data indicate the relative and increasing importance of that market.

TABLE 12.5

IMPORTS AND EXPORTS OF RAINBOW TROUT, FRANCE, 1968-73

Year	Imports (MT)	Value in Dollars (000's)	Price (per kg)	Price (per lb)	Exports (MT)	Value in Price Dollars (000's)	Price (per kg)	Price (per lb)
1968	1287.2	1622	1.26	0.57	60.7	106	1.74	0.79
1969	1423.7	2093	1.47	0.67	44.0	90	2.04	0.93
1970	1023.7	1781	1.74	0.79	379.0	680	1.79	0.81
1971	1037.7	1785	1.72	0.78	333.3	520	1.56	0.71
1972	1982.2	2815	1.42	0.65	406.9	521	1.28	0.58
1973	1986.4	3456	1.74	0.79	466.0	867	1.86	0.85
Totals and Averages	8740.9	13,522	1.55	0.70	1690.8	2784	1.65	0.75

Source: Anon. (1974). Original data secured from Customs Service.

TABLE 12.6

EXPORTS AND IMPORTS OF RAINBOW TROUT, FRANCE, 1973 AND 1974

Country	1973 (MT)	1974 (MT)
Exports from France		
Belgium-Luxembourg	407	1161
West Germany	9	62
Andorra	26	21
Others	24	46
Totals	466	1290
Imports to France		
Italy	1480	1742
Denmark	262	283
Belgium-Luxembourg	170	92
Norway	18	22
Ireland	17	12
Others	32	16
Totals	1979	2167

Source: Anon. (1971-74).

In southern France during 1975 farmer-jobbers were buying trout at $2.20 per kg ($1.00 per lb) delivered to their ponds. They, in turn, sold about 80% of their finished fish to hotels and restaurants, and 20% to retail stores in small quantities a few weeks later at about $3.33 per kg ($1.51 per lb) for dressed fish. On a heads-on, dressed-out basis, this compares to $2.73 per kg ($1.24 per lb).

The retailers, in turn, sell the $2.13 per kg ($1.24 per lb) dressed fish to restaurants or housewives at an average price of $4.13 per kg ($1.88 per lb).

There is a small but growing fee fish-out market for producers. Most fee fish-out enterprises are located in a general area south of Paris, but some are scattered throughout central France. It is estimated that 5 to 10% of total production is sold to fee ponds for recreational fishing. Trout producers' prices for these fish averaged about $2.65 per kg ($1.20 per lb) in 1975. Catch-out price at the fee fish-out pond was $4.35 per kg ($1.98 per lb), which is higher than retail prices of $4.13 per kg ($1.88 per lb). In addition, the fee fish-out price was for fish in the round, whereas retail price was for dressed fish.

Marketing Channels

Marketing channels for rainbow trout sold in France are remarkably different from those found for most products. The typical picture is farmer to wholesaler to retailer to consumer. Producers sell French-

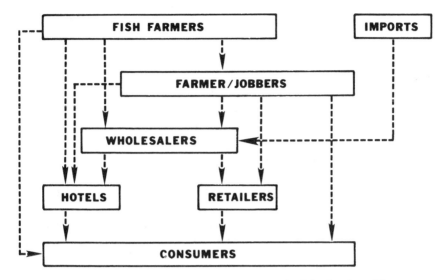

FIG. 12.5.   MARKETING CHANNELS FOR RAINBOW TROUT IN FRANCE

produced trout directly to farmer-jobbers, wholesalers, hotels and restaurants, and the ultimate consumer (Fig. 12.5). The farmer-jobbers also sell direct to all levels, including wholesalers, hotels, restaurants and the final consumer. No information is available concerning relative volumes sold at each step. Imports usually go directly to French wholesalers.

Farm Examples

Case 1.—Of the approximately 14,000 MT of rainbow trout produced in France, only about 200 MT are produced in brackish water. About one-half of total brackish production comes from the Méditerranée Pisciculture S.A. farm at Salses near the Spanish border. A brief discussion of this operation will enable the reader to understand some problems faced by French producers in using brackish water for rainbow trout production. Mr. Jouy, the owner, started this farm in 1972 and is still working at removing obstacles and problems associated with the site.

Fresh water supply is from a stream which goes underground about 10 km away and then surfaces at the fish farm from the bottom of a cliff. The oxygen content of the resurfaced water is only 5.5 ppm. Hence, artificial oxygenation is necessary to raise oxygen content. The terrain is so level that mechanical rather than gravity aeration must be used. Maximum use of the fresh water flow has not yet been attained. Existing plans call for a substantial increase in production. The fresh water flow is 1000 liters per sec. (264 gal. per sec) with a constant temperature of 17.5°C (63.5°F). Thus, there is a temperature limitation on use of the water. In summer months, production is limited by only being able to use the water until the temperature increases 4° to 5°C (8°F). Thus, year-round production is restrictive, being able to stock only limited volumes of fish in the summer.

Some of the fresh water is used to dilute salt water brought inland from the bay area. The salt water reaches a temperature of 26° to 27°C (80°F) in summer. Fresh water is added so that the brackish water has a salt content of 20 to 21 ppm. The mixed brackish water is cold enough for production only between December 15 and May 15, or for five months.

Trout can be raised from hatching to market sizes of 200 g (7 oz) in about 10 to 12 months. On a 12 month schedule of growth, eggs must be hatched about May 1, reared in the newest fresh water until December 15, and then transferred to the brackish water for growing-out before the next May.

All trout are sold in the round (undressed) to retailers within a 100 km (62 mi.) radius. There is no relationship to the central market price in Paris. Prices are fairly stable throughout the year.

Another enterprise producing rainbow trout in brackish water is found just south of Bordeaux on the Atlantic Ocean side of France. As is true for the Mediterranean side, the brackish water is too warm for trout production during the months of June through September.

FIG. 12.6.   TIDAL WATER GATE FOR SALTWATER CULTURE OF FISH

Hence, 100 fingerlings (3.5 oz) are brought in to be fattened to a market size of 250 g (8 oz). The production period is about three months. Salinity of the water varies from 18 to 30 ppm. These fish are produced in old salt beds which have been in fish culture since 1750.

Production costs are higher in brackish water than in fresh water. For example, feed costs increase about 40% due to the need to add a binder and to greater loss of feed, which dissolves faster in brackish water. This additional feed cost amounts to about $0.17 per kg ($0.08 per lb). The technicians working with brackish water culture believe that higher production cost can be recovered in the marketing system by emphasizing the pink, fleshy color and the firmer texture of the product. There is also the possibility of differentiating the product by selling different sizes of fish. These might be retail sizes of 200 g (7 oz) for portions, family sizes of 400 g (14 oz) for restaurants, and 600 g sizes (21 oz) for smoking.

There are two ways of maintaining oxygen levels in brackish water culture. With either method, tides are relied upon to exchange the water twice daily. If fish are not intensively stocked, no supplement aeration may be necessary. One way of providing necessary supplementary oxygenation is with mechanical aerators. The second method is to provide for a reservoir which is higher than the fish cultural ponds. At high tides the reservoir is filled. Between high tides, water is released from the reservoir into the cultured ponds. This system requires an extensive, expensive system of levees and water gates, however. In some parts of France such systems already exist.

Case 2.—One of the largest, if not the largest, trout enterprises in France is near Mezos, France, south of Bordeaux. The enterprise consists of two different production units. One unit hatches the eggs and raises the fry to about 50 g each (nearly 2 oz). Then the 50 g fingerlings are transported to the grow-out unit several kilometers away.

Trout eggs for hatching are imported from different countries in order to spread out the production cycle so that harvesting and marketing will be more uniform throughout the year. Eggs are imported from the USA, Australia and Denmark. Eggs produced in France are also used. Hatching takes place every month but June and July. The water source for hatching the eggs is a spring having a water flow of 350 liters per sec (92.5 gal. per sec). The spring water is a constant 13°C (55°F).

One month after hatching, the fry are moved to rearing raceways. These raceways are supplied by a separate spring-fed stream; the temperature of raceway water varies from a low of 9°C to a high of 15°C (48° to 59°F). In 6 to 8 months the fry reach an average size of 50 g (nearly 2 oz). The fish are graded for size twice during this period. The fry and fingerlings are fed 6 to 8 times daily by mechanical feeders. All work is performed by four workers.

After the fish reach 50 g they are transferred to the grow-out unit several kilometers away. Water supply for this operation comes from a large spring having a flow of 3000 to 6000 liters per sec (792.7 to 1585.5 gal. per sec). The water flows through 13 lines of raceways. Each line is divided into 5 segments, or 65 segments *in toto*. Each segment is 1 m deep, 8 m wide and 60 m long (3 ft deep × 26 ft wide × 194 ft long). The fish are grown from 50 g to 200 to 250 g (from 2 to 7-8 oz) in 4 to 4.5 months. During this time they are mechanically graded twice by size. At harvest they are sold in the round or dressed and packed in ice. Sales are made throughout France to wholesalers. Direct export sales are also made to supermarkets in Belgium and Switzerland.

Production efficiency is very high at this farm. Output of trout per worker per year ranges between 70 and 80 MT.

At this facility a quick-freezing plant was constructed. It became operational in 1976 and was the first specialized quick-freezing trout operation in France. The plant is capable of processing, storing and marketing several thousand metric tons of trout annually.

Case 3.—One rather typical enterprise was visited in central France. This fish farmer had three different farms: one to hatch eggs and

FIG. 12.7.   RAINBOW TROUT EGG HATCHING EQUIPMENT

produce fry; one to rear fry to fingerling size; and one to complete the grow-out process.

At one time this farmer raised his own brood stock and produced his own eggs. He then shifted to importing one-half of his eggs from Denmark. He recently shifted entirely to imported eggs from Denmark. He remarked that it was cheaper to buy all eggs than to produce his own. He buys eyed eggs annually during the January 15-May 5 period. Eggs are purchased in units of 200,000 at a time. A hatching tray is divided into 4 compartments, each holding 10,000 eggs, for a total of 40,000 eggs per tray. After hatching, the yolk fry go by gravity to the holding tank under the tray.

From the holding tank, fry are put into rearing tanks, each 7 m long, 1.5 m wide and 0.5 m deep (22 ft X 5 ft X 18 in.). Each rearing tank holds 30,000 fry. These fry are then grown to a maximum length of 6 cm (2.5 in.).

At 6 cm the fry are transported to the second farm. This farm has sufficient raceways to rear the fry to 10 cm (4 in.). A raceway is divided into 3 segments, each of which is 30 m long, 3 m wide and 0.5 m deep (92 ft long X 3 ft wide X 18 in. deep).

After reaching 10 cm fingerlings are transported to the grow-out farm for rearing to market sizes varying between 150 and 220 g (5.5 to 7.5 oz).

This farm (all three units) is designated as a disease-free certified

FIG. 12.8.   FRY REARING TROUGHS

FIG. 12.9.   NEW FROZEN TROUT PROCESSING PLANT UNDER CONSTRUCTION

farm. Hence, sales prices for fingerlings are higher than noncertified farm prices. However, prices for food fish remain competitive with other producers. Sales of fingerlings from this farm during September, 1975 were priced at $8.70 per kg ($3.85 per lb) for 7 cm (2.75 in.) fish. This amounted to about $0.033 each. The price at noncertified farms was $5.16 per kg of fish ($2.35 per lb), costing about $0.02 per fingerling.

Feed conversion from time of hatching to market size was 1.8 to 1. Feed cost was $0.69 per kg ($0.31 per lb).

Case 4.—One of the nicest, but by no means the largest, fish farms visited by the author was in Brittany. This farm was owned by an elderly gentleman more than 80 years of age. Daily work was performed by one hired man, but during busy times such as grading an additional man was hired. Production was 50 MT (55 ST) per year.

The farmer had no pollution, had disease-free spawning stock, produced all his own eggs and raised only his own fry and fingerlings. He had no disease problems and only minimal parasite problems.

This farmer had his own private spring which had a normal water flow of 300 liters per sec (79.3 gal. per sec). The lowest flow ever encountered was 180 liters per sec (47.6 gal. per sec).

Water temperature at the spring was 9°C (48°F). The temperatures of the discharge waters after being used in the fish farm were never lower than 6° or 7°C (44°F) in winter and never exceeded 21°C 70°F) during the worst heat wave.

Eggs are produced from January 15 to February 15 from 700 to 800 female spawners and 200 to 300 male breeders. Egg output averages 1.5 million annually. From these he sells between 200,000 and 400,000 fingerlings annually and 750,000 food-size trout. The survival rate is between 65 and 76%. His 12 to 15 cm (5 to 6 in.) fingerlings in 1975 sold for $0.08 to $0.09 each. The remainder of the fish, about 50 MT, were sold at 200 to 250 g sizes (7 to 8 oz). About 50% were sold locally through fish markets or to a nearby fish freezing plant. This plant freezes mostly saltwater fish. The remainder were sold through the central market in Paris.

Case 5.—This involves what is perhaps one of the most interesting trout enterprises in France. The spring water is a constant 10.5°C (51°F). Two parts of recirculated water are used for each part of new spring water.

Eggs are purchased during the November to February period. Production time, from hatching to market size, averages 16 months. Survivability during the rearing and growing-out stages to market weights averages 56% of the eggs. The fish are marketed between 160 and 200 g (6 to 7 oz) to fish jobbers who add 20 to 40 g (1 to 1.5 oz) of weight per fish. Eighty percent of their production is marketed between April and August to coincide with the tourist season in the area. In this area, producer trout prices do not follow the seasonal low prices of April-June of the National Market in Paris.

FIG. 12.10.   OVERHEAD AERATION FROM CANAL LEADING FROM A RIVER

FIG. 12.11.   INTENSIVE AERATION OF TROUT WATER

Maximum use is made of aeration, and production per unit of water flow was the highest found on any trout farm visited in western Europe by the writer. Production was 555 kg of fish per liter of fresh water per sec (76.8 lb per gal. per min). Disease problems were minimal and mortality was less than that of many farms where only a fraction of his output per unit of fresh water was attained.

Additional labor efficiency was high. Production per worker averaged between 70 and 80 MT per year.

Production Costs

As is true in nearly all, if not all, trout producing countries, the major cost is for feed. This is true for France. In 1975, feed costs for different feeds in 20,000 kg (44,000 lb) trucklots varied from a low of $333 per MT ($367 per ST) for 45% protein feeds, for fish between 100 to 300 g (4 to 11 oz), to a high of $378 per MT ($417 per ST) for 50% protein feeds for young fry. Fish pellets for other stages of growth ranged between these two figures. A feed conversion of 1.7 kg of feed per kg of growth resulted in an average cost of about $0.58 per kg ($0.26 per lb) for feed. This amounted to about 36% of total production cost.

From discussion and data secured from interviewing numerous trout producers, rather consistent cost of production figures by areas appeared. In 1975 total costs of producing trout in France, including feed, labor, fingerlings and fixed overhead costs, ranged between $1.38 and $1.83 per kg ($0.63 to $0.83 per lb). The lower figure of $1.38 per kg was representative of larger, more efficient producers and average-size producers in Brittany. The higher figure of $1.83 per kg was for the smaller, less efficient producers outside Brittany. A mean cost of $1.60 per kg ($0.73 per lb) would probably be representative of the average producer.

OUTLOOK

In 1975 rainbow trout production was hovering around 15,000 MT annually (16,530 ST). Estimates by French trout producers and specialists suggest production could increase by 15 to 30% before all existing fresh water of the proper temperature and purity, including seasonal production, would be used.

This projected increase does not take into account possible increases from fish disease control. The worst disease is viral hemorrhagic septicemia (VHS) which may be restricting total production by 15%. In addition, some farmers have problems with infectious pancreatic necrosis (IPN) which affects young trout.

Moreover, if all producers were able to increase output per unit of fresh water flow to the maximum possible, production might well in-

FIG. 12.12.   FEED STORAGE BINS FOR TROUT FEED

FIG. 12.13.   MECHANICAL FEEDING OF TROUT FROM BOTH SIDES OF FEEDER

crease by 50 to 60% (see section on Hydrologic Information). In addition, if brackish water culture expands, there may be no foreseeable limit on this type of production. There are thousands of hectares of brackish water area in France which could conceivably be converted to production. Present techniques result in 4 to 20 MT per ha (1800 to 9000 lb per acre). Among these four possibilities, it would appear possible for trout production to at least double in the foreseeable future.

BROWN TROUT *(Salmo trutta)*

Brown trout are not commonly grown in France as a food fish. However, there is a large demand for brown trout fingerlings for restocking rivers and streams. The number of fingerlings stocked could not be determined. However, there are approximately 4.2 million licensed anglers. Many of these people fish primarily for brown trout. Hence the number of fish stocked could be somewhere between 50 and 100 million fish.

CARP *(Cyprinus carpio)*

There is one intensive carp producer in France. He feeds his fish and fertilizes his ponds and is reputed to be the largest producer of carp in the country. His total production of 225 to 250 MT is taken from 200 ha of water. Average production is 1.25 MT per ha (2200 lb per acre).

The remainder of carp producers are extensive enterprises, using no feed and little or no pond fertilization. The number of these producers is unknown but may number several hundred. On the extensive-type operations, production is estimated to be about 180 kg per ha (160 lb per acre). Total cultured carp production is difficult to determine. No government studies or data are available. Discussions with different individuals who are extremely knowledgeable about France's cultured fish industry resulted in widely varying estimates of total production, hectares of production and area of consumption. The most pessimistic estimate of carp production was 5000 MT (5510 ST). The most optimistic estimate indicated 14,000 MT (15,428 ST), and the most common estimate was 8000 to 10,000 MT (8816 to 11,020 ST). Since total production is a function of total hectares of water area and yields per hectare, it was only natural that these estimates also varied widely. Estimated consumption ranged from 80% of some unknown quantity exported to West Germany and 20% consumed in France, to 80% consumed in France and 20% exported to West Germany. Data secured in West Germany for carp imports from France for the 1970-75 period indicate that even the most conservative

FIG. 12.14.   RAINBOW TROUT PRODUCED IN BRACKISH WATER
ON THE WAY TO MARKET

FIG. 12.15.   CARP *(CYPRINUS CARPIO L)* AREAS IN FRANCE, 1975

estimate of production and exports to West Germany was too high. Annual imports from France during these years varied from low of 300 MT to a high of 792 MT (see Chapter 8). Apparently, in order to cut through some of the confusion, the French government is planning to do a thorough study of the carp industry. Areas of production are shown in Fig. 12.15.

More solid estimates indicate that very few farmers make their living solely from carp production. Production and sale of carp is a supplementary farm enterprise. There was general agreement that many fingerlings were sold for sports fishing and that nearly all carp food fish were sold at the end of three summers of growth at slightly more than 1 kg of weight.

EUROPEAN EELS *(Anguilla anguilla)*

There were only 3 intensive eel farms in France in 1975, and all 3 were discontinuing operations. Reported reasons for discontinuance are relatively low domestic eel price and technical production problems, such as high mortality during the 0.2 to 20 g growth period.

A general statement is that the French have no preference for eels.

FIG. 12.16.   HARVESTED EEL BEING TRANSPORTED TO HOLDING TANKS

Most of the wild catch is exported to Italy, Belgium, Holland and West Germany. Approximately 90% of the total catch is exported. Total catch is about 8000 MT (8816 ST). This is divided into 4000 MT caught in fresh water, 3000 MT from brackish waters of the French Mediterranean, and 1000 MT from brackish waters on the Atlantic side of France. Because of the abundance of wild eel, prices are low. In spite of these factors, there are several enterprises either planned or in process to culture eel extensively (no feed). One such project is near Guerande where extensive culture is planned on old salt beds. Eel culture is also conducted at Croisic near Nantes.

Culturing techniques call for feeding freshly ground redworms during the first three weeks after the elvers are captured. Then in brackish or salt water, the eels are gradually converted to fish fillets. In fresh water a binder can be added to artificial foods, and after the first three weeks finely ground fish meal can be fed. Feed conversion with mackerel is about 5 to 1, while feeding other raw fish results in about a 7 to 1 ratio. During the first season of growth the eels may grow from 0.2 to 0.4 g elvers to 20 to 30 g black eels. At the end of the second season, eels of 150 to 200 g can be expected.

One intensive farm visited used salt water in its intensive operation. The ponds were 1000 m² and 1 m deep (10,760 ft² x 39 in. deep). Each pond produced 1.2 to 1.5 MT (1.3 to 1.4 ST), for a total output of 10 to 12 MT per ha (4.5 to 5.3 ST per acre). Each pond has 5 m³

FIG. 12.17.   HOLDING TANKS FOR HARVESTED EEL

of salt water flow per hour (1322 gal.). Each pond also was stocked with a few sea bass (*Dicentrachus labrax*) and mullet (*Mugil cephalus*). These fish feed on the eel excreta.

COHO SALMON *(Oncorhynchus kisutch)*

There is only one coho salmon farm in France. It is operated by the federal government as a pilot demonstration farm. Production in 1975 was between 40 and 50 MT. In 1976 production was forecast at 120 MT. It is believed that between government and private individuals, production could reach 3000 MT. Part of existing production is used for stocking streams and part for food fish.

CRAYFISH *(Astacus leptoctylus)*

Crayfish are a gourmet item in France and command a premium price. In addition to the capture of an unknown quantity of wild crayfish, about 1300 MT are imported. About 80% of this volume is imported from Turkey and 20% from Poland and eastern Europe. Domestic production from 6 culturing enterprises is estimated at 10 MT annually.

The largest crayfish enterprise in France was visited. This farm has 4 ha of water divided into 32 ponds. About 10 MT (11 ST) of crayfish are in various stages of culture.

FIG. 12.18.   SALTWATER CULTURING PONDS AND FEEDING STATIONS
FOR EEL

FIG. 12.19.   SEINING HARVESTING CHANNEL IN EARTHEN EEL
CULTURING POND

FIG. 12.20.   MECHANICAL LOADING OF EEL CATCH FROM HARVESTING
CHANNEL IN EARTHEN EEL POND

This farmer cultures Turkish crayfish which mature in 2 to 3 years, as compared to 5 to 6 years for the Polish crayfish. Several hundred thousand juvenile crayfish are produced annually. Each female produces 50 to 100 eggs. Shrimp pellets are fed exclusively. Average weight of food-size crayfish sold is 40 g or 25 per kg (11 per lb). The producer price in 1975 was about $6.50 per kg ($3.00 per lb), or $0.26 per crayfish. Crayfish sold for propagation purposes are, of course, higher priced. With imports of 1300 MT and a price of $6.50 per kg, the minimum amount spent for imports is about $8.5 million annually.

It is essential that each crayfish has its own quarters. Hence each pond is furnished with thousands of terra cotta pipes about 20 cm long and 6 cm in diameter (8 in. × 2.5 in.). In the more intensively cultured ponds there can be as many as 40 crayfish and 40 terra cotta pipes per m² (10 ft²), which necessitates stacking the pipes 3 and 4 high. These resemble a hotel with each dweller looking out of his private picture window. With production of this intensity, there can be an annual gross return of $10.80 per m² every 2 or 3 years. This calculates to $108,-000 per ha ($43,725 per acre).

HYDROLOGIC INFORMATION

One measurement of management effectiveness is the volume of

FIG. 12.21.   DIP NET FOR HARVESTING CRAYFISH

FIG. 12.22.   INTENSIVE CULTURE OF CRAYFISH IN TERRA COTTA PIPES

FIG. 12.23.   CULTURED CRAYFISH, THE END PRODUCT

trout which can be produced per unit of flowing water. In the USA Buss and Miller state that in one heavily aerated hatchery in Pennsylvania production of more than 82 lb per gal. of water flow per min (589.6 kg per liter per sec) was attained. They said that this was far above the accepted maximum of 50 to 55 lb per gal. of water (360 to 396 kg per liter per sec). While there is no recommended minimum, it would appear that a production rate of one-half of the maximum could be attained to increase the effectiveness of a given water flow.

In West Germany the commonly accepted standard is that 100 kg of production should result from a flow of 1 liter per sec (13.8 lb per gal. per min). It would appear to be about 28% of the maximum potential.

In Denmark the production rate is normally about 200 kg of production per liter of water flow per sec (27.8 lb per gal. per min). This result is 100% better than the West German level, but is still only 56% of the maximum potential. Growers in Denmark admit that they do not stock their waters at the maximum rate, but understock because sometimes in winter the streams feeding the ponds freeze over and water flow and oxygen levels drop.

In Italy where the trout waters never freeze and minimum temperature is 12°C (54°F), a high production rate is 143 kg of production per liter per sec (19.9 lb per gal. per min). While this rate is better than for West Germany it is only 40% of the maximum possible.

In southern France one average trout enterprise producing 50 MT of trout annually uses only spring water which does not freeze in winter. This farmer produced at the rate of 278 kg per liter per sec of water flow (38.6 lb per gal. per min).

At a second trout farm in central France, approximately one-third of total water in the system was fresh water and two-thirds was recirculated. Maximum aeration was used. Production rate was 555 kg of fish production per liter of fresh water per sec (76.5 lb per gal. per min). Disease problems were minimal, and mortality was less than that of many farms where only a fraction of this output per unit of water was attained.

These latter examples indicate that production of rainbow trout in western European countries, where stream flow is not restricted by freezing in winter and water recirculation in summer still permits water temperature to remain low enough, has not attained its maximum production volume. If maximum use is made of trout waters, production could double in Italy and increase by significant amounts in France and Spain.

SPECIAL ACKNOWLEDGEMENTS

M. OLIVER CLEMENT, 50, Avenue de Verdun, Gazinet

M. PHILLIPPE FERLIN, Chef de la Division Aménagements, Littoraux et Aquaculture, Boîte Postale N3-33610 Cestas Principal, 50 Avenue de Verdun, Gazinet

M. MICHAEL FOURNIS, 44290 Guemene-Penfao

M. PIERRE GAILLARD, Pêches Maritimes et Continentales, 3 Rue Milton, 75009, Paris

M. OLIVER LEDOUX, 7 RD Louis Pasteur, 34470 Perols.

DR. JEAN-JACQUES SABAUT, G.I.E.R.N.A., Domaine du Roulet, 33240 Saint-André-De-Cubzac

M. ANDRE TESSIER, Syndacat des Pisciculteurs, Salmoniculteurs de France, 28, Rue Milton, 75009, Paris

M. WILLIOT, 50 Avenue de Verdun, Gazinet

REFERENCES

ANON. 1971-74. Export and import data. European Trout Federation Office, Treviso, Italy.

ANON. 1974. Customs Data. GTGREF, Technical Center for Rural Engineering of Waters and Forests, Division of Coastal Management and Aquaculture, Report 4, Sept., 53. (French)

DESPLANQUES, L., GAILLARD, P., and LIGIER, Y. 1972. Marketing of trout. French Fish Culture No. 31, Third Quarter, 18, 20, 127. (French)

TESSIER, A. 1975. Salmonoid fish culture. French Fish Culture. No. 43, Third Quarter, 14. (French)

Chapter 13

Spain

RAINBOW TROUT *(Salmo gairdneri)*

Cultured fish destined for sale for food use is a relatively recent development in Spain. There was little or no production before 1965. In 1975 there were 112 private fish farmers producing between 5000 and 6000 MT (5510 to 6612 ST) of rainbow trout. The total volume of rainbow trout eggs produced for all purposes was 100 million compared to 3 million Atlantic salmon *(Salmo salar)* and 6 million brown trout *(Salmo trutta)* eggs. All of the production of rainbow trout is consumed domestically. There are practically no exports or imports of commonly cultured freshwater fish.

Rainbow trout spawn in the October-March period and after reaching about 6 cm (2.5 to 3 in.) fingerlings are placed in concrete raceways for growing-out. A small proportion of the trout are grown in earthen ponds.

All fish are fed pelletized foods. Fish are harvested at about 200 g (7 oz). A few are raised to 2 or 3 kg sizes (4.5 to 6.5 lb). These latter are for smoked filets. Nearly all of the 200 g fish are sold in the round, packed in ice. Average production time from egg laying to the 200 g size is about 15 months.

Most rainbow trout are sold by the farmer directly to buyers for fish markets. The fish market in turn sells directly to the final consumer.

Fish farms exist in all parts of the Iberian Peninsula but the majority are located in the northern half of the country. The ideal farm is located close to a large spring but many farms are using river water. In 1975 there was little or no fee recreation fishing, but some farmers were in the planning stages of initiating fee fishing outlets for the public.

OTHER FRESHWATER FISH

Other than rainbow trout, there is no commercial production of freshwater cultured fish destined for immediate food consumption. However, the state has 23 fish hatcheries in production. These hatcheries produce a variety of fish for restocking river and stream beds. In 1974 the restocking program was as follows:

Species	Number of Fingerlings
Brown Trout *(Salmo trutta)*	4,648,150
Rainbow Trout *(Salmo gairdneri)*	3,707,562
Largemouth Bass *(Micropterus salmoides)*	1,005,000
Carp *(Cyprinus carpio)*	838,500
Atlantic Salmon *(Salmo salar)*	363,800
Northern Pike *(Esox lucius)*	162,000
Tench *(Tinca tinca)*	7,000
Other	36,790
Total	10,768,802

MARINE CULTURE

Cultivation of marine fish in Spain is only in its beginning phases. In 1975 there were two private groups experimenting with Atlantic salmon culture *(Salmo salar)*. One of these was located at the mouth of the river Arosa in Villagarcía province and one at the mouth of the river Ortigueira in La Coruña province. Both of these projects were still in the experimental phases and salmon culture can be said to be still non-commercial. Salmon at these two sites are produced in saltwater net enclosures. Eggs and fingerlings are produced in fresh water for stocking in these pens.

EUROPEAN EEL *(Anguilla anguilla)*

There is no culture of the European eel in Spain. An undisclosed volume of wild eel is caught annually. This is consumed domestically. There is no exportation of elvers or adult eel and no importation.

OUTLOOK

Production of rainbow trout for food started around 1965 in Spain. In 10 years production reached between 5000 and 6000 MT. In the foreseeable future production is estimated to keep increasing at about a

10% annual rate. Water limitations may check production at about the 10,000 MT limit. Additional production can also be attained by intensifying production per unit of flowing water. The domestic market has readily accepted rainbow trout and can probably expand to accommodate increased production.

There is also a predicted future for salmon culture. However, commercial production may never exceed several hundred metric tons annually.

SPECIAL ACKNOWLEDGEMENTS

MR. JOSÉ VERA KIRCHER, Direccion General De Pesca Maritima, Ministerio De Comercio, Madrid, Spain
DIRECTOR GENERAL, Ministerio De Agriculture, Instituto Nacional Para La Conservacion De La Naturaleza, Madrid 3, Spain

Chapter 14

Portugal

RAINBOW TROUT *(Salmo gairdneri)*
and BROWN TROUT *(Salmo trutta)*

As in Spain, culturing fish for food use is a relatively recent development. All production began within the past ten years. In 1976 there were only four private individuals culturing rainbow and brown trout. The area of production is in northern Portugal in Paredes de Coura, Vila Conde, Manteigas and Visev. About 250 MT (276 ST) of trout, nearly all of them rainbow, are produced annually. All production of trout is consumed domestically. There are practically no exports or imports of freshwater cultured fish of food sizes.

The trout spawn in the November-February period. In addition to domestic egg production, about 1.5 million rainbow trout eggs are imported. The fry are raised in concrete troughs until they reach about 6 cm (2.5 to 3 in.). They are then stocked in concrete raceways for growing-out.

All fish are fed pelletized foods. Trout are harvested at about 200 g (7 oz). Nearly all the 200 g fish are sold in the round, packed in ice. Average production time from hatching to the 200 g size is from 12 to 15 months, depending on water temperatures.

Nearly all trout is sold by the farmer directly to retail fish markets which sell to the final consumer. Hence the marketing channel is very direct.

In 1976 there were few or no fee fish-out operations. However, this type of outlet is being discussed.

CARP *(Cyprinus carpio)*

In 1976 there were several carp farms in Portugal. Total production was estimated at 50 MT (55 ST).

Carp spawn in the March-May period and are then raised in earthen ponds. Farming is extensive, which means that little or no feeds are fed. After about 2 to 3 years the carp reach their market sizes of about 1 kg (2.2 lb) and are sold directly to housewives or to retail fish markets.

MARINE CULTURE

Cultivation of marine fish centers on steelhead trout *(Salmo irideus)* and Atlantic salmon *(Salmo salar)*. However, production cannot be said to be truly commercial. About 500,000 steelhead trout eggs and 300,000 Atlantic salmon eggs are imported yearly for restocking streams leading to the ocean. Production of both species combined is less than 100 MT (110 ST).

OUTLOOK

It is predicted that production of cultured fish will increase at least 20% in the next 10 years. This increase will be essentially in rainbow trout. It is difficult to predict the expansion of marine fish culture since it is not really commercial at present. With marine cultured fish, much depends on price relationships and price levels of the same and other species of wild marine fish. An increase in the price of salmon and steelhead trout would be conducive to expansion of culturing.

SPECIAL ACKNOWLEDGEMENTS

MR. T.J. HOWELLS, Sociedade Continental De Alimentacéo, L., Lisbon, Portugal

MINISTRY OF FORESTRY AND AGRICULTURE, Lisbon, Portugal

Chapter 15

United Kingdom

RAINBOW TROUT *(Salmo gairdneri)*

In the United Kingdom the only freshwater fish farmed in significant quantities is the rainbow trout, with a production of 1000 to 2000 MT (1100 to 2200 ST). This volume is produced by about 150 farmers. About 2000 MT (2200 ST) of imports come into the country each year (Tables 15.1 and 15.2). Most of the imported fish are frozen and come from Denmark and Japan. Fresh iced fish also are imported from Ireland as well as Denmark. The three main weight classes imported are: 140 to 168 g (5 to 6 oz), 168 to 224 g (6 to 8 oz) and 196 to 252 g (7 to 9 oz).

The main trout products for table market in order of popularity are: (1) frozen gutted; (2) fresh ungutted (in the round); and (3) smoked. Most of the frozen trout (perhaps 90%) is sold to hotels and restaurants. Most fresh trout passes through the wholesale fish markets, notably Billingsgate (London), Manchester and Birmingham and is retailed by fishmongers.

Production

There has been increasing diversification of production techniques. These have probably occurred largely as a prevention for diseases and parasites. More and more farmers are producing their own eggs to avoid bringing in diseased stock. Rearing facilities for fry have gone to concrete or fiberglass tanks until the fry reach 7 cm (3 in.) to prevent clinical outbreaks of whirling disease. Automated feeders are used to aid in labor efficiency. For grow-out facilities the farmers use

TABLE 15.1

IMPORTS OF RAINBOW TROUT, UNITED KINGDOM, 1967-70

| | Metric Tons | | | |
	1967	1968	1969	1970
Fresh or iced	817	905	725	613
Frozen				
Denmark	499	701	433	828
Japan	436	512	617	617
Total	1752	2118	1775	2058

Source: Shepherd, (1973). Dr. Shepherd's source for fresh and iced trout was H.M. Customs and Excise and for frozen trout the sources were Export Companies and Japanese Fisheries, etc.

narrow raceways constructed of concrete or brick. These raceways may be only 2 m (6 ft) wide.

Dried pellet food is used exclusively. Protein content varies with fish sizes. Lower-protein feeds are fed during the grow-out stages. In 1975 feed cost accounted for about two-thirds of total production cost.

Production cost of 170 to 250 g (6 to 8 oz) fish was estimated at $0.62 per 454 g (1 lb). Feed constituted $0.41. Wholesale prices paid to farmers averaged $1.54 per kg ($0.70 per lb), leaving a profit margin of $0.81 per kg ($0.08 per lb). This amounted to 13% of production cost. Fish weighing between 170 and 250 g (6 to 8 oz) in the round are in demand. The production cycle from hatching to market size required 10 months in the southern United Kingdom and up to 18 months in the north.

TABLE 15.2

IMPORTS OF RAINBOW TROUT, UNITED KINGDOM, 1971-75

| Country of | | Metric Tons | | |
Origin	1971	1972	1973	1974
Denmark[1]	1459	1485	1164	991
Ireland[2]	200	300	400	500
Japan[3]	401	534	553	540
Total	2060	2319	2117	2031

[1] Anon. (1971-1974B). Data for 1973 and 1974 corresponds to information from Anon. (1975).

[2] Estimated by author from data secured in Ireland.

[3] Obtained from Japan Fisheries Agency, Tokyo, Japan.

CARP *(Cyprinus carpio)*

During 1976 it was reported to the author that there was commercial production of carp in Great Britain. Attempts to verify this statement were fruitless. However, production must be less than 100 MT annually.

MARINE CULTURE

Experimental development work is being conducted by the Ministry of Agriculture, Fisheries and Food and the White Fish Authority. This includes Atlantic salmon *(Salmo salar)*, plaice, turbot of the Pleuronectidae family and Dover sole *(Microstomus pacificus)*. Of these the Atlantic salmon is farmed commercially in Scotland. Production was reported to be about 200 MT (220 ST) in 1974 with attempts to double production in 1975. However, the author could not verify these reports of production and was unable to obtain further data on volumes, techniques, feeds, etc.

OUTLOOK

The outlook for continued expansion of trout farming appears bright. From 1965 to 1975 production increased from 800 to nearly 2000 MT. In addition, 2000 MT were imported. Production could increase by at least the present volume of imports. In the opinion of some leaders the United Kingdom can absorb up to 50,000 MT. While this figure may be optimistic it would appear that production may be limited by profit margins and water availability. No forecast of marine production is possible because of limited information. Suffice it to mention that production of cultured salmon has been increasing rapidly in recent years.

SPECIAL ACKNOWLEDGEMENTS

DR. C.J. SHEPHERD, University of Stirling, Stirling, Scotland

REFERENCES

ANON. 1971-74A. Customs data. Japan Fisheries Agency, Tokyo, Japan.

ANON, 1971-74B. European Federation of Salmonoids—Breeders (FES), Treviso, Italy.

ANON. 1975. The Journal of the Fresh Fishing Industry, Bulletin for Members of the Freshwater Fishing Association, Danish Association, No. 3. Viborg, Denmark. (Danish)

SHEPHERD, C.J., 1973. Commercial aspects of trout farming in Europe. Two Lakes Fifth Fishery Management Training Course Report. Two Lakes, Ramsey, England, Oct. 5-7.

SOLOMON, D.J., STOTT, B., LLOYD, R., and SHEPHERD, C.J., 1975. Limiting factors in freshwater fish farming: A report on a two day discussion meeting held on July 8 and 9, 1975 at Queen Elizabeth College, London. J. Inst. Fish Mgmt. 6, No. 4, 95-98.

Chapter 16

Ireland

RAINBOW TROUT *(Salmo gairdneri)*
and BROWN TROUT *(Salmo trutta)*

Culturing of rainbow trout for food consumption in Ireland began about 1965. Since then the industry has grown slowly. In 1975 there were only 6 fish farms with total production between 500 and 600 MT (550 to 660 ST). Of this amount only 50 MT were consumed domestically and nearly all the remainder was exported. Of the estimated 500 MT exported, over 95% went to Great Britain and limited quantities to France.

The reason so little trout is consumed domestically is that traditionally the Irish are a meat-eating people. Beef, mutton and lamb constitute major portions of the diet. When the Irish eat fish it is a saltwater species, the cost of which is only about one-half the rainbow trout price.

The typical trout farm is fed by river or stream waters rather than from springs. The rivers are described as "racy". This means that the water flow varies considerably, depending on rainfall. The water flows vary from lethargic to torrid Niagara currents. This makes it difficult to locate a farm in order to take advantage of gravity flows without subjecting the farm to periodic flooding.

Each farm generally performs all steps in the fish culturing system. They raise spawners, produce eggs, hatch them, raise the fry to fingerlings and grow them out to food market sizes. Water temperatures vary from a low of 5° to 7°C (41° to 45°F) in winter to a seasonal high of 14° to 16°C (67° to 71°F) in summer. The normal spawning period is from November through February. The normal hatching month is March. After hatching, the fry are grown to about 7 cm

(2.5 in.) in concrete tanks and are then stocked for grow-out in Danish type earthen ponds. No mechanical aeration is used. After hatching, a period of about 18 months is required for the fish to reach market size because of the cold water. Since most fish are exported this means market sizes are essentially those demanded in Great Britain (170 to 220 g, or 6 to 8 oz). This size is also in demand in France. The fish are sold in the round (whole and ungutted) and are iced for delivery. They are sold directly from farms to importers, who are usually wholesalers in London, Manchester or Wales.

All fish are fed artificial foods. Fry are fed pellets having 48 to 50% protein content. However, during the grow-out process a much cheaper feed containing only about 35% protein is fed.

Considerable effort is made to ensure low cost production because of low prices. This is shown by the type of earthen ponds used and cheaper feeds. In 1975 the farm price was estimated at $1.54 per kg ($0.70 per lb).

In addition to the 500 to 600 MT, there are some rainbow trout stocked in enclosed public ponds for sports fishing. In 1975 there were three such ponds or lakes stocking a total of only 50,000 fish.

The Irish government is conducting some experimental raising of rainbow trout and Atlantic salmon *(Salmo salar)* in brackish water and seawater. However, there is no commercial production of these two species by private growers at present.

In addition to rainbow trout production, there is a governmental restocking program for brown trout. An estimated 1 million one-year-old fish averaging about 110 g (4 oz) are stocked annually in public waters and by angling clubs. Anyone can fish free of charge without a license in either the public or angling club waters for freshwater fish. However, salmon fishing requires a license.

Because of the good fishing available in public waters, there is no demand for fee fish-out facilities. Hence there are no fee fish-out lakes or ponds in Ireland.

EUROPEAN EEL *(Anguilla anguilla)*

There is no culturing of eel in Ireland. However, there is commercial fishing for this species. The amount of catch is unknown, but may total 1000 MT (1110 ST). Certain inland waters have very high eel production. For example, Lough Neagh, which covers 35,000 ha (87,500 acres), produces 700 MT per year. This amounts to about 20 kg per ha (20 lb per acre).

The government, in order to maintain the commercial catch of eels, has a restocking program. Elvers (migrating young eels) are captured in some of the infertile waters and transported to more fertile inland waters.

Adult eels are commonly caught in nets as they migrate downstream. Domestic consumption of eels is small and nearly all of the estimated 1000 MT are exported. Exports go mainly to France, Great Britain and West Germany, with minor quantities to Holland and Denmark.

OUTLOOK

The outlook for the Irish cultured fish industry is for continued growth in production of rainbow trout. Within 10 years, production is estimated to nearly double to 1000 to 1200 MT from the present 500 to 600 MT. While it is possible for some salmon production to occur in seawater, the outlook is for very minor quantities of production. This is largely due to the lack of trash fish for feeding.

SPECIAL ACKNOWLEDGEMENTS

MR. P.C. DOLAN, Department of Agriculture and Fisheries, Dublin
MISS JACQUELINE DOYLE, Fisheries Division, Dublin
MR. C.J. MACGRATH, Department of Agriculture and Fisheries, Dublin

Chapter 17

European USSR

Akiro Homma and Hiroshi Shibukawa

Estimates of the freshwater catch from inland fisheries in the USSR range from 500,000 to 850,000 MT annually (551,000 to 937,000 ST). The Caspian Sea area accounts for about 390,000 MT (430,000 ST) of the higher estimate. The catch is increasing each year. Most of the catch is from natural waters. However, to maintain and increase the catch has meant limiting catches by species and sizes, creation of fish ladders around dams, creation of spawning areas and artificial hatching, and release of young fry and mature fish.

Sturgeon, trout and carp are the major species being managed. Silver carp *(Hypothalmichthys molitrix)* and grass carp *(Ctenopharyngodon idellus)* from China have been introduced and are cultured in a limited way. Fish farming is generally limited to pond culture, mainly in the Ukrainian Republic. Carps are the major species. These include common carp *(Cyprinus carpio)*, bleak carp *(Alburnus alburnus)*, bream *(Abramis brama)* and crucian carp *(Carassius carassius)*. The main species for culturing is probably the Ukranian or bleak carp. There are plans to greatly increase cultured production, which presently is in the neighborhood of 200,000 MT (220,000 ST). In addition some rainbow trout *(Salmo gairdneri)*, native catfish (not identified) and a hybrid of sturgeon called Bester are cultured.

About 70% of all pond production involves carp. There is considerable emphasis on creating new hybrid strains of carp for culturing and managing by crossing native and foreign species. Some of this work dates back 300 years. Research is aimed at increasing production by breeding for better growth rates, resistance to diseases, adaptability to cold water and higher survivability. These are goals for silver and grass carp also.

During the fifth 5-year plan it was decided to culture fish intensively in the heated water affluents from thermal electrical power plants. There were 10 to 12 of these possible production areas in 1974. Carp are grown in the summer and rainbow trout and sturgeon in the winter. Feeds fed to carp in intensive culture contain no animal protein and 3 to 4% fat content. Feed conversion is 4 units of feed to 1 of fish weight. Even with this conversion rate, production cost is less than in Japan, where the feed conversion is better but the feed ingredients are more expensive.

Production of carp in ponds is about 2 MT per ha (1787 lb per acre). Plans call for increasing yields to 5 MT per ha (4468 lb per acre) by using fertilizers, minerals and artificial food supplements. In one area, yields of 3 MT per ha are reported.

The Ukrainian or bleak carp *(Alburnus alburnus)* has superior growth in the colder waters. At the end of the first year the fingerlings range between 70 and 100 g (3 to 4 oz); the average is 1500 g (54 oz) after 2 years, 2500 to 2700 g (89 to 96 oz) after 3 years, and 3500 g (125 oz) after 4 years. The males mature in 3 years and the females in 4 years.

To achieve maximum production per hectare of pond water, poly-culture is being promoted. This means that top water, bottom feeders and mid-water feeding fish are raised together. A typical stocking density would be 7000 carp, 2000 to 2500 grass carp, 50 silver carp, a few pike-perch *(Lucioperca lucioperca)* and 30 to 50 catfish per hec-tare. It can be seen from these stocking figures that grass carp are playing an important role in cultured production.

Rainbow trout play only a minor role in fish farming. Production is not large enough to be listed in official statistics. They are cultured in the winter discharge waters of several electrical generation plants. They are grown in netted boxes 3×4×1.5 m (10×13×5 ft). In 10 months they reach market sizes of 150 to 200 g (5.5 to 7 oz). Produc-tion may be 100 MT (110 ST) annually. There is also one experimental hatchery using concrete raceways. This hatchery has 1200 m² (about 12,000 ft²) and production is 20 kg per m² (4.1 lb per ft²). Total production of this hatchery is 24 MT (26 ST). The goal is to double production per unit of water surface.

The inland fresh water catch in 1973 is shown in Table 17.1. The catch of carp is important, as well as herring. World sturgeon pro-duction, of which the USSR accounts for 95%, totals about 20,000 MT (22,000 ST). About 90% of the USSR catch comes from tributaries of the Caspian Sea. The sturgeons taken by fishermen are great stur-geon *(Huso huso)*, Baltic sturgeon *(Acipenser sturio)*, starred sturgeon *(Acipenser stellatus)*, sterlet sturgeon *(Acipenser ruthenus)*, and Bes-ter, a hybrid cross.

TABLE 17.1

INLAND FRESHWATER CATCH, USSR, 1973

Fish	Metric Tons		
	Caspian Sea (000's)	Other Areas (000's)	Total (000's)
Cyprinidae	8	181	189
Common or Northern pike (*Esox lucius*)	1	14	15
Catfish	3	11	14
Bass	1	29	30
Sturgeon	—	20	20
Herring	373	—	373
Salmon	—	28	28
Others	3	179	182
Total	388	462	851

Biological minimum sizes for *Huso huso* are between 100 and 150 kg (220 to 330 lb). Adult females contain 500,000 to 800,000 eggs. These fish are taken in the Don and Volga rivers and tributaries. However, after maturing sexually the sturgeons grow much larger.

Biological minimum-sized Baltic sturgeon *(Acipenser sturio)* weigh between 15 and 40 kg (33 to 88 lb) and produce 150,000 to 300,000 eggs. They are caught in the Volga and Don rivers and in the Black and Mediterranean seas.

Biological minimum-sized starred sturgeon *(Acipenser stellatus)* weigh between 10 and 20 kg (22 to 44 lb) and female adults produce 150,000 to 200,000 eggs. They are found in the Volga and Don rivers and tributaries.

Because of plans to build four hydroelectric dams on the Volga (one already completed) and also dams on the Don River, scientists have concentrated on the methods of maintaining and increasing fishing populations, particularly of sturgeons. To do this means not only construction of fish ladders to enable fish to by-pass the dams, but also knowledge of the reproduction cycles and maintenance of hatcheries. They found that the *Huso huso* lays eggs in April and the small fish hatched go downriver in June. The *Acipenser sturio* lays eggs in April and goes downriver in September. The *Acipenser stellatus* lays eggs in June and the small fish go downriver in October.

At the sturgeon hatchery near Volgograd, which was founded in 1960 and contains 50 three-hectare ponds (371 acres), eggs are taken from natural sturgeon, hatched, raised to 13 to 15 cm (5 to 6 in.) and released. In 1972, 6 million sturgeon were released. These were 60% *Acipenser sturio*, 40% *Acipenser stellatus* and a few *Huso huso*. After stripping the fish, using pressure only, the eggs are placed in hatching trays and at temperatures of 8° to 10°C (46° to 50°F) are hatched in 9 to 10 days. After hatching, the fry are placed in the hatchery ponds where they feed on natural foods. Stocking density of fry varies from 40,000 per ha (16,000 per acre) for *Huso huso* to 60,000 to 70,000 for the other 2 species (24,000 to 28,000 per acre). From the egg to 7 to 8 g sizes (0.25 oz) survivability of the *Acipenser stellatus* is between 55 and 60%. For *Huso huso* survivability is 40 to 42%. Survivability of the *Acipenser sturio* was not obtained but it must be high because the return rate from natural waters is 50%, compared to 20 to 30% for the *Acipenser stellatus* and 10% for the *Huso huso*. When the fingerlings reach a length of 13 to 15 cm (5 to 6 in.) they are released. This is called the "cigarette" stage and they weigh only 6 to 10 g (0.25 to 0.33 oz). It takes 42 to 45 days from spawning to release.

The number of sturgeon passing the Volga River dam in 1972 was 800,000. Plans call for an increase to 1,600,000. By the year 2000 production is planned to increase from 20,000 MT in the Caspian Sea to 50,000 MT (22,040 to 55,100 ST). Of these *Huso huso* will account for 20% with each of the other 2 species accounting for 40%. To achieve this goal plans call for hatching and releasing 10 to 15 million *Huso huso* one-year-old fish, 130 to 135 million *Acipenser sturio* and 290 to 300 million *Acipenser stellatus*. The sturgeon reach biological minimum size in 15 years and have a life span of 30 to 50 years.

For about 15 years a hybrid cross called Bester has been produced. Adult female *Huso huso* eggs are fertilized with milt of male sterlet sturgeon *Acipenser ruthenus*, a sturgeon which is smaller and does not migrate down the rivers to the Caspian Sea. The advantage of these is the reduction of maturing from 15 to 8 years. In 1972 production of Bester was still in a pilot stage. In 1974 production was 225,000 fish.

Sturgeon fishing is prohibited in both the Caspian Sea and the waters of the Volga dam. Fishing is permitted only in other parts of the main rivers and tributaries.

Other species being cultured for release in natural waters are: (1) Inconnu, *Stenodus leucichthys*, which lives only in the Caspian Sea and the Volga River and nearly became extinct before scientists aided in its recovery; (2) Atlantic salmon *(Salmo salar)*—about 3 million are raised and released annually into the Black Sea and Baltic Sea; and (3) *Salmo iridius*, which is released in the Black and Asov seas. About 80,000 are released annually.

Experiments are being conducted with steelhead *(Salmo iridius)* imported from Canada, chum salmon *(Oncorhynchus keta)*, and pink salmon *(Oncorhynchus gorbuscha)*.

Chapter 18

Hungary

Dr. Shigeru Arai

The Hungarian Peoples' Republic is an inland country of 93,000 m² (35,900 mi.²). The country is nearly 75% agricultural land and has a population of 10,500,000.

The area in natural waters of rivers and lakes is 122,000 ha (301,000 acres), while cultured pond area for fish is 21,000 ha (51,870 acres). With 5.8 ha of natural waters for each hectare of cultured waters, the natural catch amounts to only 20 to 25% of total fish production. In 1958 total fish catch and production was 12,976 MT (14,300 ST) with nearly 71% of this amount, 9152 MT (10,086 ST), coming from pond cultured fish. By 1970 total production and catch reached 25,998 MT (28,650 ST) with cultured fish accounting for 19,197 MT (21,155 ST) or 74%. In 1976 catch and production were estimated at nearly 30,000 MT (33,060 ST) with cultured production accounting for 80% (24,000 MT) of total supply.

State farms control 61,000 ha (150,670 acres) of natural waters; fishery associations account for 51,000 ha (126,000 acres) and fishing or angling clubs and associations account for 10,000 ha (24,700 acres). These clubs and associations issue special permits to anglers for the right to fish in club or association waters. The important species in natural waters are:

Common carp (*Cyprinus carpio)*
Barbel *(Barbus barbus)*
Sturgeon (*Acipenser*)
Tench (*Tinca tinca*)
Bream (*Abramis brama*)

Danubian wels (*Silurus glanis*)
Pike-perch (*Lucioperca lucioperca*)
Common pike (*Esox lucius*)
Asp (*Aspius aspius*)
Brown trout (*Salmo trutta*)
Perch (*Perca fluviatilis*)

All fish cultured are raised in ponds. In 1974 a total of 21,000 ha (51,870 acres) was cultured. Of this area, 15,000 ha were in state farms and 6000 ha in fisheries cooperatives. Carps are raised almost exclusively. Polyculture is practiced. The usual stocking is 85% common carp (*Cyprinus carpio*) and 15% other carps, such as grass carp (*Ctenopharingdon idellus*), silver carp (*Hypothalmichthys molitrix*) and bighead carp (*Aristichthys nobilis*). Average production per hectare is about 1 MT (892 lb per acre). This is not intensive production per unit of water area as is found in other countries such as Japan and Israel.

The annual per capita consumption of fish before World War II was about 0.5 g (1 lb). In 1976 fish consumption amounted to about 2.7 g (5.9 lb) per person. This amounts to slightly less than 5% of total consumption of animal meats. By 1980 the government plans to increase production and consumption to 5 g (11 lb) per person. Because of water pollution problems the catch of wild fish from natural waters is not expected to increase. Plans call for increasing the numbers and areas of cultured ponds as well as output per hectare of water area.

To accomplish these plans research efforts have been intensified at the Fish Culture Research Institute at Szarvas. This institute has 700 ha with 400 of them (988 acres) under water. Researchers are placing emphasis on combination production of ducks and fish using the same waters. In this plain area of Hungary there are many areas called Sodic soil, which is very poor agricultural land commonly used for meadows. At this institute they have been conducting research into a 10 year cycle of land and water use. During the first five years ducks are raised with carp. In the sixth and seventh years the ponds are drained and alfalfa is raised. In the eighth, ninth and tenth years rice is grown and then the 10 year cycle is repeated. Annual production by this method per hectare is 4 MT of ducks and fish annually during the first five years, 8 MT of alfalfa in the sixth year, and 6 MT of alfalfa in the seventh year. Rice production during the last 3 years of the cycle varies between 3 and 3.5 MT per ha (2670 to 3123 lb per acre). The idea is to increase the productivity of the land with the duck-fish rotation. This scheme is projected to be extended to other farms and to increase fish production. The ducks at this institute are reported to reach 2.6 kg (5.7 lb) in only a few months time. Feed conversion is 3.27:1,

Courtesy of Dr. Shigeru Arai

FIG. 18.1.   COMBINATION FISH WITH DUCK CULTURE

which is extremely good. The ducks are intended to raise the natural fertility of the water and lower the production costs of the feed by replacing some of the grain fed at other farms.

As can be seen in the following example, fish feeding costs are relatively high, amounting to nearly 55% of production costs. This example was for Hartobagy State Farm in 1973 where polyculture of carps was practiced.

	Million Forints
Cost of feed	13.0
Water management	1.6
Fertilizers	0.5
Fry cost	0.6
Labor	5.0
Maintenance	1.0
Other costs	2.0
Total	23.7

It costs 23,700,000 forints to produce 1500 MT (1653 ST) of fish. With forints worth 4.28 cents each, each kilogram of production costs $0.676 or $0.307 per lb. The selling price was $0.749 per kg ($0.34 per lb) so the net profit was $0.073 per kg ($0.033 per lb). Even though unprocessed grain was fed to the fish the cost of feeding was nearly 55% of production costs.

FRY PRODUCTION

Nearly all fry are produced at specialized hatcheries. The largest is near Szazhalombatla not far from Budapest. This farm supplies about 60% of all fry in the country from a pond area of less than 50 ha (124 acres). This hatchery uses the warm water discharge of a nearby power plant to heat waters year round. During certain times of the year live pressure steam is used.

TYPICAL STATE FARM

A state farm which might be considered typical is Bikal, southeast of Lake Balaton. The total area of this farm is 8300 ha (20,500 acres). The farm employs 1050 people. Grain, fruit, hops and meats are produced. Among the meats are 2200 MT of pork, 1700 MT of fish, 750 MT of rabbits, 500 MT of ducks, 350 MT of beef and 190 MT of turkeys. The 1700 MT (1873 ST) of fish produced are common carp, grass carp and bighead carp. Small amounts of pike-perch and wels are also cultured. This production is from 960 ha (2371 acres) of water area. Production is 1.8 MT per ha (1604 lb per acre) which is considerably above the state average.

Chapter 19

Asian USSR

Clinton E. Atkinson

Aquaculture in the Soviet Far East is completely different than that taking place in European Russia. The area itself is extremely broad, extending from east to west for about 7500 km (5000 mi.). Although there are a number of large modern cities (Vladivostok, Khabarovsk, Irkutsk, etc.) the population density as a whole is low — probably comparable to that of the state of Alaska. However, the climate is the factor that has discouraged the development of pond culture in Siberia and the Soviet Far East. The winters are severe and, except for the Sakhalin and Primora regions, the ponds are covered with ice for 4 to 6 months of the year and the growing season is short.

The marine environment is similarly severe and it is difficult to develop an extensive marine aquaculture program. Again with the exception of Sakhalin and the Primora regions, the entire coast of the Soviet Far East is affected by the flow of cold currents along the coast, originating in the northern Bering Sea (and some from the Arctic Ocean) to form the Oyashio, Okhotsk and Liman currents. Generally the water is too cold to grow the common forms of fish and shellfish cultured in the more temperate waters of other countries.

In very recent years the Soviet scientists have begun a series of studies on the development of mariculture along the Pacific coast of the USSR, especially in the warmer waters of Primora (e.g., Peter the Great Bay) and Sakhalin. Although much valuable work has been done by the scientists of the Pacific Scientific Institute of Fisheries and Oceanography (TINRO) at Vladivostok and five branch laboratories, recent success of the Japanese and others in the culture of scallops, abalone, king and Tanner crab, the northern seaweeds, etc. has done much to encourage a re-examination of the potential of mariculture in the Soviet Far East. Although a final decision has not been reached on the appropriate species to culture and the methods to use, plans have been developed to establish production farms in the next year or two for kelp and other seaweed and probably for scallops and other shellfish as well (Ayushin and Chigirinskii 1976).

Although mariculture and pond culture have not been developed in the eastern part of the Soviet Union, the results from "fish ranching" (i.e., the taking of spawn, incubation and releasing the young in the rivers and lakes to grow) has been outstanding. Also associated with the work of the hatcheries have been innumerable experiments on the transplantation and acclimatization of various species of fish to new waters. Although a list of species used in these transplants is not available, much of the effort has centered around sturgeon, whitefish and salmon. Perhaps the most spectacular of all these experiments has been the mass transplant of millions of pink and chum salmon eggs to the Murmansk region and the establishment of a new run of Pacific salmon to the eastern Atlantic and northern seas.

SALMON[1]

Although several earlier investigators had expressed concern for the future of the salmon runs to the Amur River, I.I. Kuznetsov was the first to take active steps to protect the natural spawning runs of salmon and to propose artificial propagation, ". . . which, when there is an excess of spawners on the grounds, can provide a real addition to natural reproduction." Mainly as a result of his efforts to maintain salmon runs in the Far East, measures were taken in 1924 to establish catch quotas, protect spawning grounds, regulate fishing seasons, and undertake artificial spawning. Thus in 1927-28 the first salmon hatcheries were built at Teplovka Lake (tributary to the Amur River) and at Ushi Lake (tributary to the Kamchatka River). Later a third hatchery was built on the Bidzhan River, tributary to the Amur River and about 55 mi. above the hatchery at Teplovka Lake (Atkinson 1960).

It is interesting to note the growth in the production record for chum salmon (*Oncorhynchus keta*) at Teplovka Lake. There has been an increase of 200 to 300% between the pre-war (1928-37) and postwar (1938-52) periods, or from an average of 12.5 to 23.2 million eggs and 8.1 to 20.4 million fry released. At the same time, however, the

---

[1]The portion of the text describing the propagation of Pacific salmon in the Soviet Far East has been taken almost *verbatum* from *Salmon Aquaculture in Japan, the Koreas and the USSR*, prepared by the author in 1976.

neighboring Bidzhan hatchery began to experience difficulty in obtaining sufficient eggs to operate. This was blamed specifically on local development and poaching.

The hatchery at Ushki Lake produced mostly sockeye (*Oncorhynchus nerka*), coho (*Oncorhynchus kisutch*) and a few chum salmon. A production level of 20 to 26 million eggs was maintained before World War II. Afterwards, however, the run suffered a catastrophic decline, dropping from 16 million eggs in 1947 to only 3.9 million in 1952. The criticism at the time was directed toward failure to maintain the facilities in proper operating condition; but whatever the cause, the hatchery has now been rebuilt and is in full operation.

The Japanese also built a number of salmon hatcheries in the southern part of Sakhalin and the Kurile Islands during the latter part of the 1920s. By the beginning of World War II a total of at least 12 hatcheries were in operation with a total capacity of 170 million eggs. The average annual take by the Japanese hatcheries, however, was only about 73 million eggs (Chernyavskaya 1964).

Between 1946 and 1960 the 12 existing Japanese hatcheries were improved, and an additional 12 new hatcheries were built between 1955 and 1960. Together they provided a total capacity of 265 million eggs for Sakhalin and 100 million eggs for the Kurile Islands. By 1964 there was a total of 25 hatcheries in operation in Sakhalin and the Kuriles — 20 in Sakhalin and 5 on Iturup Island (Kuriles).

The relative size of the hatcheries was described by Chernyavskaya (1964) as follows:

No.	Capacity
7	Less than 10 million eggs
12	10-20 million eggs
6	Over 20 million eggs

In 1959 all hatcheries (except for three on Iturup Island) had rearing ponds. In 1958 the young were fed for 2 or 3 months at 10 of the hatcheries before release. The food was ground, frozen fish waste (pollock, cod, etc.), supplemented at times with fish meal; no meat products were used. There is no information on the type of food now being used in the Soviet salmon hatcheries.

Both Tables 19.1 and 19.2 show the rapid growth in production by the USSR hatcheries since 1962. In Sakhalin, for example, the numbers of pink (*Oncorhynchus gorbuscha*) and chum salmon fry released almost doubled in the 10 year period — 257 million fry released in 1962 and 468 million in 1971 (Rukhlov 1973A). Further, according to Doroshev, the total number of salmon fry released by the hatcheries in

TABLE 19.1

NUMBERS OF YOUNG SALMON RELEASED FROM SAKHALIN HATCHERIES, 1962-71

Species	Region	1962	1963	1964	1965	1966	1967	1968	1969	1970	1971
						(in millions)					
Pink salmon	East Sakhalin	44.7	15.7	81.1	38.4	88.1	78.3	191.1	108.2	191.1	87.3
	Southwest Sakhalin	23.5	11.4	43.3	10.2	32.6	30.1	47.9	18.8	48.7	7.5
	Gulf of Aniva	29.2	17.3	36.7	31.6	17.6	36.0	42.9	16.7	40.5	17.2
	Total	97.4	44.4	161.1	80.2	138.3	144.4	281.9	143.7	280.3	112.0
Chum salmon	East Sakhalin	80.0	108.1	105.1	201.3	185.5	169.9	93.3	186.9	116.9	188.3
	Southwest Sakhalin	75.2	78.3	69.5	112.6	93.9	102.6	97.7	115.3	100.3	143.3
	Gulf of Aniva	4.5	12.7	0.2	4.4	18.6	3.1	—	20.0	1.2	24.7
	Total	159.7	199.1	174.8	318.3	298.0	275.6	191.0	322.2	218.4	356.3
Total, pink and chum salmon		257.1	243.5	335.9	398.5	436.3	420.0	472.9	465.9	498.7	468.3
Percentage chum salmon of total		62	82	52	80	68	65	40	60	44	76

Source: Rukhlov (1973A).
Note: Minor discrepancies noted in chum salmon totals for 1963 and 1964.

TABLE 19.2

TOTAL NUMBER OF SALMON RELEASED FROM HATCHERIES IN THE
SOVIET FAR EAST, 1970-74

Species	1970	1971	1972 (in millions)	1973	1974
Pink salmon (*Oncorhynchus gorbuscha*)	423.1	206.3	398.8	269.4	457.1
Chum salmon (*Oncorhynchus keta*)	218.4	446.6	351.6	413.7	336.8
Sockeye salmon (*Oncorhynchus nerka*)	—	—	9.3	—	9.1
Coho (*Oncorhynchus kisutch*)	—	—	3.2	10.3	4.8
Total	641.5	652.9	762.9	693.4	807.8

Source: McNeil (1976) — Personal communication from Dr. S.I. Doroshev, Head, Laboratory of Acclimatization and Aquaculture, VNIRO, Moscow.

the Soviet Far East increased from some 642 million in 1970 to 808 million in 1974.

Note that the salmon hatchery production of the USSR and Japan are of about the same magnitude. The real difference is that the Japanese effort is almost exclusively on chum salmon while that of the Soviets is almost equally divided between pink and chum salmon. Further, the combined hatchery production for the Soviet Union and Japan could well approach 2.5 or even 3 billion fry for release in the next 2 or 3 years.

The Soviet scientists have made a number of studies of the rate of adult return from fry released, both by extensive marking programs and by simple comparison of the numbers of fry released to actual count of the numbers of adult returns to the hatchery. Although the most recent comprehensive program by Kanid'yev et al. (1970) gives a coefficient of return for the 1956-63 brood years of 0.21 to 0.62%, these are returns to the hatchery and do not include the high seas catch of the Japanese or the more distant Japanese and Soviet coastal fisheries. It would appear that the estimate of 1.3% for autumn chum salmon at Teplovka hatchery (Levanidov 1954), or the estimate of 1 to 3% (rarely 5%) for Sakhalin pink and chum salmon (Chernyavskaya 1964), would still be a more accurate estimate to use at this time. Such a rate of return would be quite comparable to the results being obtained from salmon hatcheries in Japan and the United States.

Based upon hatchery costs given by Kanid'yev et al. (1970), some attempt has been made to estimate Soviet costs of salmon hatchery production. For example, we may consider the cost of 197,800 rubles to produce 131,900,000 fry for release in 1959-63, a return of 1 to 3%. At the present U.S. exchange rate (U.S. $0.29 = 1 ruble) the average cost per adult return would be from $0.01 to $0.04 per fish. The cost return ratio for 1968 (i.e., 1 million fry for release at a cost of 1506 rubles) gives an even more favorable production cost. There are problems here, however, in trying to compare costs between the two economic systems. Until more information is available on just what is inluded in the hatchery costs, these figures should be used with caution.

Kanid'yev *et al.* (1970) also give the results of an interesting study by V. Ya. Levanidov on the relationship between the size of young chum salmon and the survival from char predation. He has been able to show that the larger the young chum salmon, the better the chance of survival. Levanidov has also been able to demonstrate a similar relationship between the survival of young chum salmon and fish-eating birds.

Although somewhat out-of-date, the following notes obtained during a visit to the Soviet Far East in 1959 may be of help in understanding some of the operational detail of the salmon hatchery program of the USSR.

Khabarovsk Region (Amur River)

There are apparently three hatcheries now operating on streams tributary to the Amur River: Teplovka Lake (Bira River), Bidzhan, and Udinsk (built in 1959-60 on the Amgun River). In 1959 the Teplovka hatchery had (in some years) handled between 40 and 55 million eggs and the Bidzhan hatchery about 12 million eggs. The new hatchery at Udinsk was being designed with a capacity of 25 million eggs.

The hatchery at Teplovka Lake is located about 138 m (450 ft) above the outlet. The area of the lake is 0.5 ha (1.3 acres), with a maximum depth of about 4 m (12 ft) and an average depth of 1.5 m (5 ft). The water supply for the hatchery comes from springs with an average monthly temperature of 3.2°C (38°F) in winter and 6.8°C (44°F) in summer (Vasil'ev 1954A).

The outlet stream is about 5 km (3 mi.) in length, flowing into the Bira River and then into the Amur. The young fish migrating out of the lake are counted by a trap placed in the outlet stream.

The collection of eggs usually begins in October at the Teplovka hatchery. The adult salmon, after removal of the eggs or sperm, are sold for human food—a practice similar to the one followed in Japan.

The eggs are incubated in trays placed in troughs. By February the eggs hatch and the young are held in the hatchery until free-swimming.

The young fish are then released into the lake where they feed upon *chironomids* and other natural food.

The Teplovka hatchery was the first hatchery established in the Soviet Far East and over the years has probably been the most successful. The average return from fry released has been about 1.3%, more than 6 times the 0.2% from natural spawning.

Sakhalin Region

The hatchery at Kalinin (southwest Sakhalin) was built in 1925 and was operated by the Japanese until 1939. After World War II the fishing industry, including hatcheries, was placed under the jurisdiction of the Sakhalin Fishing Authority (SAKHALIN-RYBPROMA). Under its direction the water supply and ponds at Kalinin were rebuilt in 1951, and new houses and a garage were added in 1952-54, as well as a new hatchery building in 1959-60. The hatchery at Kalinin has probably been one of the most successful hatcheries of the Sakhalin group and is a favorite site for tests and scientific studies.

In 1959 the hatchery was operating on an annual budget of 300,000 to 400,000 rubles a year. The supervisor was a trained economist and the hatchery technicians were all university graduates trained in fish culture. The results of the hatchery operations are not usually published but are reported only to the Sakhalin Fishing Authority.

The Kalinin hatchery handles both chum and pink salmon. In the first year of operation after the war (1951) the hatchery took 5 million eggs. In 1958 the quota was set at 21 million eggs; the hatchery took 28 million eggs. The new hatchery building increased the capacity of the hatchery to 33 million eggs in 1959. The capacity was further increased in the latter part of the 1960s. For example, in 1967 a total of 48.6 million fry were released from Kalinin and in 1968, 55.8 million.

The water supply for the Kalinin hatchery comes from a spring, is filtered through sand and gravel, and is carried into the hatchery through a covered flume. The water temperature is about 4°C (39°F) in the winter and 9°C (48°F) in the early fall. The water supply does not freeze in winter.

The salmon are trapped at a weir located a short distance above salt water. Chum salmon are taken from August until the beginning of October, and pink salmon slightly earlier (i.e., from early August until the latter part of September). The fish are spawned at the weir and the eggs are washed and taken to the hatchery to water-harden.

The eggs are picked after one or two days and placed on standard hatchery trays (about a foot square and 0.375 in. deep) with about 1500 chum eggs or 2000 pink eggs on each tray and 10 trays stacked together. The stacks of trays are placed in concrete troughs, built into the floor of the hatchery but similar in design to those used in Japan and the United States. The Soviet technologists reported an average of 1.5% egg mortality at time of "pick-off."

After the eggs are eyed and just before hatching, the stacks of trays are transferred to raceways (about 10 m long, 1.34 m wide and a water depth of 25 cm) (37 ft, 4.5 ft and 10 in.). The bottom of the raceways are covered with about 6 cm (2 in.) of gravel. When the young hatch they drop through the screens on the bottom of the trays and enter the gravel.

Several weeks before transferring the eggs to the raceways, the gravel is washed and sterilized with calcium chloride at the rate of 10 kg per m^2 (2 lb per ft^2).

At the time of the visit in 1959, the young fish were fed ground fish waste (pollock, cod, etc.) by placing the food on shallow trays suspended about 5 cm from the bottom. The trays were made of wood and the food placed on the tray about 1.5 to 2 cm deep. About 1.5 to 2 kg of food were placed on each tray twice a day, or at a rate of 5 mg per fish at the beginning of feeding to about 20 mg per fish at the end. Some fish meal is used to supplement the ground fish diet.

Although unconfirmed, it is believed that the Soviet hatchery technologists have now developed a more efficient way of feeding the young fish and a better formulated food (Frolenko 1964; Kanid'yev et al. 1970).

Cooperative Programs

Because of the decline in the salmon runs and the growing restrictions on the salmon fisheries adopted by the Japan-USSR Northwest Pacific Fisheries Commission, in 1962 Japan proposed to establish in the Soviet Far East a series of salmon hatcheries operated jointly by the two countries. Finally, on June 8, 1975, the two governments agreed to establish such a station on an appropriate river in southern Sakhalin. Subsequently there has been a series of meetings between Japanese and Soviet hatchery experts, and they now have agreed to construct a joint salmon hatchery on the Pioner River (southwest Sakhalin).

Although a number of details must still be worked out, tentative plans call for the construction of a hatchery in 1977 and for completion and operation in 1978. The hatchery will have a capacity of 30 million eggs (25 million chum, 3 million pinks, 1 million silver, and 1 million other). The estimated cost of about six million dollars will be shared equally by the two countries.

Also, in December 1972 the first Japan-USSR Joint Symposium on Aquaculture of the Pacific Ocean was held in Tokyo, and subsequently annual symposia have been held alternately in Japan and the USSR. These symposia have been organized by Tokai University (Shimizu) and VNIRO (All-Union Scientific Research Institute of Marine Fisheries and Oceanography) (Moscow). Although the papers cover a broad field of subjects related to fish culture and ocean farming, many are either directly or indirectly related to salmon aquaculture. For example, it is in these seminars that problems of disease and genetics have been discussed (Shikama 1973; Altukhov 1973).

Finally, there have been continuing exchanges of experts, data and other materials related to salmon propagation between the countries. Kanid'yev *et al.* (1970) refer to such an exchange.

REFERENCES

ALTUKHOV, YU. P. 1973. Rational management of marine biological resources in the light of population genetics. *In* Propagation of Marine Resources of the Pacific Ocean. First Japan-USSR Joint Symposium of Aquaculture of the Pacific Ocean (Tokai Univ. and VNIRO). Sept. 1973. pp. 31-40. Tokai Univ., Tokyo. (Russian)

ANDREEVA, M.A. 1954. Fish culture and fish preservation methods relating to the conservation and utilization of salmon in Kamchatkan waters. *In* Transactions of a Conference on Questions of the Salmon Fisheries of the Far East. NAUK SSSR (USSR Academy of Science), pp. 70-77. (Russian)

ANON. 1976. Japan-Soviet salmon propagation meeting. Fisheries Economics No. 6960, 1. (Japanese)

ATKINSON, C.E. 1960. Fisheries Research: Its Organization and Program within the USSR. Biological Lab., U.S. Bureau Fisheries, Seattle.

ATKINSON, C.E. 1964. The salmon fisheries of the Soviet Far East. M.S. Thesis. Univ. of Washington.

ATKINSON, C.E. 1976. Salmon aquaculture in Japan, the Koreas and the USSR. Proceedings of the Conference on Salmon, Aquaculture and the Alaskan Fishing Community. January 9, 10 and 11, 1976. Cordova, Alaska. Univ. of Alaska Sea Grant Rep. 76-2, pp. 79-154.

AYUSHIN, B.N., and CHIGIRINSKII, A.I. 1976. Bioecological basis and trends of mariculture development in the Far-East coast. Proceedings of the Fifth Japan-Soviet Joint Symposium on Aquaculture of the Pacific Region, Tokyo and Sapporo. Sept. 1976. (Russian)

BOGDONOVA, E.A. 1964. Diseases of young salmon and means of control in the fish hatcheries of Sakhalin and the Amur. *In* Salmon Fisheries of the Far East. NAUK SSSR (USSR Academy of Science), pp. 186-188. (Russian)

CHERNYAVSKAYA, I.K. 1964. The state of artificial rearing of salmon in the Sakhalin region. *In* Salmon Fisheries of the Far East. NAUK SSSR (USSR Academy of Science), pp. 175-178. (Russian)

DISLER, N.N. 1954. The development of Amur River fall chum salmon. *In* Transactions of a Conference on Questions of the Salmon Fisheries of the Far East. NAUK SSSR (USSR Academy of Science), pp. 129-143. (Russian)

DVININ, P.A. 1954. A survey of the salmon fisheries of Sakhalin and an analysis of the work of the fish hatcheries of the Sakhalin Fisheries Authority. *In* Transactions of a Conference on Questions of the Salmon Fisheries of the Far East. NAUK SSSR (USSR Academy of Science), pp. 78-86. (Russian)

FROLENKO, L.A. 1964. The rearing of young chum and pink salmon in Sakhalin hatcheries with artificial food. *In* Salmon Fisheries of the Far East. NAUK SSSR (USSR Academy of Science), pp. 184-185. (Russian)

KANID'YEV, A.N., KOSTYUNIN, G.M., and SALMIN, S.A. 1970. Hatchery propagation of the pink and chum salmons as a means of increasing the salmon stocks of Sakhalin. J. Ichthyology *10*, No. 2, 249-259.

KOKHMENKO, L.V. 1962. Food organisms for fall chum salmon and freshwater salmonoids in the nursery waters of Teplovka Hatchery. Bulletin, Pacific Research Institute of Fisheries and Oceanography. TINRO *48*, 67-84. (Russian)

KOKHMENKO, L.V. 1972. The food of the mykiss (*Salmo mykiss*) [Walb] in Kamchatka. J. Ichthyology *12*, No. 2, 282-291.

KOPOSOVA, A.F. 1964. Some questions on the effectiveness and the economics of artificial propagation of salmon in Sakhalin. *In* Salmon Fisheries of the Far East. NAUKA (USSR Academy of Science), pp. 179-183. (Russian)

KRYKHTIN, M.L. 1962. Material on the early life history of masu. Bulletin, Pacific Research Institute of Fisheries and Oceanography. TINRO *48*, 84-132. (Russian)

KUDRYAVTSEV, V.V. 1954. Discussion of the conference papers. *In* Transactions of a Conference on Questions of the Salmon Fisheries of the Far East. NAUK SSSR (USSR Academy of Science), pp. 174-175. (Russian)

LANDYSHEVSKAYA, A.E. 1965. Increasing the effectiveness of the work of the Sakhalin Fish Hatcheries. Commercial Fisheries *41*, No. 10, 9-10. (Russian)

LAZEREV, M.S. 1954. On the work of the Sokolovo Fish Hatchery. *In* Transactions of a Conference on Questions of the Salmon Fisheries of the Far East. NAUK SSSR (USSR Academy of Science), pp. 87-93. (Russian)

LEVANIDOV, V.YA. 1954. The way to increase the reproduction of the chum salmon of the Amur. *In* Transactions of a Conference on Questions of the Salmon Fisheries of the Far East. NAUK SSSR (USSR Academy of Science), pp. 120-128. (Russian)

LEVANIDOV, V.YA., and LEVANIDOVA, I.M. 1962. The spawning-rearing waters of Teplovka Hatchery and their biological productivity. Bulletin, Pacific Research Institute of Fisheries and Oceanography. TINRO *48*, 3-66. (Russian)

McNEIL, W.J. 1976. Personal communication, Juneau, Alaska.

OSTROYUMOV, A.G. 1964. Dynamics of the numbers of salmon in the Kamchatka rivers. *In* Salmon Fisheries of the Far East. NAUKA (USSR Academy of Science), pp. 69-72. (Russian)

RUKHLOV, F.N. 1973A. A contribution to the question of the method of artificial reproduction of the Pacific salmon. Bulletin, Pacific Research Institute of Fisheries and Oceanography. TINRO *91*, 11-17. (Russian)

RUKHLOV, F.N. 1973B. The freshwater period of life of the spawning fall chum and pink salmon of Sakhalin. Bulletin, Pacific Research Institute of Fisheries and Oceanography. TINRO *91*, 25-30. (Russian)

RUKHLOV, F.N. 1973C. Peculiarities of the downstream migrants of artificially-reared pink salmon. Bulletin, Pacific Research Institute of Fisheries and Oceanography. TINRO *91*, 31-36. (Russian)

SHIDLOVSKIY, A.L. 1954. Discussion of the conference papers. *In* Transactions of a Conference on Questions of the Salmon Fisheries of the Far East. NAUK SSSR (USSR Academy of Science), pp. 154-185. (Russian)

SHIKAMA, YASUMASA. 1973. Pathological Study of Fish Diseases in Propagation of Marine Resources of the Pacific Ocean. First Japan-USSR Joint Symposium of Aquaculture of the Pacific Ocean (Tokai Univ. and VNIRO). Sept. 1973, pp. 41-49. Tokai Univ., Tokyo. (Russian)

SMIRNOV, A.I. 1954A. Rationalization of the biotechnology of the culture of salmon in Sakhalin. *In* Transactions of a Conference on Questions of the Salmon Fisheries of the Far East. NAUK SSSR (USSR Academy of Science), pp. 94-110. (Russian)

SMIRNOV, A.I. 1954B. Discussion of the conference papers. *In* Transactions of a Conference on Questions of the Salmon Fisheries of the Far East. NAUK SSSR (USSR Academy of Science), pp. 195-196. (Russian)

SMIRNOV, A.I. 1960. Salmon culture in the Far East. Commercial Fisheries *36*, No. 10, 30-37. (Russian)

VANYAEV, N.A. 1954. Discussion of the conference papers. *In* Transactions of a Conference on Questions of the Salmon Fisheries of the Far East. NAUK SSSR (USSR Academy of Science), pp. 173-174. (Russian)

VASIL'EV, I.M. 1954A. Experiences of the work at the Teplovka Fish Hatchery of the Amur Authority. *In* Transactions of a Conference on Questions of the Salmon Fisheries of the Far East. NAUK SSSR (USSR Academy of Science), pp. 11-119. (Russian)

VASIL'EV, I.M. 1954B. Discussion of the conference papers. *In* Transactions of a Conference on Questions of the Salmon Fisheries of the Far East. NAUK SSSR (USSR Academy of Science), pp. 187-189. (Russian)

ZHUYKOVA, L.I. 1973. The effect of the degree of ripeness of the eggs on the embryological development of fall chum salmon. Bulletin, Pacific Research Institute of Fisheries and Oceanography. TINRO *91*, 18-24. (Russian)

Chapter 20

Republic of Korea

Koo-Byong Park

In recent years aquaculture production in Korea has shown a phenomenal increase. In the decade from 1965 to 1974, production increased from 73,705 MT (81,223 ST) to 340,324 MT (375,250 ST) or by 362%. Aquaculture, particularly shallow sea aquaculture, is contributing significantly to food production and the earning of foreign exchange in Korea.

This rapid increase in the aquacultural production, however, was not achieved by a rapid expansion of fish culture but by a rapid increase in the production of seaweeds and mollusks cultured in coastal shallow seawaters. The most important products in terms of quantity are seaweeds such as laver, dulse, agar agar and kelp, being followed by mollusks such as oysters, hard clams, shortnecked clams, cockles, sea mussels, abalones, pearl mother shells and octopuses. In 1974 the production of seaweeds amounted to 244,795.4 MT, accounting for 71.9% of the total aquacultural production. The production of mollusks was 95,353.2 MT, accounting for 28.0%. The remaining aquacultural production consisted of 168.9 MT (186 ST) of fish and 5.4 MT of crustaceans, such as shrimp and blue crab.

These figures show that the production of cultured fish is negligible in Korea, although a number of species of fish have been and are being cultured (Table 20.1).

History of fish farming in Korea, with the exception of common carp farming, is short, and most fish farming is still in the pilot stages. Freshwater fish such as carp, eel and rainbow trout are species cultured in Korea (Table 20.1). Farming of these species and some others will be briefly discussed.

TABLE 20.1

PRODUCTION OF CULTURED FISH BY SPECIES, KOREA

Species	Metric Tons				
	1970	1971	1972	1973	1974
Carp	3.0	6.4	24.5	29.3	44.5
Eel	9.4	157.6	5.9	35.7	85.0
Loach	—	—	—	—	15.7
Rainbow trout	3.4	7.1	15.4	6.1	—
Grass carp	—	0.6	1.8	0.3	0.8
Goldfish	0.2	1.6	1.5	2.4	4.7
Others	22.9	21.0	17.1	13.7	18.2
Total	38.9	194.3	66.2	87.5	168.9

Source: Anon. (1971-75).

CARP (Cyprinus carpio)

Carp culture in Korea has a history of hundreds of years. There is a historical record which suggests that carp were cultured as early as in the beginning of the third century. However, carp farming for commercial sale began about 50 years ago under the Japanese regime with the establishment of a public hatchery.

At the present time, carp farming is practiced all over the country. Although many carp farms are scattered throughout the country, the scale of most farms is small as reflected in the small total production of carp. In 1974 the total production of carp, including wild carp caught in natural waters, was 149.8 MT, of which 105.3 MT or 70.3% were wild. It should be noted that a considerable part of the wild carp catch was produced in government-run public hatcheries and released in natural waters as fry.

Three different methods are presently used in culturing carp in Korea, namely: (1) still-water culture; (2) running water culture; and (3) cage culture. Among these, method (1) is the oldest and most common method. Only two carp farmers use method (2). Method (3) was introduced recently, and a couple of carp farmers adopted this method.

In the still-water method ponds and reservoirs are used. Ponds are built with concrete or earthen walls. In carp farms with adjacent streams or rivers, ponds are designed so as to keep a small quantity of water flowing through ponds by gravity.

Since there is a large number of reservoirs for irrigation of rice fields, and many other reservoirs for power generation or multipurpose uses which can be rewardingly stocked with carp, Korea has favorable conditions for adopting still-water carp culture methods. These reservoirs, however, are not fully utilized yet. In many instances carp farming in reservoirs is extremely extensive, allowing the carp to feed on natural feeds.

In recent years, under the UN Korean Upland Development and Watershed Management Project, a large number of irrigation-fish ponds have been newly constructed, and some of them have already been stocked with carp fry. In addition, stocking reservoirs and ponds with carp has been encouraged by strong government support. Fry produced in public hatcheries and purchased from private hatcheries have been supplied free of charge to carp farmers. The government purchases fry of 3 to 4 cm (1 to 2 in.) for about $0.08 each to deliver to carp farmers. The Choseon Daily is, at the present moment, undertaking a "Raise Fish Where There Is Water" campaign, supplying free carp fry under the auspices of the Office of Fisheries.

Some carp farmers specialize in producing seedlings. But many carp farmers produce eggs and fry for their own use. In regard to the stocking rate, carp fry of 3 cm (1 in.), for example, are released in still-water ponds at the rate of about 6 per m^2 (less than 1 per ft^2). In running water ponds, carp yearlings of 10 to 13 cm (4 to 5 in.) are stocked with a higher stocking density.

In feeding carp, finely hulled wheat, trash bread, oil cake, silkworm pupae, fish meal and vegetables are widely used in Korea. Doughy, pasty feed compound made of these stuffs is prepared at each carp farm. No mass-produced commercial feeds are available.

There is a good market for carp in Korea. No carp farmer suffers from overproduction. Instead, carp production lags behind growing demand. Prices of carp received by farmers per kilogram range from about $1 to $3 ($0.45 to $1.35 per lb), whereas the average price of sea fish does not exceed $0.20 per kg ($0.09 per lb). This means that carp are valued at more than five times the average value of captured sea fish. Market sizes of carp vary from 500 g to 1 kg (1 lb to 2.2 lb). It takes about two years from hatching to produce market size carp.

Primary consumers of carp are sightseers who visit the fish farms. Carp are cooked and served at these farms. Sliced raw fish or pepper-pot soup of carp is popular with sightseers, but most sightseers prefer the former to the latter. For Korean tastes, raw fish goes well with beer or white liquor. Since, however, it is believed that carp as well as many other freshwater fish are intermediate hosts of the liver-fluke, many people do not eat raw fish. A recent study by Dr. Seh Kyu Chun of the Pusan Fisheries College indicated that carp are not an intermediate host of liver-fluke. If this is proved to be true through further studies, the demand for raw carp will increase.

Carp are also demanded for medicinal uses in Korea. It is said that when pregnant women eat carp they can have babies with pretty white skin, and that eating carp is conducive to the recovery of health after delivery. According to one of the most famous Oriental medical books, Huh Joon's *Tongeui-Bokam* (Oriental Medical Thesaurus) published in 1613 in Korea, the flesh of carp is efficacious for jaundice, morbid thirstiness, dropsy, etc., and the gall bladders of carps heal amaurosis and strengthen one's sight.

The demand for carp as sport fish has been growing rapidly. Carp stocked in fee-fishing ponds are caught with tackle and bait by the consumer. These carp are free when the consumer pays an entrance fee of $2 to $4 although the total catch is limited to 1 to 2 kg (2.2 to 4.4 lb). At fee-fishing ponds where the entrance fee is not paid or only a small sum is paid, the consumer has to pay a fee of $2 to $3 per kg ($0.91 to $1.36) for carp caught. The entrance fee and the fee for carp caught vary, depending on the distance between fishing ponds and population centers. Carp are often co-stocked with crucian carp in fee-fishing ponds.

In view of conserving the stock of freshwater fish, the Office of Fisheries recently prohibited fishing of wild freshwater fish in many natural inland waters. This has increased and will increase the demand for cultured carp as sport fish.

Profitability of carp farming in Korea varies greatly from farm to farm. Although it is difficult to obtain reliable cost-benefit data on many carp farms, it seems that there are carp farmings which are relatively profitable. For example, data obtained from a carp farm in Yangbuk-myon, Yangsan-kun, Kyongsang-nam-do which adopted the running water culture method showed the profitability of carp farming presented in Table 20.2.

The full cost of production per kilogram can be calculated by adding 15% for interest on the capital investment. The price received was $2.80 per kg ($1.27 per lb). Thus a relatively high profit on investment can be obtained. If facilities were fully utilized more profit could be obtained. This carp farm produced only about 2 MT in 1975, but it can produce as much as 20 MT per annum.

However, this is not to suggest that all the carp farms in Korea are invariably profitable. Profitability varies between carp farms and from farmer to farmer according to the suitability of the site and efficiency of operations. There are many unprofitable carp farms. For example, data obtained from a carp farm in a suburb of Pusan, which produced 2.5 MT of market-size carp in 1975 in a 5035 m² (1.5 acre²) still-water pond built of concrete, showed that the cost of production (including the depreciation cost and interest for

TABLE 20.2

COST-BENEFIT DATA ON CARP FARMING IN A 693 M² RUNNING WATER POND,
YANGSAN-MYON, YANGBUK-KUN, KYONGSANG-NAM-DO, KOREA, 1975

Item	($)	(%)
Capital Investment		
Value of land	210	14.2
Construction of pond	1260	85.8
Total	1470	100.0
Operating Costs		
Carp yearlings	200	10.3
Feeds	1120	57.5
Labor	480	24.6
Maintenance and repairs	6	0.3
Other	100	5.1
Depreciation[1]	49	2.2
Total	1955	100.0
Income	4800	
Profit	2845	
Ratio of profit to:		
Operating costs		1.46
Operating costs — depreciation excluded		1.50
Gross Income		0.59

[1] Life of 30 years for pond.

capital investment) exceded $4 per kg ($1.82 per lb), but the average price received was $2.60 per kg ($1.18 per lb). It seems, however, that if carp farms are properly managed, most carp farmings can obtain favorable returns on investment.

JAPANESE EEL (Anguilla japonica)

Eel farming in Korea began in the late 1960s with experimental culture. Since the beginning of the 1970s heavy demands from Japan for elvers stimulated an increased interest in eel farming and brought an eel farming boom.

Eels are widely cultured in Korea at present, but the annual production of cultured eels is highly variable, reflecting that eel farming in Korea is not as yet an established industry (Table 20.3).

In 1974 the total production of eels, including wild eels caught in natural waters, was 145.8 MT, of which the cultured eels accounted for 85 MT or 58%. Eel farming in Korea is characterized by the production of seedlings for exports. Culture of eels through to market size

TABLE 20.3

PRODUCTION OF CULTURED EELS BY PROVINCE AND YEAR, KOREA

Province	Metric Tons				
	1970	1971	1972	1973	1974
Kyongki-do	—	—	0.6	9.5	8.1
Kangwon-do	—	—	—	—	—
Chungcheon-buk-do	—	—	—	—	—
Chungcheon-nam-do	—	—	—	1.5	15.3
Kyongsang-buk-do	—	—	1.0	1.4	1.8
Kyongsang-nam-do	—	—	—	—	2.0
Cheonra-buk-do	—	—	3.3	20.3	51.6
Cheonra-nam-do	9.4	157.6	1.0	3.0	6.2
Cheju-do	—	—	—	—	—
Total	9.4	157.6	5.9	35.7	85.0

Source: Anon. (1971-75).

is practiced rather incidentally in some eel farms. Since no technique of artificial breeding of eels has been developed, eel farming has to be based on young elvers collected when they reach river mouths. These young elvers are domesticated to eat artificial feeds and reared for 2 to 3 months and then exported as seedlings.

In the southern part of South Korea, young elvers averaging 0.12 g and 4 cm (1.5 in.) are caught in river mouths from February to May, while in the northern part smaller ones are caught. These elvers are stocked in ponds for 2 or 3 months as already mentioned. During this feeding period, they reach 10 cm (4 in.) in length. On the other hand, the time required for eels to reach market size of 200 to 300 g (7 to 11 oz) is about 2 years.

Two different methods of raising eels (i.e., still-water culture and circulating filter system) are used in Korea. The still-water culture is the most common method in raising both elvers and market-size eels.

In still-water ponds water is aerated by splasher paddles (water wheel). During the spring and early summer when the water temperature is not high enough, eels are raised in covered ponds filled with heated water.

Eel farming by the circulating filter system was introduced in the early 1970s, and only a few eel farmers adopted this method. This is the most intensive fish farming method in Korea. To enhance the growth rate of elvers, it is necessary to keep a high average water temperature. Hence in the circulating filter system the pond is covered with a greenhouse built of angle iron frames and polyethylene or canvas covers. Pond water is heated by pipes, which are laid on the pond bottom, through which hot water is circulated. Dirty water in the culture pond is pumped into the filter pond to be cleansed, and then returned to the culture pond. Culture pond water is constantly aerated by an aerator and splasher paddles.

In the height of summer, when the water temperature outside the greenhouse rises above 25°C (77°F), the warm-water circulating filter system is transferred into the running water system. In some eel farms, elvers in the greenhouse pond are moved into still-water ponds, which are constructed on the same farm, during summer.

Main feeds used by Korean eel farmers are flesh of fresh mackerel and compound feed imported from Japan. Powdered compound feed is mixed with finely chopped flesh of mackerel to make a thick paste feed. In the early stage of feeding, other feeds can be fed to elvers and these include oysters and small earthworms.

There is a large export market for both elvers and market-size eels, and the eel has a very high commercial value. Until the end of the last century, Koreans had an aversion to eels because of their snake-shaped unacceptable appearance, and no Korean tried to catch eels for food use. Consequently rivers of Korea teemed with wild eels at that time. It was reported that the stock of eels was so abundant in those days that Japanese fishermen who first engaged in eel fishing in Korea could easily catch as much as 300 kg (660 lb) each in a day. After the turn of the century, however, Koreans came to like eels as their tastes were gradually changed by association with Japanese who liked eels. This led to an overexploitation of wild eels. Demand for eel as a food fish in domestic and foreign markets has steadily grown, and eels are highly prized, not only as a delicacy, but also as excellent health food. These combined effects have made eels the most expensive fish. The farm gate prices of cultured eels of market size reach about $8 per kg ($3.64 per lb) in the domestic market. Compared with the price of beef, it is about twice as expensive. Prices of cultured elvers exported to Japan and Taiwan are much higher. The ruling prices in 1976 range from about $44 (an average weight of 10 g each) to $80 (an average weight of 20 g each) per kg ($20 to $30 per lb). Quantities and values of exported cultured eels in the last three years are shown in Table 20.4.

Until 1971 Japan was the only importing country of Korean eels, but in 1972 Taiwan began to import them also. Eel exports to Taiwan have sharply increased in the last two years. The quantity and value of eels exported to Taiwan in 1975 amounted to 71,047 kg (156 ST) (54% of the total export) and 51% of the total export value of eels.

A cost-benefit ratio is calculated in Table 20.5 based on data obtained from an eel farm in Roksan-myon, Kimhae-hun, Kyongsang-nam-do, which produced about 2 MT of elvers by the circulating

TABLE 20.4

QUANTITY AND VALUE OF CULTURED EEL EXPORTS BY YEAR, SIZE
AND KIND, KOREA

Size and Kind	Quantity (kg) and Value ($)	1973	Metric Tons 1974	1975
Market size[1] (Under 10 per kg)	Quantity	2,830	5,263	2,386
	Value	11,278	27,175	16,640
	(Unit price)	(3.99)	(5.16)	(6.97)
Under 100 to 400 per kg	Quantity	67,215	109,279	71,328
	Value	4,318,436	2,563,815	1,463,129
	(Unit price)	(64.25)	(23.46)	(20.51)
500 to 2,500 per kg	Quantity	11,379	12,726	55,403
	Value	1,847,411	878,573	2,004,298
	(Unit price)	(162.35)	(69.04)	(36.18)
3,000 to 6,000 per kg[2]	Quantity	239	—	—
	Value	127,886	—	—
	(Unit price)	(535.09)	—	—
Processed	Quantity	13,465	2,500	1,500
	Value	83,943	15,230	10,500
	(Unit price)	(6.23)	(6.09)	(7.00)
Others	Value	—	150,000	—
Total	Quantity	95,128	129,768	130,617
	Value	6,388,954	3,634,793	3,494,567
	(Unit price)	(67.16)	(28.01)	(26.75)

Source: Anon. (1973-75).
[1] A small quantity of wild eels caught in natural waters is included.
[2] Exports of elvers under certain size have been prohibited by the government since 1974 for better use of the limited elver stock.

filter system in 1975. Although a relative bumper crop was harvested on this farm, the cost-benefit ratio was not high. This was mainly due to the unusually low export price of eels in 1975.

Cost of production per kilogram can be calculated in Table 20.5 by computing interest at 15% per annum for capital investment. Cost of production was about $19 per kg ($8.64 per lb). The average price received per kg was about $25 ($11.36 per lb), showing a favorable comparison.

The previous figures, however, by no means indicate that all eel farming in Korea is always profitable. The profitability of eel farming varies greatly from year to year. This is due mainly to wide annual variations in export prices and operating costs. Export prices of eels have shown extremely wide fluctuations, as can be seen in Table 20.4. In addition to this, operating costs vary annually to a considerable extent due to the fluctuation in quantity of annual production and total

TABLE 20.5

COST-BENEFIT DATA ON CULTURING 2 MT (2.2 ST) OF ELVERS IN A
660 M^2 (7102 FT^2) POND USING THE CIRCULATING FILTER SYSTEM,
ROKSAN-MYON, KIMHAE-KUN, KYONGSANG-NAM-DO, KOREA, 1975

Item	($)	(%)
Capital investment		
Construction of pond	7060	—
Construction of greenhouse	4800	—
Construction of cold storehouse	1200	—
Refrigerator	2400	—
Water heating system	6000	—
Generator	5000	—
Feed cooking equipment	1100	—
Aerator	1300	—
Total	28,860	—
Operating costs		
Young elvers	8000	23.8
Feeds	3320	9.9
Labor	9840	29.3
Maintenance and repairs	1286	3.8
Fuel	2000	5.9
Electric power	1200	3.6
Chemicals	600	1.8
Rental fee for land	400	1.2
Taxes	3200	9.5
Miscellaneous	1200	3.6
Depreciation[1]	2572	7.6
Total	33,618	100.0
Income	49,150	
Profit	15,532	
Ratio of profit to:		
Operating costs	0.46	
Operating costs—		
depreciation excluded	0.50	
Gross income	0.32	

[1] Lives of pond, cold storehouse and greenhouse are regarded as 30 years, 35 years and 8 years, respectively, and that of other equipment varies between 5 and 10 years. Depreciation costs for pond and cold storehouse were figured out by dividing their initial costs by the number of years of their lives; for the greenhouse and other equipment, 10% of initial costs was subtracted before calculations.

catch of elvers for stocking. Although no published data on the prices of elvers are available, it is known among eel farmers that prices vary greatly from year to year according to the catch. One eel farmer said that the average price of elvers per kg was about $700 in 1973, $300 in 1974, $30 in 1975 and $400 in 1976 ($318 per lb in 1973, $136 in 1974, $13.60 in 1975 and $182 in 1976). Annual production of eels varies greatly, too, and is subject to a number of factors. A main risk involved in the production of eel is loss through various diseases and parasites, and in a serious case all the eels under culture may die. In many cases eel farming is uneconomical, and it is regarded as a highly risky industry.

The market demand for elvers and market-size eels seems to continue to increase. Unfortunately, however, there is a narrow limitation in the expansion of eel farming since the arrivals of elvers to river mouths are limited. The exact quantity of elvers collected around Korean waters is unknown, but it is estimated that it is, on the average, only about 10 MT (11 ST) per annum. Culture of market-size eels should be encouraged rather than simply exporting elvers as seedlings.

RAINBOW TROUT (Salmo gairdneri)

Rainbow trout were imported from the United States in the mid-1960s. In 1965 an experimental rainbow trout culture was begun in Kangwon-do with 10,000 eyed eggs brought from the United States. In 1966 a public hatchery run by the local government was established in Pyongchang-myon, Pyongchang-kun, Kangwon-do, and the hatchery started production of rainbow trout fry to deliver to would-be rainbow trout farmers. Rainbow trout eggs used in producing fry were brought from the United States and Japan (Table 20.6).

Following a set of experiments, artificial spawning and larval rearing was successful in 1970, providing a strong base for rainbow trout farming in Korea.

The major eel farming province is Kangwon-do. Most rainbow trout farms are also concentrated in this most mountainous province in Korea.

The best water temperature for raising rainbow trout is between 10° and 18°C (50° to 65°F); hence the water temperature should not go much over 20°C (68°F) in summer nor fall too low in winter. Sites which satisfy this condition and at the same time assure adequate supplies of water can be found only in limited places in Korea. Because of this constraint, the total area of rainbow trout farming was only 3.19 ha (7.9 acres) in 1973. The number of farmers engaging in rainbow trout farming is limited to about 18 at present. Annual average production of rainbow trout is less than 10 MT (11 ST) (Table 20.1). Total production in 1974 was so small that it was not shown in official statistics.

Running water culture and still-water culture methods are used in raising rainbow trout in Korea. In both cases, an important source of water is spring water. Using spring water, extremely high or low wa-

TABLE 20.6

IMPORTED NUMBER OF EYED EGGS OF RAINBOW TROUT BY
YEAR, KOREA

Year	USA	Japan	Total
1965	10,000	—	10,000
1966	200,000	—	200,000
1967	800,000	30,000	830,000
1968	500,000	—	500,000
1969	1,972,000	100,000	2,072,000
1970	800,000	100,000	900,000
Total	4,282,000	230,000	4,512,000

Source: Sun Tae Kim et al. (1971).

ter temperatures can be avoided. In still-water ponds a small quantity of pond water is continuously replaced by spring water. In some rainbow trout farms where running water culture is practiced, pumped-ùp well water is used.

In feeding rainbow trout, farm-made feeds are used. Ingredients of feed include silkworm pupae, fish meal (made of Alaska pollacks in the main), rice-bran, flour, oil cake and various vegetables. Specially prepared pelleted feed is also used in some farms.

Rainbow trout is an unfamiliar fish for most Koreans. Those who have tasted it soon come to like it. Rainbow trout are consumed by sightseers who visit rainbow trout farms, as was the case with carp. Raw rainbow trout are a most popular dish. They are also demanded as sport fish, and some fee-fishing ponds are stocked to meet this demand.

Rainbow trout is a high-priced luxury fish in Korea. The price of rainbow trout received by farmers was about $3 per kg in 1975 ($1.36 per lb). In spite of the high price, demand for rainbow trout has increased and exceeds the present level of production, particularly with those farms situated near population centers.

It is probably too early to talk about the profitability of rainbow trout farming in Korea since such farming is still largely in the experimental stage. Though a few farmers are still skeptical, there is little doubt that it is a promising industry. Its profitability is suggested by a hypothetical cost-of-production schedule prepared by a rainbow trout farmer who is raising rainbow trout in still-water ponds in Nam-myon, Cheongseon-kun, Kangwon-do and produced about 2000 market-size rainbow trout in 1975 (Table 20.7).

Feed constitutes the major item of cost in rainbow trout farming. The cost of feed accounted for about 92% of the total operating costs in

TABLE 20.7

HYPOTHETICAL COST OF RAISING 100,000 MARKET-SIZE RAINBOW TROUT VALUED AT
75 MILLION WON[1] ($150,000) — TIME REQUIRED IS 18 MONTHS; KOREA, 1975

Item	Cost (won)
Fry (150,000 — 5 won each)	750,000
Feed (7,500,000 won for raising fry for 6 months, and 30,000,000 won for raising 100,000 rainbow trout for 12 months)	37,500,000
Wage (one manager, 900,000 won, and 2 assistants, 1,080,000 won)	1,980,000
Electric power and fuel	72,000
Maintenance costs of ponds and hatchery	120,000
Miscellaneous (firewood, chemicals and others)	360,000
Total[2]	40,782,000

[1] 500 won equal approximately $1.00 (U.S.).
[2] Total won equals $81,564.

the case shown in Table 20.7. Adequate supplies of suitable feeds at reasonable prices are of special importance in raising rainbow trout.

Full utilization of existing culture facilities and intensification of culture will alleviate the narrow limit on production of rainbow trout from the lack of suitable sites, although a significant expansion of the industry cannot be expected even in the future.

OTHERS

Loach (Misgurnus anguillicaudatus)

Korea has native loach with which all Koreans are familiar. A considerable amount of wild loach is caught annually. In 1974 the catch amounted to 214.6 MT (236 ST). Loach farming has been practiced for years, although its contribution to fish farming has been negligible except for 1974, when 15.7 MT (17 ST) of loach were produced under cultured conditions.

Loach are raised in small still-water ponds. The breeders are collected from wild stock. Techniques of intramuscular injection of frogs' pituitaries are often applied to loaches to induce maturation and spawning, but they are practiced only on an experimental small scale.

Ponds are fertilized with compost or cattle-dung to help produce natural feeds. Supplementary artificial feeds such as fish meal, flour, oil cake and barley-bran are sometimes fed.

Loaches are in demand as a food fish as well as bait fish for commercial fisheries, and they are expensive in Korea.

Grass carp (Ctenopharyngodon idellus)

Grass carp were introduced with silver carp from Japan by Dr. In-Bae Kim of the Pusan Fisheries College in 1963. Experiments with the artificial spawning of grass carp using hormone injection techniques, conducted at the Pusan Fisheries College and the public hatchery in Cheongpyong, Kyongki-do, were successful in 1970. This paved the way for grass carp farming. Since then pilot-scale grass carp farming has been practiced on a few fish farms.

Special consideration should be given to culture of grass carp, for they are herbivorous animals and there is virtually no limitation in the Korean feed supply.

Catfish (Ictalurus punctatus)

Channel catfish were imported from the United States in 1972. They have been experimentally cultured in the ponds of the Pusan Fisheries College and public hatcheries.

The low water temperature which lasts for months in the winter season is the greatest difficulty encountered with catfish farming in Korea. Dr. In-Bae Kim says that if channel catfish can survive through the severe winter season in Korea, special attention should be paid to their acclimatization and the popularization of their culture.

Although there is an indigenous catfish (Parasilurus asotus) in Korea and Koreans admire it as food fish, no attempt to raise it has been made as yet.

Blue gill (Lepomis macrochirus)

Blue gills were introduced from Japan in 1970. Some fish farmers are raising them and they are becoming popular as sport fish.

Bass (Micropterus salmoides)

Largemouth bass were imported from the United States in 1973. They are presently under experimental culture at the public hatchery in Cheongpyong.

Sea fish

No appreciable sea fish culture has yet been practiced in Korea. However, seedlings of a few species of sea fish have been produced under culture.

Fry of a couple of species of sea breams of 2 to 3 cm in length are caught in coastal waters and raised in net enclosures for 2 to 3 months. They are fed with minced low-priced fish. All of them are exported to Japan.

Yellowtail fry were raised by a similar method for the same purpose in the past. However, in 1975 the government prohibited the catching of yellowtail fry with a view to conserving the natural stock of yellowtail.

OUTLOOK

The fact that fish farming requires relatively expensive feeds is one of the most important reasons that it has not achieved a rapid expansion in Korea, while rapid development has taken place in shallow sea culture of seaweeds and mollusks which do not require feeding. In most fish farming, feed constitutes the major item of operating costs, accounting for more than 50%. Furthermore, it is impossible at present to obtain adequate quantities of suitable commercial feeds produced by a centralized mass production system. This is a serious constraint on large-scale fish farming. Therefore, sufficient supplies of commercial feeds fitted for each species at reasonable prices are of special importance in the expansion of fish farming.

Another major constraint on the expansion of fish farming is related to the construction of ponds. The climate of Korea is characterized by a rainy warm summer and a cold winter which lasts for about three months. It is necessary therefore to build deep ponds with strong walls to prevent flood damages as well as winter kills. Because of this, coupled with high costs incurred in purchasing land for pond construction, heavy capital investments are involved in fish farming. On the other hand, time required to obtain a return on investment is long. In Korea, where high interest rates prevail, this is one of the most serious constraints. Sufficient supplies of government grants or low interest loans could alleviate this constraint to a great extent. However, these are not likely to be realized in the near future since fish farming, which produces luxury food items or sport fish for the limited demands of a small number of people, cannot receive high priority in the aquacultural development program.

With these constraints, a significant expansion of fish farming in Korea cannot take place in the foreseeable future. This, however, does not rule out the possibility of slow but steady development of fish farming in years to come. It can be stated with certainty that increasing per capita incomes and leisure, which will result from the high growth rate of the national economy, will create new demands for cultured fish as both luxury food fish and sport fish, thus stimulating investors' interests in fish farming.

SPECIAL ACKNOWLEDGEMENTS

DR. SEH KYU CHUN, Pusan Fisheries College, Pusan, Korea
DR. IN-BAE KIM, Pusan Fisheries College, Pusan, Korea

REFERENCES

ANON. 1971-75. Yearbook of Fisheries Statistics, 1971, 1972, 1973, 1974 and 1975. Office of Fisheries, Seoul, Korea. (Korean)

ANON. 1973-75. Misc. data. Korean Inland Water Export Promotion Association, Seoul, Korea.

SUN TAE KIM *et al.* 1971. Spawning and hatching of the rainbow trout. Report of Freshwater Culture Research No. 8 (August). (Korean)

Chapter 21

JAPAN

E. Evan Brown and S. Nishimura

Fish have been raised by Japanese farmers for hundreds of years. However, production for commercial sale began about 150 years ago. The industry grew slowly until the 1930s when the government placed emphasis on marine harvest. During World War II the government emphasized freshwater culture and production of freshwater fish expanded. After the war the cultured fish industry, though largely ignored by the government, continued to grow. In 1950 output of shallow-sea cultured fish reached 48,000 MT (52,896 ST), including mussels, mollusks and seaweeds, and freshwater cultured fish reached 5000 MT (5510 ST). From 1950 through 1974 the output of cultured marine fish, mussels, mollusks and seaweeds expanded from 48,000 MT to 880,000 MT (969,760 ST) or by 1733% (Table 21.1). This was at an annual rate of more than 69%. Inland freshwater cultured fish production increased from 5000 MT (5510 ST) in 1950 to 67,000 MT (73,834 ST) in 1974. This was a 1240% increase, or an annual rate of nearly 50% (Table 21.1). Marine harvest of wild fish increased 199% during these 25 years and the inland freshwater wild harvest increased only 77%. In 1950 the cultured fish industry accounted for 1.6% of total production. In 1974 the cultured fish industry accounted for 8.8% of total production. At present, the government is making renewed efforts to stimulate marine and freshwater culture, including sea ranching, in anticipation that the results of new international 200 mi. zones will reduce the Japanese marine catch by several million tons annually.

The 880,000 MT of shallow-sea culture in 1974 represented only 8.2% of total supply, but accounted for 13.6% of total value (Table 21.2).

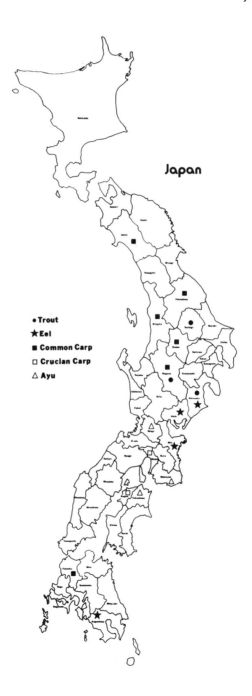

FIG. 21.1.   MAIN AREAS OF PRODUCTION BY SPECIES OF FRESHWATER
CULTURED FISH, 1976

TABLE 21.1

FISH PRODUCTION[1] (NOT INCLUDING WHALES), JAPAN

Year	Marine Harvest	Shallow Sea Cultured Fish Harvest	Inland Fresh-water Harvest (Wild)	Inland Cultured Freshwater Harvest	Totals
1950	3255	48	63	5	3371
1951	3774	88	60	6	3928
1952	4646	113	53	9	4823
1953	4387	144	57	8	4596
1954	4303	145	82	9	4539
1955	4658	154	82	11	4905
1956	4487	180	90	13	4770
1957	5067	244	81	14	5406
1958	5197	214	78	15	5504
1959	5567	225	75	15	5882
1960	5817	284	74	15	6190
1961[2]	6287	322	81	18	6708
1962	6346	362	84	20	6812
1963	6200	390	85	23	6698
1964	5869	363	89	30	6351
1965	6382	380	113	33	6908
1966	6558	405	103	37	7103
1967	7241	470	97	42	7850
1968	7993	522	103	52	8670
1969	7976	473	112	52	8613
1970	8598	549	119	49	9315
1971	9149	609	101	50	9909
1972	9400	648	109	56	10,213
1973	9722	791	114	64	10,691
1974	9749	880	112	67	10,808

Source: Anon. (1976C).
[1] All figures indicate thousands of metric tons.
[2] Between 1961 and 1965, 87% of the marine harvest came from domestic waters.

TABLE 21.2

DISTRIBUTION AND VALUE OF THE VARIOUS TYPES OF MARINE AND FRESHWATER FISH HARVEST, JAPAN, 1974

Distribution of Harvest	Harvest (000 MT)	Volume (%)	Harvest ($ million)	Value (%)
Marine harvest (wild)	9749	90.2	4595	62.8
Shallow sea culture	880	8.2	775	13.6
Inland freshwater (wild)	112	1.0	329	5.8
Inland cultured	67	0.6	213	3.8
Totals	10,808	100.0	5699	100.0

Source: Anon. (1976C).

The freshwater cultured fish accounted for 0.6% of supply but amounted to 3.8% of total value. This means that each unit of cultured marine fish was valued at nearly twice the average fish value, while each unit of freshwater cultured fish was valued at nearly six times the average value. These data indicate the emphasis on culturing highly valued species.

FRESHWATER CULTURED FISH

In Japan there are five major species of freshwater fish produced by culturing methods, i.e., in ponds or running water enclosures. The pond classification in government statistics is separated into: (1) agricultural ponds, used mainly for agriculture and where the water levels vary considerably; (2) specially built ponds of concrete and/or stone with some running water; (3) still-water ponds, used primarily for fish culturing and with nearly constant water levels; and (4) running water enclosures. In a few cases pens, cages and nets are used. The five major species raised in the systems are: (1) rainbow trout *(Salmo gairdneri)*; (2) eel *(Anguilla japonica)*; (3) common carp *(Cyprinus carpio)*; (4) crucian carp *(Carassius carassius)*; and (5) ayu or sweetfish *(Plecoglossus altivelis)*. In addition to these five there is some culturing of: (1) loach *(Misgurnus anguillicaudata)*; (2) gray or jumping mullet *(Mugil cephalus)*; and (3) so-called native trout which are comprised of *Salvelinus pluvius, Oncorhynchus masu* and *Oncorhynchus rhodurus*. These minor species will be discussed in separate sections.

In 1974, of the 5 major cultured species, a total of 66,584 MT (73,376 ST) were cultured and 31,063 MT (34,231 ST) of wild fish were caught (Table 21.3). Of the total supply of 97,647 MT (107,607 ST), 68% were cultured and raised to market sizes for food. Ninety-one per cent of the trout were cultured, 89% of the eel, 82% of the common carp, 28% of the ayu or sweetfish and 8% of the crucian carp. By volume of production, common carp ranked first with 26,323 MT (29,-008 ST), trout second with 17,631 MT (19,429 ST), eel third with 17,077 MT (18,819 ST), ayu or sweetfish fourth with 4712 MT (5193 ST), and crucian carp fifth with 841 MT (927 ST). The remaining volume of freshwater fish was harvested as wild fish caught in inland fresh waters, although many of these were produced in hatcheries before being released in native waters. Hence to some extent these wild fish were partially cultured, even if they had not been fed to a finished weight.

In recent years the volume of freshwater fish cultured has been increasing. For example, common carp production between 1965 and 1974 increased from 7973 MT (8786 ST) to 26,323 MT (29,008 ST)

TABLE 21.3

VARIETIES AND VOLUMES OF FISH PRODUCED BY INLAND FRESHWATER
FISHING AND FROM CULTURED IMPOUNDED WATER AREAS, JAPAN, 1974

Variety	Freshwater Fishing (MT)	Cultured Fish (MT)	Totals (MT)	Cultured (%)
Trout				
(Salmo gairdneri)	1,667	17,631	19,298	91
Eel				
(Anguilla japonica)	2,083	17,077	19,160	89
Common carp				
(Cyprinus carpio)	5,698	26,323	32,021	82
Crucian carp				
(Carassius carassius)	9,347	841	10,188	8
Ayu (sweetfish)				
(Plecoglossus altivelis)	12,268	4,712	16,980	28
Total	31,063	66,584	97,647	68

Source: Anon. (1976C), p. 228.

for an average yearly increase of 23% (Table 21.4). Trout during the same ten years increased from 5747 MT (6333 ST) to 17,631 MT (18,386 ST) for an average yearly increase of 21%. Ayu also increased by a total of 26% annually, while crucian carp and eel declined.

Production Methods

Official government statistics classify production facilities by four methods. These are: (1) running water ponds, (2) still-water ponds, (3) agricultural or farm ponds and (4) nets, pens or cages (Table 21.5). As can be seen in Table 21.5, running water culture is used for trout, ayu and common carp. However, there is a world of difference in this classification. The trout and ayu enclosures or raceways may turn over or replace the water every hour while in the carp ponds the water may be replaced only once daily.

The still-water ponds have only enough fresh water coming in to replace losses. With carp these are simple ponds that may be earthen. With eel they are circulating-filter systems, comparable to running water culture for trout, except the running water is used over and over again. These eel ponds may be inside buildings that resemble greenhouses. The water goes through mechanical and biofilters and is recirculated back to the pond. During circulation the water is aerated to increase its oxygen content. The major disadvantage of this sys-

TABLE 21.4

VOLUMES OF CULTURED FRESHWATER FISH PRODUCTION BY SPECIES, JAPAN

Species	1965	1966	1967	1968	Metric Tons 1969	1970	1971	1972	1973	1974
Eel (Anguilla japonica)	16,021	17,020	19,605	23,640	23,276	16,730	14,233	13,355	14,862	17,077
Carp (Cyprinus carpio)	7,973	9,829	10,886	14,460	13,971	15,865	17,840	23,037	26,406	26,323
Crucian carp (Carassius carassius)	1,504	1,441	1,622	1,615	1,776	1,391	1,209	1,183	1,369	841
Trout[1] (Salmo gairdneri)	5,747	6,232	7,882	9,454	10,254	10,632	12,749	13,515	15,707	17,631
Ayu (sweetfish) (Plecoglossus altivelis)	1,320	1,701	N.A.	2,343	N.A.	3,411	3,941	4,317	4,428	4,712
Total	32,565	36,223	39,995	51,512	49,277	48,029	49,972	55,407	62,776	65,637

Source: Anon. (1976C).
[1] Includes small amount of trout other than Salmo gairdneri.

TABLE 21.5

NUMBER AND TYPES OF MANAGEMENTS, NUMBER AND TYPES OF PRODUCTION PLACES, SURFACE AREA AND HARVEST VOLUMES BY SPECIES OF CULTURED FRESHWATER FISH, JAPAN, 1974

Species	Manage-ments (No.)	Total Managements				Total Places	Total Places or Ponds			
		Running Water Ponds	Still-water Ponds	Farm Ponds	Nets or Cages		Running Water Ponds	Still-water Ponds	Farm Ponds	Nets or Cages
Rainbow trout	1,375	1,371	—	—	4	1,600	1,596	—	—	4
Common carp	4,517[2]	1,244	1,802	1,218	399	6,177[2]	1,646	2,641	1,603	455
Crucian carp	299	—	1	298	—	603	—	1	602	—
Eel	2,774[2]	155	2,632	—	1	3,064[2]	159	2,914	—	2
Ayu	317	315	—	—	2	337	335	—	—	2

Species	Total Area	Surface Area (1000 m^2)				Total[1]	Harvest Vol (MT)			
		Running Water Ponds	Still-water Ponds	Farm Ponds	Nets or Cages		Running Water Ponds	Still-water Ponds	Farm Ponds	Nets or Cages
Rainbow trout	1,708	1,704	—	—	4	17,636	—	—	—	—
Common carp	35,861	2,039	6,037	27,236	549	23,902	4,125	3,164	10,002	6,611
Crucian carp	19,682	—	1	19,681	—	717	—	—	717	—
Eel	27,027	389	26,637	—	—	16,075	455	15,617	—	2
Ayu	486	486	—	—	—	4,114	4,114	—	—	—

Source: Anon. (1976C).
[1] Doesn't include production figures of minor prefectures.
[2] A management sometimes has both running water and still-water facilities.

tem is that it has a tendency to encourage the spread of disease. If fish in one production unit become diseased, the system spreads the infection to other production units in which the same water is circulated.

Agricultural or farm ponds are primarily used for rice production. Crucian carp and some common carp are grown in these ponds. When the rice is nearly mature the rice fields are flooded prior to harvesting. When harvesting begins the fields are drained. During this time the flow of water into these ponds may cease. In spite of this limitation, farm ponds are often used for fish production because: (1) investment is minimized compared to other methods, (2) water temperatures are conducive to fast growth of warm water fish, such as common and crucian carp, and (3) feeding is relatively easy. Disadvantages include: (1) oxygen supply depends on plankton and it is difficult to manage the ponds to maximize plankton growth. Compared to other methods of production, output is low per unit of water. To maximize production the water supply during the low water season must be at least one-half of the flow of other seasons. As water flow increases in the spring and fall it must be possible to control water flows so that the plankton growth is not excessively decreased. Also, the mud layer at the bottom of the pond must not be excessive or harvesting becomes difficult unless complete draining can be practiced. Water should be between 2 and 3 m (6 to 10 ft) deep so that in hot weather some cooler water can be found at the bottom of the pond. Otherwise feeding may have to be curtailed, thus decreasing the rate of gain. The pond should be subject to sunlight for a considerable part of the day so that plankton growth is encouraged. In Japan desirable pond sizes range from 1 to 3 ha (2.5 to 7.5 acres).

With pond culture, feeding should take place near the center of the pond. The water depth should be at least 1 m (3 ft) even in extremely dry weather. The feeding area should be sunny and not subject to strong winds. When the pond is initially stocked, the fish have a tendency to gather near the bank. Hence feeding should begin the day after stocking and in an area 2 to 3 m (6 to 10 ft) from the bank. Over time the distance can be extended so that the fish are gradually moved to the center of the pond for feeding. The usual practice is to have a walkway extending into the pond with some provision made for stocking a minimum amount of feed at the end of the pier. Special care needs to be given to scattering the feed evenly over the feeding area. In Japan, the usual feeding practice is to feed about 10 times daily from 7 a.m. until 5 or 6 p.m. in running water culture, and about 4 times in pond culture.

Most varieties of cultured fish will stay near the area where they are fed. If they are unable to secure feed at that spot they have a tendency to move to another part of the pond and may never return to the feeding area, but rely on natural foods. When this situation occurs, growth of fish may be uneven and poor, sometimes with 20 to 40% of the fish being produced without supplementary feeding. This can be a major problem to pond culturists; and the greater the stocking rate (such as in larger ponds), the greater the possibility of this problem arising. Thus, the rate of gain per kilogram of fish stocked may be decreased considerably.

Net, pen or cage culture is found in many lakes and some rivers. Fish are grown in enclosures anchored in the larger body of water. Usually no effort is made to aerate the water. Instead water currents within the larger body of water are relied upon to bring a constant supply of fresh water to the net, pen or cage.

As can be seen in Table 21.5, trout are nearly always cultured in running water enclosures called raceways. Common carp are produced with all systems or methods. Crucian carp are produced in farm ponds, eel in still water (recirculating) ponds and ayu in raceways.

Rainbow Trout (Salmo gairdneri)

In 1974 cultured production of trout was 17,631 MT[1] (19,429 ST). The wild catch was 1667 MT (1837 ST). Hence of the total 19,298 MT (21,266 ST), 91% was cultured. Of the total cultured production, Nagano Prefecture accounted for 23%, Shizuoka Prefecture for 18%, Gifu for 7%, and Tochigi for 6%. The remaining production, 46%, is accounted for in other prefectures. Rainbow trout are widespread in Japan, being produced on all four main islands. Rainbow trout production is expanding rapidly. For example, in 1965 production was only 5747 MT (6333 ST). Thus in 10 years production expanded 207%, for an average annual increase of nearly 21% (Table 21.4).

In 1974 there were 1375 separate managements in Japan. All of these but four used running water culture methods. Four producers used either cages or nets (Table 21.5). There were 1,708,000 m² or 170.8 ha (422 acres) of water area in rainbow trout production. Production per m² was 10.33 kg (2.1 lb per ft²). This is very intensive production and denotes good management, feeding practices and proper use of high quality and large quantities of water per farm.

Trout require a much higher oxygen content in the water than most other fish, so a plentiful supply of cold water with high oxygen content is essential. In general, production is carried out in small concrete-

---

[1] Includes small amounts of trout other than rainbow. Rainbow trout totaled 16,684 MT (18,386 ST).

lined units called raceways. Water is secured from deep wells or from mountain streams. Rainbow trout producers utilize running water culture almost exclusively. The advantages and disadvantages of this type of culture follow respectively:

(1) The volume of fish produced is generally higher, and more fish can be produced per given area of water surface.

(2) Ponds or races are smaller in contrast to pond culture, and thus easier to manage.

(3) Fewer people are required per unit of production.

(4) Harvesting is easier compared with larger ponds.

and:

(1) A continuous volume of water flow is required during the growing season.

(2) Essentially the same equivalent volume of water is required during each time period.

(3) Care must be exercised at all times to ensure that debris carried by the water does not interfere with the water flow.

(4) The heavy flow of water requires the sides and bottom of the small ponds or races to be of durable material, such as concrete. Thus construction costs are liable to be higher in comparison to other methods of culture.

In 1974, 26,168 MT (28,837 ST) of feed were fed. Nearly all of this was pelletized food (Table 21.6). The average feed conversion for the entire industry was 1.48, meaning 1.48 units of feed fed per unit of output.

TABLE 21.6

TYPE AND AMOUNTS OF FEED FED, BY SPECIES OF FRESHWATER
CULTURED FISH, JAPAN, 1974

Type of Feed	Rainbow Trout (MT)	Ayu (MT)	Common Carp (MT)	Crucian Carp (MT)	Eel (MT)
Fresh and frozen fish	133	267	297	1	15,916
Fresh silkworm pupae	—	—	2,698	7	7
Dry silkworm pupae	20	3	7,147	303	20
Pellets	25,568	7,063	33,589	170	28,076
Other	447	17	1,824	457	785
Total	26,168	7,360	45,555	938	44,804

Source: Anon. (1976C), pp. 229-235.

The quantity of feed fed to trout varies by size of the fish and temperature of the water. Regardless of the country of production, feed-

TABLE 21.7

QUANTITY OF FEED TO BE FED TO RAINBOW TROUT IN RELATION TO WATER TEMPERATURE
AND SIZE OF FISH, JAPAN

Length of Fish		Water Temperature									
(cm)	(in.)	(C) 2°(F)36°	4°40°	6°45°	8°49°	10°53°	12°57°	14°63°	16°68°	18°71°	20°74°
		(Percent of live weight to be fed)									
less than 2.5	1 or less	2.0	2.5	3.1	3.5	4.2	4.6	5.5	6.3	7.1	7.9
2.5- 5.0	1 -2	1.7	2.2	2.6	3.0	3.4	3.9	4.4	5.2	5.9	7.5
5.0- 7.5	2 -3	1.4	1.7	1.9	2.3	2.7	3.1	3.6	4.3	4.8	5.5
7.5-10.0	3 -4	1.1	1.3	1.5	1.7	2.2	2.4	2.8	3.3	3.9	4.4
10.0-13.0	4 -5	0.8	1.0	1.1	1.4	1.6	1.8	2.1	2.5	2.8	3.2
13.0-15.0	5 -6	0.7	0.8	0.9	1.1	1.4	1.5	1.7	2.0	2.2	2.6
15.0-18.0	6 -7.5	0.6	0.7	0.8	0.9	1.1	1.3	1.4	1.6	1.8	2.0
18.0-21.0	7.5-8.5	0.5	0.6	0.7	0.8	1.0	1.1	1.2	1.4	1.5	1.7
21.0-23.0	8.5-9.25	0.5	0.6	0.6	0.7	0.9	1.0	1.1	1.3	1.4	1.7
23.0-26.0	9.25-10.5	0.4	0.5	0.6	0.6	0.8	0.9	1.0	1.1	1.3	1.6
26 and over	10.5 or more	0.4	0.4	0.5	0.6	0.7	0.8	0.9	1.1	1.3	1.7

Source: Chiba (1968), Vol. II (Trout), p. 145.

ing charts are relatively uniform. A typical example of feeding guides is shown in Table 21.7.

Production Costs.—Production costs for cultured rainbow trout are similar to those of western Europe and the USA. However, wages are usually a more significant cost because output per worker is lower. Only on the very large trout farms does production per person reach the output found elsewhere. There are many small producers who produce less than 20 MT (22 ST). Thus, on these farms productivity may be only one-third of the output per person found on the larger farms in other countries. An example of the typical distribution of costs and returns is found in Table 21.8.

Marketing.—In general, rainbow trout as a food fish are not highly regarded in Japan. Even after some 70 to 80 years of culturing, the people still prefer native marine and freshwater species. However, because of inflation and rapidly increasing consumer prices, many Japanese have in recent years adapted their diets to include the relatively inexpensive rainbow trout. For example, in 1965, of the total produc-

TABLE 21.8

PRODUCTION COSTS, SELLING PRICE AND NET RETURNS FOR RAINBOW TROUT[1], FUJINOMIYA TROUT COOPERATIVE, SHIZUOKA PREFECTURE, JAPAN, 1975

Item	(cents per kg)	Production Costs (cents per lb)	(%)
Fry[2]	14.7	6.7	12.2
Feed[3]	69.0	31.4	57.1
Wages, including bonus[4]	16.0	7.3	13.2
Pumping costs[5]	9.7	4.4	8.0
Repairs and maintenance	3.0	1.4	2.5
Depreciation[6]	2.0	0.9	1.7
Insurance and taxes	2.0	0.9	1.7
Office and transportation	2.0	0.9	1.7
Miscellaneous	2.3	1.0	1.9
Total costs[7]	120.7	54.9	100.0
Selling price	150.0	68.2	—
Net returns[8]	29.3	13.3	19.5

Source: Fujinomiya Trout Cooperative, Shizuoka Prefecture, Japan.
[1] This represents 75 MT (82.7 ST) of production.
[2] 670,000 fry purchased at 1.3 cents each. Survival to 10 g was 82%. Survival to market sizes of 150 g was 75%.
[3] 116.25 MT of feed purchased at $444 per MT ($403 per ST). Feed conversion was 1.55 to 1.
[4] Three men employed at $4000 each.
[5] Pumping at the rate of 450 liters per sec, or 119 gal. per sec.
[6] Total investment of $333,000.
[7] Rental costs of the 1600 to 1700 m² of land would be additional.
[8] Returns to land and management was $39,067.

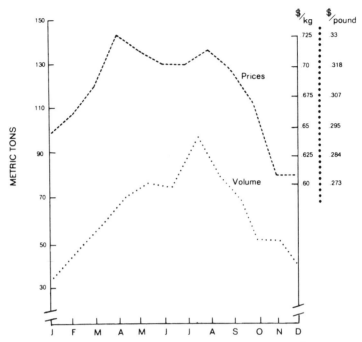

FIG. 21.2.   COMPARISON OF AVERAGE MONTHLY VOLUMES OF RAINBOW
TROUT SOLD AND PRICES PAID, TOKYO CENTRAL WHOLESALE FISH
MARKET, 1974

tion of 5747 MT (6333 ST), 2122 MT (2339 ST) or 37% were exported. At that time trout was very cheap, wholesaling for $0.75 per kg ($0.339 per lb) in Tokyo. The export price for frozen and dressed trout was only $0.81 per kg ($0.369 per lb) FOB Tokyo. After allowing for processing costs, packaging, and dress-out loss the export price was less than the domestic price. Nevertheless, the export market had to be used to sell all the trout produced. However, by 1974 the situation had turned around. Production increased by 207% to 17,631 MT (19,429 ST). However, exports increased to only 2518 MT (2775 ST) or by 19%. Domestic consumption increased from 3625 MT (3995 ST) to 15,113 MT (16,654 ST) or by 317%. Wholesale prices in Tokyo had increased to $1.51 per kg ($0.69 per lb), or more than double the prices of ten years before. Trout were still relatively inexpensive compared to other marine and freshwater cultured fish.

In 1974, 853 MT (940 ST) of rainbow trout were sold through the Tokyo Central Wholesale Fish Market (Table 21.9). This amounted to nearly 5% of total production, down from 8% in 1972. This indicates that the central wholesale fish markets located in the large cities may handle as much as 25% of total production. The major markets for trout

TABLE 21.9

VOLUME OF RAINBOW TROUT SOLD AND PRICES BY MONTHS, TOKYO CENTRAL WHOLESALE FISH MARKET, JAPAN

Month	1972			1973			1974		
	kg	$ per kg	$ per lb	kg	$ per kg	$ per lb	kg	$ per kg	$ per lb
January	45,075	1.00	0.45	64,151	1.01	0.46	36,922	1.40	0.64
February	67,839	0.98	0.45	87,551	0.98	0.45	54,993	1.45	0.66
March	91,375	0.95	0.43	124,561	1.05	0.48	62,289	1.50	0.68
April	98,753	0.92	0.42	106,917	1.11	0.50	77,723	1.61	0.73
May	102,653	0.93	0.42	114,599	1.20	0.55	83,437	1.58	0.72
June	120,631	0.92	0.42	102,154	1.19	0.54	81,246	1.57	0.71
July	148,076	0.97	0.44	95,013	1.26	0.57	104,748	1.56	0.71
August	99,352	1.08	0.49	76,817	1.36	0.62	87,669	1.58	0.72
September	74,616	1.02	0.46	68,021	1.25	0.57	83,057	1.54	0.70
October	61,602	0.96	0.44	52,485	1.33	0.60	62,217	1.45	0.66
November	65,338	0.93	0.42	44,124	1.26	0.57	64,379	1.34	0.61
December	67,858	0.92	0.42	47,312	1.28	0.58	54,751	1.35	0.61
Totals and Averages	1,043,168	0.96	0.44	983,705	1.17	0.53	853,431	1.51	0.69

Source: Anon. (1972-74).

TABLE 21.10

RAINBOW TROUT EXPORTS FROM JAPAN BY COUNTRIES, 1967-74

Importing Country	1967	1968	1969	Metric Tons 1970	1971	1972	1973	1974
USA	1149.3	1385.0	991.1	1118.1	1800.9	990.8	807.5	848.6
Canada	202.1	247.2	167.7	143.4	577.4	525.3	426.6	623.5
Great Britain	480.6	563.7	680.0	680.3	400.9	533.7	522.6	539.8
Australia	49.4	63.4	25.4	50.4	80.0	64.7	74.2	212.3
Belgium	348.0	186.1	230.9	111.5	65.0	142.7	221.8	152.6
South Africa	3.4	10.6	43.3	45.1	38.3	16.4	38.3	53.1
Sweden	17.0	13.1	19.4	21.7	34.7	15.9	2.3	24.9
West Germany	118.1	50.6	652.6	786.1	39.6	131.5	168.9	22.7
Holland	13.3	6.8	42.2	13.7	20.4	27.7	15.9	20.2
Others[1]	12.6	36.0	24.8	13.7	19.3	70.8	67.0	20.5
Total	2393.8	2562.5	2877.4	2984.0	3076.5	2519.5	2345.1	2518.2

Source: Anon. (1967-74).

[1] Includes, during various years, Hong Kong, Singapore, Philippines, Norway, Switzerland, Italy, Gibraltar, Samoa, New Hebrides.

are in the large central cities for restaurant use and for sale in supermarkets. The secondary markets are in mountainous recreational areas and for fee fish-out ponds. While prices have increased rapidly in the past few years, the fish are relatively inexpensive (Fig. 21.2) and sell for less than comparable fish in western Europe and North America. See these sections for comparisons.

While domestic consumption now accounts for 86% of production, compared to 63% ten years ago, the export market is still important as a stabilizing influence on domestic prices. Exports are relatively stable, ranging from 2345 MT (2584 ST) to 3077 MT (3391 ST) between 1967 and 1974 (Table 21.10). Trout are exported to between 15 and 18 countries annually. The major exports markets in 1974 were: (1) USA, (2) Canada, (3) Great Britain, (4) Australia and (5) Belgium. Exports to the USA are declining, while exports to Canada and Australia are increasing rapidly.

Japanese Eel (Anguilla japonica)

Eels are considered a gourmet fish in Japan and command premium prices. For these reasons culturing of eels began at an early date. During the 1950s and early 1960s, as Japan industrialized rapidly, the natural catching waters were restricted by dams and pollution and the wild catch declined. As prices rose and culturing techniques became known, eel farming intensified by increased number of producers and increased production per surface unit of water. In 1955, 4000 MT (4400 ST) were

FIG. 21.3.   INTENSIVE INDOOR EEL CULTURING FACILITY

cultured. From then until 1968 production increased rapidly. In 1968 production reached its high mark of 24,000 MT (26,400 ST). In the 1970-72 period, production declined drastically as a result of various diseases among the fry and the concurrent decline in the catch of natural elvers or glass eels (juvenile stage). In 1972 production reached its lowest level with only 13,000 MT (14,300 ST) produced. During the following years production increased slightly as elvers from Europe as well as those from many other areas were imported.

Market prices followed the supply schedule. Following are the yearly average prices for live eel of 150 to 200 g (5.5 to 7.0 oz) from 1966 through 1974 at the Tokyo Central Wholesale Fish Market. Price is in U.S. dollars per kilogram and pound.

	$ per kg	$ per lb
1966	2.00	0.91
1967	1.78	0.81
1968	2.01	0.91
1969	2.96	1.35
1970	4.00	1.82
1971	5.03	2.29
1972	6.36	2.89
1973	5.17	2.35
1974	6.56	2.98

In addition to smaller supplies affecting market prices between 1969 and 1972, real per capita incomes increased and there was a rapid expansion in demand for frozen and fresh ice-packed fish rather than for the more costly live fish. With rapidly increasing prices between 1970 and 1973 interest in eel culture intensified. Existing producers tried to expand production, new producers entered the business and industrial firms started culturing in warm water effluents. Production units became more capital intensive. The important problem was the supply and price of elvers. This problem was difficult because the supply of elvers is determined by the catch under natural conditions. Before 1968, if more elvers were needed they were imported from Taiwan, but in that year Taiwan began culturing eel and exported food size eels to Japan rather than elvers. In addition some Japanese firms began exporting Japanese elvers to Taiwan to take advantage of lower production costs. Hence Japanese elver prices increased rapidly. In 1969 elver prices were only $33 per kg. In the spring of 1973 the price of elvers in Japan reached a phenomenal figure of $1300 per kg ($590 per lb), or about $0.43 each.

Import firms went around the world to secure alternative species. France became the main source of supply and accounted for 217 MT (238 ST), or 90% of the 1973 imports. Most imported elvers died because of differences in water temperatures and disease. The retail prices of imported cultured eels was lower than the native species; hence the imported price was as low as $30 per kg ($14 per lb). By 1976 eel diseases were widespread in Japan as a result of introduction of new species from more than ten countries. The major problems became disease control and high feed prices.

Because of problems in obtaining enough elvers and then the introduction of new diseases, domestic production of food-sized eel has been erratic. Production since 1965 has varied from a low of 13,355 MT (14,717 ST) to a high of 23,640 MT (26,051 ST) (Table 21.4). Since total domestic supply is 89% cultured, the supply of wild eels has not been sufficient to moderate rapid price changes brought about by erratic shifts in production volumes.

In 1974 there were 2788 separate managements in the major producing prefectures (Table 21.5). These managements were classified as using essentially still-water ponds. While this classification is technically correct, there is a world of difference between carp and eel still-water ponds. The carp pond is exposed to the elements and is fed by a stream or well having sufficient water flow to maintain the pond level with some water flowing out of the pond. Of the eel ponds, many are enclosed in plastic-type greenhouses. The water is usually from wells, heated to the proper temperature, pumped into the

FIG. 21.4.   TYPICAL EEL CULTURING FARM, WITH INDOOR AND OUTDOOR
PONDS, SHIZUOKA PREFECTURE

greenhouse-type pond, and is recirculated over and over again. Bio-filters are used to remove some of the waste materials, and mechanical aerators are utilized extensively. This system is explained in a little more detail under production costs in this chapter.

A small number of eel farmers use running water culture, similar to trout raceways. In 1974 about 5% of the eel farms used running water and 95% used still-water recirculation systems.

The still-water farms had a total surface area of 26,637,000 m² or 2664 ha (6583 acres). The farms using running water culture had 389,-000 m² of surface area (23.1 ha or 57.1 acres). Production per m² for the still-water ponds was 0.6 kg (0.12 lb per ft²). Production per m² of running water surface was 1.2 kg (0.26 lb per ft²).

Of the 17,077 MT (18,819 ST) of production in 1974, Shizuoka Prefecture accounted for 38%, Aichi Prefecture for 23%, Kagoshima Prefecture for over 6%, and Mie Prefecture for 6%. These 4 prefectures and 8 other major producing prefectures made up 94% of production. The remainder was produced by numerous other minor areas.

A variety of food is fed to eels. However, fish mash makes up 63% of the total, followed by fresh or frozen trash fish with 36%. Other miscellaneous foods make up the other 1% (Table 21.6). In 1974 the feed conversion for all of Japan was 2.6 to 1, or 2.6 units of food per unit

FIG. 21.5.   EEL FEEDING STATION, SHIZUOKA PREFECTURE

FIG. 21.6.   FEEDING EEL

of gain. This may appear to be excessive compared to many other cultured fish, but it must be remembered that more than one-third of the diet was other fish. Feed conversion for these other fish may have been 6 or 7 to 1.

Production Costs.—The average eel farmer in Japan has 9743 m² of water area (0.97 ha or 2.41 acres) and produces 5.8 MT (6.4 ST) of eel yearly. Hence production per establishment and per person is low. Facilities are costly since land may sell for $1 million per ha (over $400,000 per acre). Production practices call for intensive methods. There are four different methods used. All of them consist of concrete or rock and concrete walled running water or pond enclosures.

The four methods in order of intensity of production are:

(1) Circulating system—Only a few of these are in use. In experiments, 200 kg have been produced per m² (41 lb per ft²). Water is heated or maintained at a constant 25°C (77°F). Only enough new water is added to replace losses. The recirculated water is filtered by both mechanical filters and biofilters.

(2) Kochi System—This system takes its name from the prefecture of Kochi on Shikoku Island. About 5% of total production comes from the Kochi system. The eels are kept inside in plastic-covered greenhouses during the entire production cycle. Very little new water is added. Temperatures are maintained at 25°C (77°F). Bacterial action is used in biofilters to clean the recirculated water. Production is 10 to 15 kg per m² (2 to 3 lb per ft²).

(3) Raceway or Semi-raceway System—This system accounts for about 2 or 3% of total production. The water is used once and discharged. Costs are high for maintaining water temperatures. The water exchange rate is about once daily. Production is slightly in excess of the pond system (4).

(4) Pond System—This is the most common method. The elvers are grown in heated 25°C (77°F) water until they weigh 10 g (0.5 oz). This indoor pond is about 50 cm deep (18 in.). After about one month they are transferred to a pond 70 cm (28 in.) deep for the second month. They are then transferred to another pond 90 cm (36 in.) deep for two more months. By this time the outdoor pond waters are suitable for stocking and they are transferred to this outdoor pond for grow-out. This pond is also 90 cm (36 in.) deep. They are kept in this pond for 14 months. They are then marketed after about 18 months at 150 to 200 g sizes (5.5 to 7 oz). In the last outdoor pond they are often wintered-over outside. The eels throughout this production cycle may be separated by size 6 or 8 times. Hence labor needs are demanding.

The elvers are fed minced worms for the first 2 or 3 weeks and then

FIG. 21.7.   SIZING AND GRADING EEL

FIG. 21.8.   SIZED EEL

are gradually shifted to a mash diet containing 55 to 60% protein. The more mature eels have a mash diet containing 43% protein. A typical diet contains:

	(%)
White fish meal	70
Potato starch	20
Yeast	5
Vitamin mixture	1
Salt	1
Minerals	1
Other	2

Since the pond system is typical, production costs are presented for this system (Table 21.11). These figures are based on 400 different farms in Shizuoka Prefecture. Of the 150,000 elvers stocked, only about 72,000 or 40% survive to market sizes. The feed conversion is 2 to 1. No raw fish are fed as in some smaller facilities. Fixed costs constitute nearly 24% of total cost. Interest accounts for nearly one-third of fixed cost. The high expenses under variable costs are feed, labor and elvers. Total production cost in 1975 was $5.42 per kg ($1.88 per lb). However, the farmer's price was $6.67 per kg ($3.03 per lb), resulting in an 18.9% return or profit on sales. For these farms producing 12 MT (13.2 ST) gave a total return to management or profit of $15,096. This is a high income in Japan.

Marketing.—Eel as a food is mentioned in archives of A.D. 718. Specialized cooking methods go back 500 years. Eel culturing began in 1879. Among the Japanese this species of cultured fish is considered highly nutritious and conducive to good health. Hence the demand is high. However, Japanese scientists have met with only limited success in reproducing the eel elvers by artificial means. The elvers are caught at only a few locations. Elvers are imported from Korea, People's Republic of China and Europe in commercial quantities. Very few eels are exported since the Japanese price is the highest in the world. The price fluctuates widely from year to year, depending on the catch of elvers. Formerly elvers were imported from Taiwan, but in recent years the Taiwanese have been raising the elvers to market sizes and exporting them to Japan. Some Japanese elvers are exported to Taiwan for rearing there. This is because production cost is less in Taiwan.

The elver catching season is from December to March. After an 18 month productive cycle the market-sized eel are ready in the June-October period. However, because of extreme variations in individual growth which may vary by 200 times, eels are harvested and sold throughout the year.

TABLE 21.11

PRODUCTION COSTS, SELLING PRICE AND NET RETURNS FOR EEL PRO-
DUCTION, SHIZUOKA PREFECTURE, JAPAN, 1975

Item	Production Costs (cents per kg)	(cents per lb)	(%)
Fixed costs			
Land and water rent	28.2	12.8	5.2
Depreciation[1]			
Ponds	6.9	3.1	1.3
House	13.9	6.3	2.6
Heating equipment	16.7	7.6	3.1
Culturing tools	6.9	3.1	1.3
Vehicles	5.6	2.5	1.0
Repairs	8.3	3.8	1.5
Interest[2]	41.5	18.9	7.6
Total	128.0	58.1	23.6
Variable costs			
Elvers[3]	100.0	45.5	18.4
Feed[4]	140.0	63.6	25.8
Heating oil[5]	40.0	18.2	7.4
Electricity	13.3	6.0	2.5
Marketing	4.2	1.9	0.8
Expendables	5.6	2.5	1.0
Labor[6]	111.1	50.5	20.5
Total	414.2	188.2	76.4
Total cost	542.2	246.3	100.0
Selling price	667.0	303.2	—
Net returns	125.8	56.9	18.9

Source: Estimated costs for 1975 from Yoshida Eel Cooperative, Shizuoka Prefecture, Japan.
[1] 4950 m² (1.22 acres) of ponds costing $25,000 annualized over 30 years; 570 m² (0.19 acre) of heated house costing $8333 annualized over 5 years; heating equipment costing $10,000 annualized over 4 years; culturing tools costing $3333 annualized over 4 years, and vehicles costing $2000 annualized over 3 years.
[2] Fifty thousand dollars at a 10% annual interest figure.
[3] Starting with 180,000 elvers weighing 36 kg (79.2 lb) and finishing with 72,000 eels weighing 12 MT (13.2 ST). Average cost of elvers for the 3 years from 1974-76 is $333 per kg ($151 per lb).
[4] Twelve thousand kilograms of feed at 70 cents per kg (13.2 ST at 31.8 cents per lb).
[5] Includes 40 kiloliters of oil at $120 per kiloliter (10,560 gal. at 45.5 cents per gallon).
[6] Two men at $6666 each.

Tokyo is the largest market for eel in Japan. About 40% of all eels are consumed in this one city. Perhaps 15 to 20% of Tokyo's eels come through the Tokyo Central Wholesale Fish Market. However, the vast majority come from eel distributors selling to fish markets, retail stalls and supermarkets. Since these sales are of a private nature and prices change rapidly, the volumes and prices at the Central Mar-

FIG. 21.9.   ELVER (EEL) SHIPPING CONTAINERS

ket are presented in Table 21.12. Prices have been relatively stable in the most recent three years for which data are available. By contrast, prices range from a low of $1.78 per kg ($0.81 per lb) in 1967 to the high of $6.56 per kg ($2.98 per lb) in 1974. In 1975 and 1976 prices were only slightly higher than in 1974.

The volume of eel sold on the Central Market rises in the summer, particularly in July (Fig. 21.10). However, prices do not decline with heavy volumes since more eels are eaten in the summer. In July one day is set aside as National Eel Day and one of the traditional meals that day is eel. Hence prices reach their peak in July in spite of sales increasing.

An unknown but significant share of eel is utilized by the better restaurants. Eel can be found frozen in most large supermarkets and in large retail fish stores. The better restaurants use fresh eel. In very recent years some eel has been sold as a precooked frozen item. This has broadened the demand base for eel. In addition some eel is canned. Of the major types of freshwater fish cultured, only eel is canned and only eel, trout and ayu are sold frozen.

TABLE 21.12
VOLUMES OF EEL SOLD AND PRICES BY MONTHS, TOKYO CENTRAL WHOLESALE FISH MARKE, 1972-74

Month	1972			1973			1974		
	kg	$ per kg	$ per lb	kg	$ per kg	$ per lb	kg	$ per kg	$ per lb
January	32,220	5.32	2.42	44,285	7.52	3.42	46,290	5.23	2.38
February	37,723	5.28	2.40	30,119	7.35	3.34	46,179	5.46	2.48
March	41,094	5.84	2.65	41,217	7.07	3.21	101,321	6.07	2.76
April	49,950	6.00	2.73	87,391	6.11	2.78	68,415	6.81	3.10
May	66,474	5.77	2.62	100,164	5.44	2.47	90,842	6.65	3.02
June	75,636	6.46	2.94	138,662	5.43	2.47	125,078	6.89	3.13
July	159,625	7.19	3.27	357,455	4.63	2.10	222,033	7.69	3.50
August	81,947	6.97	3.17	171,690	4.78	2.17	138,008	6.72	3.05
September	45,536	6.21	2.82	86,526	4.61	2.10	74,941	6.08	2.76
October	42,843	5.52	2.51	85,866	4.58	2.08	46,237	5.88	2.67
November	39,703	6.13	2.79	52,171	4.40	2.00	41,951	5.67	2.58
December	58,690	6.61	3.00	63,031	5.20	2.36	66,657	5.45	2.48
Totals and Averages	731,441	6.36	2.89	1,258,577	5.17	2.35	1,066,952	6.56	2.98

Source: Anon. (1972-74).

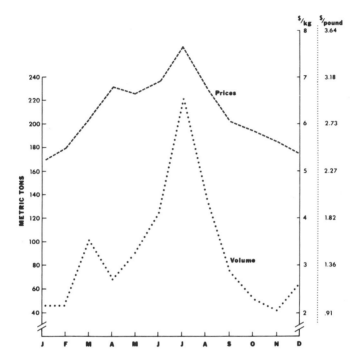

FIG. 21.10.   COMPARISON OF AVERAGE MONTHLY VOLUMES OF EEL SOLD
AND PRICES PAID, TOKYO CENTRAL WHOLESALE FISH MARKET, 1974

Carp (Cyprinus carpio)

Common carp is the leading freshwater species cultured in Japan. In 1974 a total of 26,323 MT (29,008 ST) were produced by fish farmers. An additional 5698 MT (6279 ST) of wild fish were caught. Eighty-two percent of the total supply was cultured (Table 21.3).

Cultured production has been increasing rapidly. In 1965 only 7973 MT (8786 ST) were cultured. For the 10 year period production grew at an average annual rate of 23% (Table 21.4).

In 1974 there were over 4663 separate managements located in the major production prefectures alone (Table 21.5). Carp are produced in nearly every type of production system. In 1974 nearly 27% of the managements used running water systems (similar to those for rainbow trout), 39% used still-water ponds, 26% used farm ponds, and 8% used cages or nets in lakes. Total surface area of water devoted to culture was 35,861,000 m² or 3586 ha (8861 acres).

Running water culture accounted for only 5.7% of the total carp area but accounted for 17.3% of production. Production per ha was 20.2 MT

FIG. 21.11.   MARKET SIZE EEL STORED IN CONTAINERS PRIOR TO PROCESSING

FIG. 21.12.   PROCESSING EEL

(9.0 ST per acre.) Production output was 2 kg per m² (0.4 lb per ft²).

Still-water ponds accounted for 16.8% of the total carp culture area, but accounted for only 13.2% of production. Production per ha was 5.2 MT (2.3 ST per acre). This rate of production, while low by Japanese standards, is two times higher than production in other countries. Production output per m² was 0.5 kg (0.1 lb per ft²).

Farm ponds accounted for 76% of the total carp area, but accounted for only 42% of production. Production per ha was 3.7 MT (1.7 ST per acre). Production output per m² was 0.4 kg, or less than 0.1 lb per ft².

Net or cage culture accounted for only 1.5% of the total carp area, but accounted for 28% of production. Production per ha was 120.4 MT (53.7 ST per acre). Production output per m² was a phenomenal 12.0 kg (2.5 lb per ft²).

Major production areas were: (1) Nagano Prefecture with 18% of production; (2) Ibaragi with 17%; (3) Gunma with 10%; (4) Akita with 10%; (5) Fukuoka with 7%; and (6) Tukushima with 6%. These six prefectures accounted for 68% of total production, while about 40 other prefectures accounted for the remainder.

Production per unit of running water surface is not as high as in trout-rearing facilities. This is due to lower water flow. An example of what can be done with adequate water flow is shown in the following example.

Undoubtedly the most intensive fish production rate in the world has been achieved by Mr. Kazuyoshi Tanaka, Annaka City, Gunma Prefecture, Japan. Mr. Tanaka has 16 different ponds and utilizes running water in each. Not all 16 ponds (races) are used to produce food carp; hence the following discussion is limited to the first pond (race) in the series of 16.

The first raceway has been in fish production for 78 years and is constructed of concrete and stone. The raceway measures 47 m² (505 ft²) of water surface, and the average depth is 1.4 m (4.5 ft). The raceway is stocked with carp fry averaging 85 g each, or less than 3 oz. Gross yield from the 8500 carp stocked is about 10,355 kg (22,738 lb) annually, or 10.3 MT. On a per hectare basis, this amounts to a harvest of 2203 MT (982 ST per acre). Deducting the initial stocking weight of 154 MT per ha (62.3 ST per acre), the net gain in weight would be 2049 MT (920 ST). Feed conversion is 1.3 units of feed fed per unit of fish gain.

The flow of river water into this raceway varies from 91 to 455 liters (24 to 120 gal.) per sec during the growing season. The raceway is stocked at the end of March and fed until early November. Harvesting is done by a net which is drawn the length of the pool, entrapping the fish. Ten people are utilized for this operation which requires three hours.

During times of high water due to rain, the flow of water entering the race is controlled so that the rapid flow does not suffocate the fish. The fish are fed a diet of silkworm pupae, boiled wheat and fish pellets which contain about 50% fish meal. Feeding commences around April 15 when the water temperature is about 18°C (64°F). The first week the feed is boiled wheat. The second week fish pellets are added to the wheat ration. After the second week silk pupae are fed along with fish pellets, with a little boiled wheat fed in July, August and September. Starting with the end of September until the end of October the feeding rate is gradually decreased. No feed is fed during the first two weeks of November until harvesting takes place.

Feeding commences at daybreak, which may occur as early as 5 a.m., and is continued at intervals of 1.5 to 2 hr until 11 p.m. Evening feeding is done under artificial lighting. If, during the hottest part of the summer the water temperature reaches 33°C (91°F), feeding is discontinued. Feeding is done by tossing the feed in the water in handfuls near the site where the water enters but where the current is not rapid. Feeding is discontinued while the fish are still actively feeding.

Mr. Chiba[1] writes that the most important factors of running water culture are: (1) the quantity of running water, which is directly related to the quantity of fish that can be produced; (2) the quality of water, essentially its oxygen content; (3) the need for constant running water, even if it must be circulated artificially; and (4) the temperature of water, which must be in a range where the species of fish will feed.

Mr. Chiba, in reporting his study of Mr. Tanaka's fish farming operation, correlated the flow of water into nine different races with the volume of fish produced per raceway. His analysis is shown graphically in Fig. 21.13. As stated previously, the volume of production is directly related to the quantity and quality of running water. The greater the volume of running water and the higher the oxygen content, the greater the possibility of increasing production. Since in this case study the water flows from one raceway to another, the total volume produced is related to the volume of water at any one point times the number of points (races) where the water is used. More than 10 MT of fish are produced at the first raceway, while total production from all 9

[1] The previous discussion of the case study of Mr. Tanaka has been based on two sources of information: Mr. Kenji Chiba's presentation in the book entitled *Fish Culture Seminar, Volume 1,* supplementary data furnished by Mr. S. Nishimura, Fishery Economist, Freshwater Fisheries Agency, Hino City, Tokyo, and personal data obtained by the authors from Mr. Tanaka in May 1976.

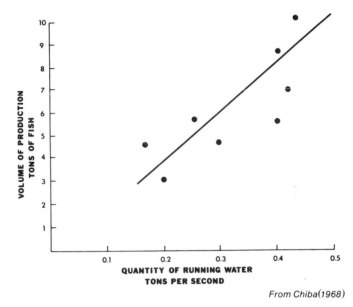

From Chiba(1968)

FIG. 21.13.   RELATIONSHIP BETWEEN RUNNING VOLUME OF WATER AND VOLUME OF FRESHWATER CARP PRODUCTION, TANAKA FISH FARM, GUNMA PREFECTURE, 1966

races is over 80 MT. Production decline after the water is used in the first raceway can be expected because of oxygen depletion and lower stocking rates in subsequent races.

Still-water ponds have only limited water flows, while farm ponds during part of the year may have none, or may even have the volume of the water in the pond reduced. As stated earlier, yields in these type ponds average 5.2 and 3.7 MT per ha (2.3 to 1.7 ST per acre), respectively. This is very high production per unit of water. The Japanese do this by managing the plankton, which increase oxygen in the water, by mechanical aeration and by proper feeding practices.

Net culture is one of the more interesting methods used to produce carp. Since this type of culture can only be done in the larger lakes, which are few in Japan, the presence of net culture is confined to the few prefectures having large lakes. Among these is Ibavagi Prefecture, which contains lakes Kasumigura and Kitaura. In 1974, 4206 MT (4635 ST) were produced by net culture in these two lakes. This volume was 16% of all cultured carp in Japan. The next highest volume of net culture is in Lake Suwa in Nagano Prefecture, where net culture was initiated in 1958.

The culturing nets are in the form of a 9 m (29.5 ft) square.The net is approximately 2 m (6.5 ft) high, with approximately 1.5 m (5 ft) submerged. On top of this net, an additional 1 m (39 in.) high net is attached to keep the fish from jumping out. A double net is used on the sides and bottom as a precaution to prevent fish from escaping in the event one net is broken or torn. The corners of the nets are anchored to the bottom of the lake and connected by a floating walkway. Nine 200 liter (50 gal.) drums are used to float the framework and each net. The usual depth of the lake where net culture is practiced averages about 3 m (10 ft).

The carp stocked are between 10 and 15 cm (4 to 6 in.) long and weigh 70 to 84 g (2.5 to 3 oz). The stocking rate is between 3500 and 5000 fingerlings per net set for an average of 320 kg (704 lb) of fingerlings per net. Stocking usually takes place in April and feeding continues for six months. Fish are fed about 4 times daily for 10 to 20 min each time. Each net, which includes approximately 81 m² (870 ft²) of water area, yields about 3508 kg (7718 lb) of fish, which is the equivalent of 431 MT per ha (158 ST per acre). The weight gain is tenfold—a gain of nine units for each unit of fish stocked. Most producers feed only pelletized fish food, which has over 39% protein content by law. Feed pellets are composed of 50% fish meal, 39% wheat flour, 6% alfalfa, 3% yeast, 1% sodium chloride and 1% vitamins.

A major finding by the Fishery Laboratory in Nagano Prefecture was that the volume of fish harvested is directly related to the number of fingerlings stocked. They found that a stocking rate of 75 fish per m² (7 fish per ft²) of water is maximum.

A second finding reported by the Fishery Laboratory in Nagano Prefecture is the need to stock uniform size fish. If the size of fingerlings stocked varies as much as 2 to 1 (e.g., 16 cm and 8 cm) the ending weight may vary as much as 3 to 1. This implies that the larger fish stocked acquire more than their proportional share of feed.

The Fisheries Agency of Nagano Prefecture has done considerable research on net pond culture during recent years. One of their most interesting and perhaps most significant findings is the correlation between average water temperatures in Lake Suwa and growth of carp as measured by the increase in the body weight of fish per day (Fig. 21.14).

Carp take 2 years to grow to market sizes of 800 to 1000 g (29 to 36 oz). In the first year spawning takes place about May 1, although this varies from place to place and from year to year. One year later the fingerlings vary from 30 to over 100 g (1 to 4 oz). In the second year the fingerlings are divided into 4 size groups: Group 1 averages 30 g (1 oz); Group 2 averages 60 g (2 oz); Group 3 averages 100 g (4 oz); and those in Group 4 are over 100 g (4 oz). They are then stocked into grow-out facilities by size about April 1. Group 4 fish (over 100 g)

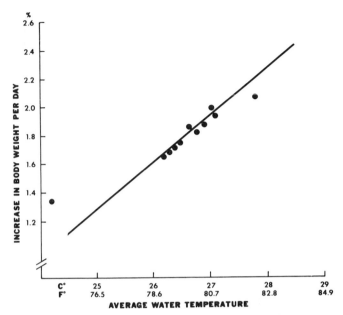

FIG. 21.14.   RELATIONSHIP BETWEEN AVERAGE WATER TEMPERATURE AND
RATE OF GROWTH OF CARP, LAKE SUWA, NAGARO PREFECTURE

reach 1 kg (2.2 lb) in the 5 months from April to August. Fish in Group
3 reach 1 kg in the 6 months from April to September; those in Group
2 reach 1 kg in the 7 months from April to October. Group 1 fish reach
800 g (28 oz) in the 8 months from April to November. The water is then
too cold for growth and the fish are wintered-over with minimum feed,
reaching the market size of 1 kg the following July. An example of typical Japanese water temperatures follows (Maebashi-shi, Gunma Prefecture):

	°C	°F
January	Below 8	Below 46
February	Below 8	Below 46
March	8	46
April	13	55
May	17	63
June	22	72
July	29	84
August	30	86
September	25	77
October	20	68
November	13	55
December	8	46

Nearly every producer produces his own eggs and hatches and raises his own fry and fingerlings. Running water producers try to stock 100 g (4 oz) fingerlings, so they often trade with pond producers. In running water the fish increase their weights by 8 to 10 times in 1 year, whereas in ponds increases of 18 to 20 times can be obtained. Hence running water culturists require bigger fingerlings to reach market sizes in one year compared to pond producers. Thus fingerling costs are higher for running water culture. This is offset, however, by the ability to partially or totally harvest a raceway when prices are high, and the average market prices are higher for raceway than for pond producers. At harvest time the pond producers concentrate their pond fish in a smaller feeding enclosure for partial harvesting. However, trapping up to 90% of the fish may require several weeks. To harvest the remaining 10% of the fish the pond must be drained. Proper sizes for these ponds are between 3 and 5 ha (7.4 to 12.3 acres) and 2 m (6 ft) deep. When ponds are over 2 m deep the water is dark and proper plankton growth is not possible. The pond culturists have the advantage in labor efficiency. Output of 60 MT (66 ST) per person is possible, while with raceway output is 50 MT (55 ST) or less.

Production Costs.—In 1974, for all of Japan, a total of 45,555 MT (50,202 ST) of feed were fed (Table 21.6). The major food was pellets, followed by dry silkworm pupae, fresh silkworm pupae, fresh and frozen fish and "other" (mainly boiled wheat).

A typical fish diet by months is:

	Wheat (%)	Pupae (%)	Pellets (%)
April	70	10	20
May	50	20	30
June	20	30	50
July	—	70	30
August	—	80	20
September	—	60	40
October	—	50	50
November	20	10	70

The wheat costs about $0.17 per kg ($0.08 per lb), the pupae about $0.37 per kg ($0.17 per lb) and the pellets about $0.33 per kg ($0.15 per lb). Proportions of feed fed vary when prices of wheat, pupae and pellets change. Pupae cannot be fed alone since they have a high phosphorus content and affect bone deposits. However, the farmers all feed some pupae since it results in a pink flesh which the Japanese consider indicative of high quality carp. The pupae-feeding of fish results in a dressed out weight of 60 to 65%, whereas pellet-fed fish dress out at only 50 to 55%. The pupae contain 58 to 60% protein, 28% fat, 4% fiber

TABLE 21.13

PRODUCTION COST PER KILOGRAM (OR PER POUND), SELLING PRICE AND RETURNS OR PROFIT,
RUNNING WATER CULTURE, GUNMA PREFECTURE, JAPAN, 1976[1]

Item	Production Costs (cents per kg)	Production Costs (cents per lb)
Fixed costs		
Land and water rent	3.3	1.5
Repairs	3.3	1.5
Taxes	3.3	1.5
Interest	5.0	2.3
Depreciation	3.3	1.5
Total	18.2	8.3
Variable costs		
Feed	51.3	23.3
Fingerlings (100 g or 4 oz)	10.0	4.5
Power	5.0	2.3
Labor	16.7	7.6
Harvesting labor	3.3	1.5
Miscellaneous	1.7	0.8
Total	88.0	40.0
Total production costs	106.2	48.3
Selling price	123.0	55.9
Returns to management or profit	16.8	7.6
Percent return	15.8	15.8

Source: Carp farmers in Gunma Prefecture.
[1] Based on 30 MT (33 ST) of production from 1.5 ha (3.71 acres).

and 3% ash, compared to 42% protein, 4% fat, 2% fiber and 12% ash
for carp pellet feed.

Production costs of common carp are relatively inexpensive compared to other species cultured. For 1976 in Gunma Prefecture, total
costs were computed at $1.06 per kg ($0.483 per lb) for a running water culturing unit requiring one person and producing 30 MT (33 ST)
from 1.5 ha (3.71 acres) (Table 21.13). Feed accounted for nearly one-
half of total expenses. Other major cost items were labor and fingerlings. Returns to management, or net profit, was $0.168 per kg
($0.076 per lb). The return above production costs was 15.8%. This
example is based on high quality management in production practices,
buying feed and selling.

Marketing.—The consumption of carp in Japan is highly diffused.
Of the total supply of 32,021 MT (35,287 ST with 82% cultured) in
1974, the Tokyo area, which accounts for a disproportionate share of
most fish, accounted for only about 10%. In some of the inland prefectures consumption of carp is often five times more per person than in

Tokyo. This is due to the inability of the inland prefectures to obtain fresh marine fish (so they turn to fresh carp) and because carp prices are lower than in the big cities. In general, carp producers sell to outlets in the local markets near the place of production. This has been the traditional method of distribution, based largely on poor transportation. As transportation improved, the local markets broadened. The major markets are tourist areas in the countryside and local people in carp production areas. Although some carp is sold at large, centralized, wholesale fish markets, the major markets are not at these centers. In general, little carp is sold in very large cities. Hence city people have not acquired a taste for carp, nor are they accustomed to using it in the home. The custom of eating carp in the home is not widespread among the younger Japanese. Hence available data suggest that demand will increase very slowly, principally in response to increases in population and rising incomes of those who have acquired the taste for carp. As in the USA, the modern Japanese consumer does not like to prepare carp because of the bony nature of the fish. Carp is not on sale at most retail markets in large cities. Hence the possibilities of rapid expansion of this segment of the industry depends on a more efficient distribution system and stimulating consumer acceptance among the younger Japanese. Carp production is widespread throughout Japan and as a result the availability of carp is good in most cities, towns and villages.

An example of the decentralized market for carp is shown by the volume handled by the Tokyo Central Wholesale Fish Market (Table 21.14). In 1974, of the total supply of 32,031 MT (35,287 ST), 0.5% was sold on this market. This is in direct contrast to shrimp, yellowtail, eel and other species, wherein the Tokyo market is the most important single market.

The consumption of carp in large cities such as Tokyo is lower per person than in rural areas largely because of price. In 1976 for example, consumer price in Gunma Prefecture was $2.00 per kg ($0.91 per lb), whereas it was over $3.00 per kg ($1.37 per lb) in Tokyo only 65 km (40 mi.) away. In the production areas producers sell direct to restaurants, fee fish-out ponds and fish dealers. The fish dealer then sells directly to the ultimate consumer. The producer receives about 62% of the consumer price. By contrast, when carp go to large centralized markets the producer sells to a local wholesaler or broker who increases the producer price of $1.00 per kg ($0.45 per lb) to $1.17 ($0.58 per lb). The local wholesaler transports and sells to a wholesaler in the city. This market increases the price from $1.17 per kg ($0.58 per lb) to $1.50 ($0.68 per lb) for sales to restaurants, other fish dealers and fee fish-out ponds. For sales to super-

TABLE 21.14

VOLUME OF COMMON CARP (*CYPRINUS CARPIO* L) SOLD AND PRICES BY MONTHS, TOKYO CENTRAL WHOLESALE FISH MARKET, 1972-74

Month	1972 kg	1972 $ per kg	1972 $ per lb	1973 kg	1973 $ per kg	1973 $ per lb	1974 kg	1974 $ per kg	1974 $ per lb
January	7,202	1.20	0.55	5,562	1.11	0.50	11,931	1.42	0.65
February	5,168	1.18	0.54	9,911	1.08	0.49	12,087	1.44	0.65
March	6,274	1.18	0.54	10,945	1.11	0.50	14,353	1.47	0.67
April	6,361	1.15	0.52	2,447	1.15	0.52	12,868	1.53	0.70
May	7,759	1.18	0.54	13,302	1.28	0.58	15,861	1.53	0.70
June	7,838	1.20	0.55	11,418	1.40	0.64	13,778	1.56	0.71
July	9,212	1.24	0.56	14,212	1.44	0.65	15,285	1.53	0.70
August	8,494	1.25	0.57	14,514	1.43	0.65	13,964	1.51	0.69
September	7,727	1.24	0.56	11,578	1.44	0.65	19,348	1.59	0.72
October	8,348	1.18	0.54	12,360	1.45	0.66	12,672	1.56	0.71
November	9,033	1.14	0.52	11,291	1.41	0.64	10,258	1.49	0.68
December	11,103	1.08	0.49	14,224	1.40	0.64	11,303	1.52	0.69
Totals and Averages	94,519	1.18	0.54	131,764	1.34	0.61	163,708	1.52	0.69

Source: Anon. (1972-74).

markets and department stores, where additional handling is required, the price goes to $1.83 per kg ($0.83 per lb). These other fish dealers, fee fishing operators, supermarkets and department stores sell the fish for about $3.33 per kg ($1.51 per lb) to the final consumer. Hence the producer receives only about 30% of the consumer dollar instead of 62%. The city dweller pays about 70 to 80% more for the same fish than does his country cousin. Even with these mark-ups, or margins, the city wholesaler cannot make a profit on carp, and always handles eel or other fish as the main source of income.

While carp prices fluctuate seasonally at the Tokyo market—with high prices in June-September and low prices in December-February —the change is not considerable. In the production areas there is the same seasonality, but prices will vary by only 25 to 30% from the low months to the high months.

Wholesalers often buy directly at the pond to eliminate some of the middleman's margins. Some marketing cooperatives have been formed to increase marketing efficiency and lower marketing margins, but they have found that, in general, they have the same distribution channels and costs faced by private businessmen.

Two possible answers suggest themselves for the low December-February prices and the high June-September prices. (1) Tourists flock to rural carp-producing areas in the summer and eat at traditional restaurants located there. (2) The volume of carp harvested in the fall increases, which may affect prices negatively.

Since there is a widespread, decentralized market for carp in small cities, towns and villages, carp are not transported long distances for sale. Nearly all of the 46 prefectures produce carp, but fish sold on the Tokyo market come from only eight prefectures. Two of these, Gunma and Ibaragi, account for 90% of sales.

Crucian Carp (Carassius carassius)

In 1974 cultured production was 841 MT (927 ST). Of this volume 34% was cultured in Osaka Prefecture and 24% in Kagawa Prefecture. The remaining 353 MT, or 42% of total production, was produced in about 20 other prefectures.

Crucian carp normally eat plankton, soft natural foods and insects. In Japan three methods are used. In the most extensive method, the fry are stocked in unfertilized farm ponds and subsist only on plankton and natural foods. In the second method the pond is fertilized to encourage the growth of natural foods. In the third method some feeding is practiced. This may be induced by household waste waters being channeled to the pond to supply nutrients and induce

FIG. 21.15.   RECREATIONAL FEE FISHING POND

plankton growth. The most important supplementary feed is dried silkworm pupae, which has a high protein content. The usual practice is to raise a few common carp and perhaps other fish, such as chubs (shiners), with the crucian carp.

It is common practice to stock rearing ponds with 16 to 20 g fish (0.5 to 0.75 oz). After about 2 years of culturing the fish are about 20 cm (8 in.) long. Between 2 and 3 years later the fish reach 30 cm (12 in.) and weigh 500 to 600 g (1 to 1.25 lb). Natural fish take 4 to 5 years to reach the same size. The 1000 g fish (2 lb) measure 35 cm (14 in.) or more in length and require 3 to 4 years. Natural fish require 5 to 6 years to reach this size.

The chief markets are for game or fee fish-out ponds, or in restaurants in the city of Osaka. Market size depends on whether the fish are used in soups, portions or as fillets. The average wholesale fish price at the Osaka Central Fish Market for the first half of 1976 was only $0.76 per kg ($0.35 per lb). Price of common carp during the same period of time was $1.42 per kg ($0.65 per lb). It is readily apparent that crucian carp have a specialized, low value market and culturing is only conducted in order to get dual use of farm pond waters.

In 1974 there were 299 managements of crucian carp with fish stocked in 603 separate ponds. About 1968 ha of water were used (4862 acres). Production was only 364 kg per ha (324 lb per acre).

Ayu or Sweetfish *(Plecoglossus altivelis)*

Ayu are native to Japan. A Dutch scientist in 1846 introduced ayu to the world's learned society. However, they did not become commercially important until about 1908. Ayu are a species of fish found in Japan, South Korea, Okinawa and the mainland Chinese coast. Eggs are laid in September, October and November, usually in a water depth of 30 to 45 cm (12 to 18 in.). Eggs number 10,000 to 100,-000 per female. Suitable water temperature for hatching is 14° to 23°C (58°to 74°F). The number of days for hatching varies from 10 to 24. The warmer the water the more rapid the hatching. Ayu are cultured most frequently by running water techniques similar to those used for trout. The fry are stocked when they are 5 to 6 cm (2 to 2.5 in.) long. The fish are sold for consumption at about 75 to 80 g (3 oz) per fish, or 17 to 22 cm (7 to 9 in.)

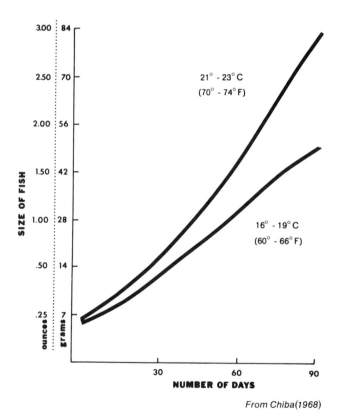

From Chiba(1968)

FIG. 21.16.   RELATIONSHIP BETWEEN GROWTH RATES OF AYU FISH AND
WATER TEMPERATURES

FIG. 21.17.   AYU POND AND FEEDING STATIONS WITH FLOOD LIGHTS FOR
NIGHTTIME FEEDING

In 1974, 4712 MT (5193 ST) were cultured. Wakayama Prefecture produced 28% of the total, Tokushima produced 20% and Shiga 14%. About 30 other prefectures produced the remaining 38%. Two-thirds of all Japanese fry cultured are captured in Lake Biwa in Shiga Prefecture. Nearly all fry cultured are wild, since it is difficult to reproduce ayu under artificial conditions.

There is a definite relationship between water temperature and growth of ayu (Fig. 21.16). The fish can be produced for sale in about 90 days from the time the fry are stocked. The ideal method of feeding is in running water raceways.

In 1974 a total of 7360 MT (8111 ST) of feed were fed to produce the 4712 MT (5193 ST) of cultured ayu (Table 21.6). The feed conversion was 1.6 units of feed per unit of gain. Nearly all the feed was artificially prepared pellets similar to trout feeds. A typical management will consist of 8 to 12 concrete enclosures with each enclosure containing about 200 m² (2152 ft²). Fresh, aerated underground water from wells 20 to 40 m (60 to 125 ft) deep is added, and the water is recirculated once. Water is exchanged every two hours. The temperature range is maintained between 18° and 22°C (64° to 72°F). In each tank or enclosure are stocked 50,000 fry. The fish are fed a 45 to 48% protein diet to which cod oil has been added at a 5% weight level. Mortality in 90 days may be only 1%. From each tank or enclosure

TABLE 21.15

PRODUCTION COSTS FOR A TYPICAL AYU CULTURING FARM PRODUCING 45.0 MT
(49.6 ST) OF FISH, TOKUSHIMA PREFECTURE, SHIKOKU ISLAND, JAPAN, 1976

Capital investment items

 11 enclosures or tanks each having 200 m²
  1 building containing 91 m²
  7 pumps
 16 aerator pumps
  6 wells, 30 m deep
  7 water wheel aerators
 12 automatic feeders
  1 electrical transformer, 85 kw
 27 flood lights

Fixed costs

Depreciation	$6,667
Interest	6,667
Repairs	1,666
Taxes and insurance	1,666

Variable costs

Fry	$26,667
Feed	35,000
Power	18,333
Wages (3 men)	10,000
Medicine	11,667
Total	$118,333
Production cost per kg	$2.63
Production cost per lb	1.20
Selling price per kg	2.87
Selling price per lb	1.30
Total profit	$10,837.00

Source: Anon. (1976F).

4 MT (4.4 ST) of marketable fish are harvested. Production per m² of water surface area is 20 kg (4.1 lb per ft²). The tanks are about 1 m (39 in.) deep.

Production Costs.—The cost of producing ayu is high. Several reasons are cited for this high cost. These are: (1) high cost of fry; (2) high costs of pumping water; (3) low productivity per person employed; and (4) high costs of preventing and treating diseases due to intensive stocking.

An example of production costs is presented in Table 21.15. The cost of fry was nearly $0.05 each. Since the ayu are sold at only about 75 g (3 oz) this makes the fry price large in relation to market weights.

TABLE 21.16

VOLUME OF AYU (SWEETFISH) SOLD AND PRICES BY MONTHS, TOKYO CENTRAL WHOLESALE FISH MARKET, 1972-74.

Month	1972 kg	1972 $ per kg	1972 $ per lb	1973 kg	1973 $ per kg	1973 $ per lb	1974 kg	1974 $ per kg	1974 $ per lb
January	3,206	2.08	0.95	9,381	2.04	0.93	7,083	3.41	1.55
February	5,633	1.80	0.82	7,898	2.71	1.23	4,871	3.46	1.57
March	11,687	2.16	0.98	20,410	2.14	0.97	4,088	3.86	1.75
April	30,039	2.64	1.20	53,806	3.16	1.44	43,726	4.49	2.04
May	66,599	3.68	1.67	116,748	3.45	1.57	98,242	4.65	2.11
June	128,094	3.52	1.60	169,644	3.23	1.47	142,330	4.54	2.06
July	161,324	3.37	1.53	183,695	3.12	1.42	144,824	4.62	2.10
August	158,795	2.88	1.31	166,730	3.08	1.40	141,193	4.23	1.92
September	149,154	1.97	0.90	110,186	2.63	1.20	104,915	3.27	1.49
October	77,568	1.48	0.67	38,942	2.40	1.09	19,187	4.12	1.87
November	14,568	1.65	0.75	10,171	1.57	0.71	7,228	4.09	1.86
December	11,195	1.97	0.90	10,578	3.51	1.60	2,559	4.18	1.90
Totals and Averages	817,862	2.77	1.26	98,189	3.03	1.38	720,246	.28	1.95

Source: Anon. (1972-74).

The proportion that fry prices constitute of total production cost was 22.5%. This is nearly double the cost of some other species. The example shows a sale price of only $2.87 per kg ($1.30 per lb), whereas the Tokyo wholesale price was over $4.00 per kg ($1.82 per lb) for the same period. However, most of the ayu on the Tokyo Central Wholesale Fish Market are fresh iced fish, while in Tokushima Prefecture over 70% are sold frozen. The frozen fish command a lower price than either live or fresh iced fish.

Marketing.—Consumption of ayu is highly seasonal. Peak prices are to be found in the April-September period. This period coincides with the tourist season, when Japanese flock to the recreation areas outside the large cities and eat at restaurants. Since most of the ayu are consumed in restaurants, this accounts for the high prices. Most

TABLE 21.17

VOLUME OF AYU SALES ON THE TOKYO CENTRAL WHOLESALE FISH MARKET

Year	Volume of Sales (MT)	Sales Price per kg ($)	Sales Price per lb ($)
1970	721.6	2.81	1.28
1971	968.9	2.45	1.11
1972	817.6	2.77	1.26
1973	898.2	3.03	1.38
1974	720.3	4.28	1.95

Source: Anon. (1972-74), pp. 300-301.

wild ayu are caught by sports fishermen and not sold. Most cultured ayu are sold live or fresh iced to these restaurants by producers or fish wholesalers.

In 1974, 16,980 MT (18,712 ST) of ayu were consumed. Of this volume 4% was sold on the Tokyo Central Wholesale Fish Market. This was the largest single market for ayu. The highest volumes reached the market during the April-September period when prices were highest (Table 21.16). The consistent high prices for ayu are shown in Table 21.17. This table also shows that prices increased considerably in 1974 with higher prices continued in 1975 and 1976. In May 1976 the wholesale price was $6.67 per kg ($3.03 per lb). These fish came from more than 30 of Japan's 46 prefectures. However, the 5 prefectures of Wakayama, Tokushima, Shizuoka, Nagano and Shiga accounted for 85% of the total.

Other Salmonidae

In Japan in this group are: (1) *Salvelinis pluvius*, commonly called Iwana or Rockling; (2) *Oncorhynchus masu*, commonly called Yamame or trout salmon; and (3) *Oncorhynchus rhodurus*, commonly called Amago or Amenouo. These three are all land-locked species.

These three species are cultured only in cold waters in mountainous areas. They have become popular only in recent years, and rapid growth is being experienced in production. They are produced in many prefectures. Unlike rainbow trout, which have never really been accepted and for which prices are relatively low, these three salmonidae have captured the Japanese imagination, are readily accepted, and command high prices. Sales are to fee fish-out operators and to restaurants and supermarkets located in recreation areas. In 1975 total production was estimated at 1000 MT (1102 ST).

In 1974 in Hiyogo Prefecture, producer price for rainbow trout was $2.63 per kg ($1.20 per lb). Producer prices for these "other salmonidae" averaged $3.93 per kg ($1.79 per lb). Returns on profits above production costs were only 1.2% for rainbow trout, and a phenomenal 26.2% for these "other salmonidae." A breakdown of production costs for Hiyogo Prefecture is shown in Table 21.18. The most

TABLE 21.18

COST AND RETURN DATA PER KILOGRAM AND PER POUND FOR *SALVELINUS PLUVIUS, ONCORHYNCHUS MASU* AND *ONCORHYNCHUS RHODURUS* (OTHER SALMONIDAE), HIYOGO PREFECTURE, 1974[1]

Item	Production Costs ($ per kg)	($ per lb)	(%)
Fry	0.98	0.45	33.9
Feed	0.98	0.45	33.9
Wages	0.45	0.20	15.5
Pond rent	—	—	—
Power	0.06	0.03	1.9
Oil and gasoline	0.03	0.01	1.1
Repairs	0.19	0.08	6.4
Transportation	0.04	0.02	1.3
Selling commission	—	—	—
Miscellaneous supplies	0.02	0.01	0.7
Depreciation			
Raceways	0.02	0.01	0.8
Materials	0.1	0.01	0.3
Office	0.03	0.01	1.0
Other	0.09	0.04	3.2
Total production cost	2.90	1.32	100.0
Selling price	3.93	1.79	—
Profit	1.03	0.47	26.2

Source: Anon. (1975A).
[1] Based on 5 MT (5.5 ST) of production.

interesting fact presented in the table is the cost of the fry: 33.9%. If it is assumed that 25% of the fry survived to market size, and that average market size was 250 g (about 9 oz), then the price per fry was over $0.06 each.

"Other" Freshwater Cultured Fish

The two cultured species under this classification are: (1) loach *(Misgurnus anguillicaudatus)* and (2) mullet *(Mugil cephalus).*

Loach are cultured in small quantities in Niigata, Aomori, and other prefectures on northeastern Honshu Island. These fish are raised in farm ponds and are similar to crucian carp in their feeding habits. Their food consists of worms, insects and plankton.

In 1941 production of loach was as high as 1600 MT (1763 ST), but declined to only 450 MT by 1962. Since then there has been a resurgence of production, and volume in 1976 was estimated at 1000 MT (1102 ST). Production declined because of the widespread use of agricultural chemicals, but with bans and strict controls by the government, production is increasing. They are small fish, seldom attaining over 40 g (1.5 oz) in size. They are sold in three sizes: (1) small, between 10 and 15 g (0.5 oz); (2) medium, 15 to 30 g (0.5 to 1 oz); and (3) large, over 30 g (over 1 oz). In 1976 the small fish were selling for $0.83 per kg ($0.38 per lb) at wholesale, the medium fish for $1.67 per kg ($0.76 per lb), and the large fish for $3.33 per kg ($1.51 per lb). They are sold alive to restaurants in large cities. The different sizes are used in different preparations and are cooked differently. The small ones are used in soup, the medium ones are grilled and the large ones are grilled and seasoned.

Gray or jumping mullet is also cultured. This fish lives in fresh, brackish and salt water, but Japanese statistics include the cultured ones under the fresh water classification.

The fry are caught in the sea and raised in the net partition ponds described under yellowtail. The cultured fish are stocked in net enclosures in the mouths of rivers or creeks. No supplementary food is fed. Natural foods consist of marine worms, insects and plankton. The mullet farmer buys captured fry about 3 cm (1 in.) long in May. In 2 to 3 years they are 30 to 40 cm (11 to 14 in.) long. They are marketed fresh on ice. This is a very inexpensive fish compared to other cultured species, but because most of these fish are marketed through game or fee fish-out ponds the authors were unable to obtain reliable price information. In 1974, 163 MT (180 ST) were cultured in Aichi and Mie Prefecture. The wild catch was 699 MT (770 ST) from rivers, 401 MT (442 ST) from lakes and 6017 MT (6631 ST) from the sea.

MARINE CULTURED FISH AND SHRIMP

At present there are seven species of marine fish and one species of shrimp cultured in Japan. These are: (1) yellowtail *(Seriola quinqueradiata)*; (2) sea bream *(Chrysophrys major)*; (3) horse mackerel *(Trachurus japonicus)*; (4) yellowjack *(Caranx delicatissmus)*; (5) filefish *(Monacanthus cirrhifer)*; (6) puffer *(Fugu rubripes)*; and (7) "other." Shrimp or prawns *(Penaeus japonica)* are not finfish, but are more important from a volume and income viewpoint than most of the cultured marine fish.

Culture of these fish and shrimp increased from 10,475 MT (11,543 ST) in 1964 to 97,996 MT (107,992 ST) in 1974 (Table 21.19). During these 11 years the annual rate of increase was 76%. This was an even more rapid expansion than for freshwater cultured species. Nearly all of this expansion was due to yellowtail and sea bream, although nearly every specie increased in importance.

While volume of production increased 812%, the number of marine fish farmers increased 341%, from 991 in 1964 to 4372 in 1974 (Table 21.20). This indicates that volume of production per farm also increased. This increase was from 10.6 MT (11.7 ST) to 22.4 MT (24.7 ST), or 112%.

Yellowtail *(Seriola quinqueradiata)*

Yellowtail is the most popular and economically significant cultured fish in Japan. Culturing is mainly along the western coast of Japan south of Tokyo, on the southeastern coast of Honshu Island, on Shikuko Island, and on the eastern coast of Kyushu.

The history of yellowtail culture dates back to 1928 when experiments were conducted at a site on Shikuko Island. However, it was not until the 1950s that the enterprise started to expand and production became commercial. When fish farming was promoted in the early 1960s by the government, yellowtail farming spread rapidly. During these years total production was about 3000 MT (3300 ST) per year.

In 1964, the earliest year for which the authors obtained detailed data, there were about 750 separate managements for yellowtail production (Table 21.20). They produced about 9200 MT (10,100 ST). Production increased rapidly from then until the present. In 1974 there were 3044 managements, producing 92,946 MT (102,241 ST) (Table 21. 19). The number of managements grew at an annual rate of 40%, while production increased at an annual rate of 100%. While this rapid increase in numbers of managements and production occurred, average

TABLE 21.19

VOLUME OF PRODUCTION, BY SPECIES OF CULTURED MARINE FISH AND SHRIMP, JAPAN, 1964-74 (MT)

Year	Yellowtail (Seriola quinqueradiata)	Horse Mackerel (Trachurus symmetricus)	Sea Bream (Chrysophrys major)	Yellowjack (Caranx delicatissmus)	Puffer (Fugu rubripes)	Filefish (Monacanthus cirrhifer)	Other Cultured Fish	Shrimp (Penaeus japonica)	Totals
1964	10,321	—	—	—	—	—	—[2]	154	10,475
1965	14,779	—	—	—	—	—	925	N.A.	15,704
1966	16,875	—	—	—	—[2]	—	815	N.A.	17,690
1967	21,169	—	—	—	46	—	313	305	21,833
1968	31,777	—	—	—	63	—	354	371	32,565
1969	32,722	—[1]	—[1]	—	49	—[1]	375	296	33,442
1970	43,354	7	454	—	23	62	11	301	44,212
1971	61,855	57	930	—	15	18	38	306	63,219
1972	77,059	127	1380	—	14	39	104	454	79,177
1973	80,439	378	2741	—[3]	16	40	150	659	84,423
1974	92,946	619	3298	48	8	25	140	912	97,996

Source: Anon. (1976C), p. 168.
[1] Under Other Cultured Fish.
[2] Under Yellowtail.
[3] Under Horse Mackerel.
N.A. - Not Available.

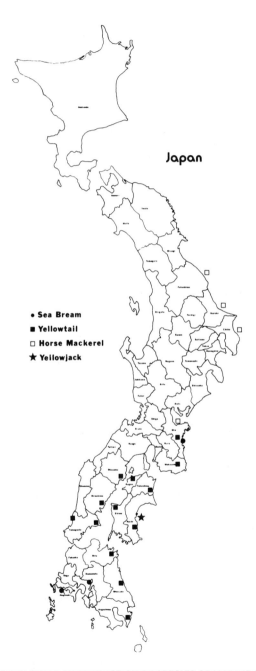

Japan

- Sea Bream
■ Yellowtail
□ Horse Mackerel
★ Yellowjack

FIG. 21.18.   MAIN AREAS OF PRODUCTION BY SPECIES OF MARINE CULTURED
FISH, 1976

TABLE 21.20

NUMBER OF MANAGEMENTS CULTURING MARINE FISH AND SHRIMP, JAPAN, 1964-74

Year	Yellowtail (Seriola quinqueradiata)	Horse Mackerel (Trachurus symmetricus)	Yellowjack (Caranx delicatissmus)	Sea Bream (Chrysophrys major)	Puffer (Fugu rubripes)	Filefish (Monacanthus cirrhifer)	Other Cultured Fish	Shrimp (Penaeus japonica)	Totals
1964	928	—	—	—	—[2]	—	—[2]	63	991
1965	786	—	—	—	40	—	126	77	1029
1966	1038	—	—	—	33	—	36	71	1178
1967	1285	—	—	—	23	—	66	76	1450
1968	1556	—[1]	—	—	14	—	103	82	1755
1969	1904	—[1]	—	—[1]	11	—[1]	167	95	2166
1970	2278	7	—	204	9	17	11	77	2603
1971	2675	27	—	278	6	14	19	75	3094
1972	3060	119	—[3]	529	6	10	23	80	3827
1973	3246	117	—[3]	745	7	8	57	86	4266
1974	3044	125	23	987	6	8	83	96	4372

Source: Anon. (1976C), pp. 168-169.
[1] Under Other Cultured Fish.
[2] Under Yellowtail.
[3] Under Horse Mackerel.

TABLE 21.21

NUMBER AND TYPES OF MANAGEMENTS, NUMBER AND TYPES OF PRODUCTION PLACES, SURFACE AREA OF WATER USED, NUMBER OF EMPLOYEES, FRY STOCKED AND TYPES OF FEED FED, BY SPECIES OF CULTURED MARINE FISH AND SHRIMP, JAPAN, 1974

Species	Managements (No.)	Types of Managements			No. Individual Places			
		On Land	Nets Extending From Shore	Marine Nets	On Land	Nets Extending From Shore	Marine Nets	Total Surface Area
Yellowtail	3,044	20	83	3,021	8	109	17,109	4,422
Horse mackerel	125			125	13		403	39
Yellowjack	23	1		22	13		34	3
Red sea bream	987	2	1	982	4	14	2,916	525
Puffer	6		3	3		3	13	11
Filefish	8			8			33	1
Other fish	83	7	1	77	55	1	178	637
Total fish	4,276	30	88	4,238	80	127	20,686	5,638
Prawns	96	87	1	11[1]	200	4	53[2]	3,221

TABLE 21.21 (Continued)

Species	Surface Area (000 m²) by Types of Mgts			Employees (No.)	Fry Stocked (No.)	Total	Food Fed (MT)	
	On Land	Nets					Fish and Other	Pellets
		Extending Drom Shore	Marine Nets					
Yellowtail	478	1,122	2,824	9,043	57,435	726,585	723,492	3,092
Horse mackerel			39	232	7,681	7,221		
Yellowjack	1	2	—	86	19	651		
Red sea bream	15	114	396	2,188	10,274	37,277		
Puffer		8	3	29	35	44		
Filefish			1	10	71	187		
Other fish	496	0	142	192	1.453	1,035		
Total fish	990	1,246	3,405	11,780	76,968	773,000	723,492	3,092
Prawns	3,004	120	97³	483	87,563	26,909	26,380	529

Source: Anon. (1976C), 170-187.

[1] Includes 10 listed as "other."

[2] Includes 40 listed as "other."

[3] Includes 84 listed as "other."

Note: There are minor discrepancies in the original data such as 83 managements for "other" fish but the number under individual headings for types of management is 85.

production per management increased from only 12.3 MT (13.5 ST) in 1964 to 30.5 MT (33.5 ST) in 1974. Hence average production per management increased 148% or 14.8% per year. Since 9043 people were employed in producing yellowtail, the average production per person was only 10.3 MT (11.3 ST). This relatively low level of output per person indicates the need for close supervision during production and the relatively high value of the species per unit of sale.

In 1966 there were 5867 different farms or places of production, operated by 1038 managements, or 5.6 places of production per management. In 1974 there were 17,226 different farms or places of production under 3044 different managements, or nearly 5.7 places of production per management (Table 21.21). In 1966 there were 3,083,000 m² (762 acres) of surface area devoted to yellowtail culture. Production was about 5.5 kg per m² (1.1 lb per ft²). In 1974 there were 4,422,000 m² (1092 acres) of surface area devoted to yellowtail culture. Production was 21.0 kg per m² (4.3 lb per ft²) (Table 21.21). Hence production increased by 3.8 times per surface area. This was possible because of changes in production techniques as explained below.

Production of yellowtail is done in a variety of ways. The oldest technique is in a land enclosure (dike) with sea gates which permit the exchange of water as the tides rise and fall. Hence there is only partial exchange of water about twice a day. Stocking levels may be only 1% that of net culture. In 1974 there were only eight of these dikes. The second technique is called net partition. With this technique at least one side of the enclosure is separated from the sea by nets, while the other sides are natural or man-made walls of concrete, stone or earth. This technique permits better exchange of fresh salt water for culturing. Stocking may be as high as 10% that of the net technique. In 1974 there were 109 of these net partition areas. The third technique is net culture. Net culture may be divided into numerous sub-types. These are nets, enclosures suspended below floating rafts, and/or buoys. Rafts may be constructed of wooden barrels, bamboo, or steel; with these systems the net enclosure is above the surface of the water. The latest method is when the net enclosures may be as much as 30 m (100 ft) under the surface buoys. This is used where there are strong surface currents caused by winds or damaging storms. With net culture, water exchange may be every few minutes and stocking may be 100 times that of the dike method and 10 times that of the net partition method. In 1974 there were 17,109 different net enclosures used. These accounted for 99% of the yellowtail farms, but only 64% of the culturing area.

FIG. 21.19.   NEWLY CONSTRUCTED YELLOWTAIL CAGE BEING LOWERED
INTO MARINE WATERS

FIG. 21.20.   FLOATING YELLOWTAIL CAGE CULTURE

In 1974 a total of 726,585 MT (799,250 ST) of food were fed to produce the 92,946 MT (102,250 ST) of production. Hence the feed conversion was 7.8 units of food to one unit of output. Ninety-six percent of the food fed was sardines, mackerel, sand eels and other trash fish, and 4% was pelletized feed.

In 1974, 321 MT (353 ST) of yellowtail fry were caught in major waters. The major catching areas in order of importance were Mie Prefecture, 38%; Kochi Prefecture, 18%; Tokushima Prefecture, 13%; and Kagoshima Prefecture, 9%. All other areas accounted for less than 22% of the stocking catch. Major areas for culturing do not correspond to catching areas since the fry are moved to places with the least pollution, the greatest culturing interest, and to areas of low-cost food. In 1974 Mie Prefecture accounted for over 19% of cultured production, and Kochi and Ehime prefectures on the island of Shikuko accounted for 16 and 17%, respectively. These three prefectures accounted for over one-half of total cultured production.

In 1974 cultured production of yellowtail was 92,946 MT (102,-241 ST) while the wild food catch was 53,000 MT (58,300 ST). Hence nearly two-thirds of the total yellowtail consumed are cultured.

Yellowtail grow to 5 to 8 kg (11 to 18 lb) under natural conditions. The market size of cultured fish is 1.0 to 1.5 kg (2.2 to 3.3 lb) in 1 year, and 2.0 to 3.0 kg (4.4 to 6.6 lb) in 2 years.

The wild fish spawn from January to August when marine water temperatures are between 16° and 29° C (61° to 84° F). The wild fry are caught in April and May when they are only about 15 mm (0.5 in.) long and weigh about 1 g. During May and June the fry are acclimatized and started on feed in floating net enclosures. They double and triple in size during this adjustment period. In June they are graded by size and stocked in floating or submerged net enclosures. Feeding rates vary from 50% of body weight for the fry to 7% for those weighing 1 kg (2.2 lb) and over. By December some culturists harvest their fish at 1.0 to 1.5 kg (2.2 to 3.3 lb) while others continue feeding into the second year.

Although spawning and fry rearing has been done on an experimental basis the cost is still prohibitive. Thus all fry are from natural stocks. A special license is required to collect and sell the fry to avoid overfishing. The fry usually follow floating seaweeds in the Kuroshio current, which flows from southern Japan to the north. The fry are captured in nets and kept in small floating net pens made of fine mesh synthetic fibers until they are acclimatized to feeding. These pens may vary up to 50 m² (500 ft²) and are from 1 to 3 m (3 to 9 ft) deep.

The most important operation initially is to separate the fry by sizes into at least small, medium and large. Without proper sizing, cannibalism takes place and mortalities of 50% may occur in a few days.

The fry are fed minced meat of sand eels, horse mackerel, shrimp and so on. During the grow-out period there must be sufficient water exchange to keep the supply of dissolved oxygen over 3 ppm and to wash out feces and uneaten food. Optimum temperatures for growth are between 24° and 29° C (75° to 84° F). The higher and lower critical temperatures are 30° C (88° F) and 9° C (48° F), respectively. Salinity should not be below 16 parts per thousand.

Floating net enclosures must be in locations where violent wave and wind action does not occur. The submersible net enclosures are more suitable to open sea areas. The basic structures of these type enclosures are shown in Fig. 21.21 and 21.22. Ten to 12 separate net enclosures are usually treated as one management unit.

As stated previously, the diet is predominantly trash fish which have low market values for human food. The small amount of pelletized food used is at least 70% white fish meal, with 5 to 10% gluten as a binder. Other ingredients include vitamin mix and minerals, especially iron and cobalt to prevent anemia. Growth with this feed

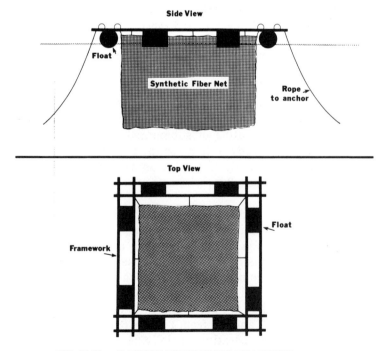

FIG. 21.21.   FLOATING NET CAGE FOR YELLOWTAIL

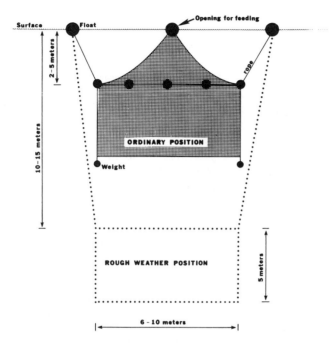

FIG. 21.22.   SUBMERGIBLE NET CAGE FOR YELLOWTAIL

is not as rapid as feeding minced fish alone, but is better when both are fed. The high cost of the pellets has prevented their widespread acceptance. The optimum feeding regime consists of a morning and afternoon feeding. Feeding rates decrease as a percentage of body weight as the fish mature.

The cultured fish ready for the market are harvested with a simple dip net. About one-half of cultured yellowtail are sold alive and are then transported to market by boat or truck in iced water. These go to cities in eastern Honshu where natural yellowtail are caught. Nearly all of these ultimately end up in restaurants. The remaining fish are sold as fresh iced fish, either in the round or gutted. These fish go to the western side of Japan or to the areas north of Tokyo where wild yellowtail cannot be caught. About 95% of them ultimately end up in restaurants while about 5% (usually the smaller ones) are sold through supermarkets.

The average producer produces between 10,000 and 20,000 fish weighing 1.0 to 1.5 kg (2.2 to 3.3 lb) each. This volume is sufficient to earn a middle class income.

It is estimated that future expansion of yellowtail culturing will be small because of government restrictions on the number of wild fry that can be harvested. Hence hope for further large-scale expansion is in reducing the costs of producing spawns and fry artificially. The main problem appears to be a shortage of suitable food for the newly hatched fish.

Production Costs.—In 1974 the average cost of producing a one-year May-December yellowtail weighing 1.2 kg (2.6 lb) was $2.47 (Table 21.22). The average selling price per fish was $4.00, giving a net return to management or net profit of $1.53 per fish. With an average production per management of 30.5 MT (33.5 ST), the average producer produced 25,407 fish and made a return for family labor and net profit of $38,873. Those farmers who can afford to do so often keep their fish for two years. During the second year growth is 2.8 kg (6.2 lb) instead of the 1.2 kg (2.6 lb) that they gain in 1 year. Hence the profit is higher for two year fish, but cost and returns data are not available.

Marketing.—Of the total fish supply of 10,807,785 MT (11,909,-872 ST) in Japan in 1974, the 6 largest wholesale fish markets handled 2,030,541 MT (2,237,656 ST) or 18.8%. These six markets in order of importance are: (1) Tokyo, (2) Yokohama, (3) Nagoya, (4) Kyoto, (5) Osaka, and (6) Kobe. The Tokyo market handled 821,714 MT (905,529 ST), or more than 40% of the volume of the 6 largest markets.

For each year from 1970 through 1974 the price of cultured yellowtail was higher on the Tokyo market than that of wild fish (Table 21.23). This premium is mostly due to the ability of the fish farmers to harvest and sell to the market when native stocks are low and prices

TABLE 21.22

COSTS OF PRODUCING CULTURED YELLOWTAIL PER 1.2 KG
(2.6 LB) FISH, JAPAN, 1974

Item	Amount ($)
Fry	0.35
Depreciation	0.23
Sea rent	0.04
Medication	0.10
Repairs and maintenance	0.02
Boat supplies	0.06
Insurance	0.10
Feed	1.30
Wages	0.23
Miscellaneous	0.04
Total	2.47
Selling price	4.00
Returns to management	1.53

Source: Anon. (1975B).

TABLE 21.23

SALES OF WILD AND CULTURED YELLOWTAIL ON THE TOKYO CENTRAL
WHOLESALE FISH MARKET, 1974

Year	Wild Yellowtail			Cultured Yellowtail		
	(MT)	($ per kg)	($ per lb)	(MT)	($ per kg)	($ per lb)
1970	9654	1.30	0.59	7,504	1.81	0.82
1971	8627	1.53	0.70	8,663	1.96	0.89
1972	8239	1.59	0.72	9,978	1.98	0.90
1973	9282	1.75	0.80	9,743	2.04	0.93
1974	6655	2.50	1.14	10,875	2.66	1.21

Source: Anon. (1972-74), pp. 116, 118.

are higher. Even though cultured yellowtail prices are higher on the Tokyo market than wild yellowtail, it is believed that prices of cultured yellowtail are lower on the Tokyo market than elsewhere. This is due to the cultured fish being fatter and containing more oil which affects the flavor.

Red Sea Bream—Red Porgy (Chrysophrys major)

Red sea bream, sometimes called red porgy, is the second most important cultured marine finfish in Japan. It is traditionally regarded as a symbol of good fortune. Thus it is not only a food material but is featured at celebration meals and is a particularly highly valued fish.

Research on the farming of this species had been carried out since 1887, and experiments on artificial propagation stem from 1902, but there were no successful results. In 1958 research on propagation began again and limited success was achieved in 1962. Since then farming has advanced quite rapidly and the output of cultured fish has expanded rapidly in recent years.

Prior to 1970 the number and volume of production of red sea bream were reported under yellowtail information. In 1970 there was a total of 204 managements producing 454 MT (500 ST). Four years later there were 987 managements producing 3298 MT (3628 ST).

The 987 managements in 1974 had a total of 2934 production places with a total area of 525,000 m² (130 acres). Hence the average production was 6.3 kg per m² (1.3 lb per ft²). Only 4 places used the dike method, 14 used net partitions and 2916 used net techniques. These were described previously under the yellowtail section of this chapter. As was true for yellowtail, the most intensive output per square meter of water surface was with net culture because of the better exchange of sea water and resulting higher water quality. The average volume of production per management was only 3.3 MT (3.6 ST). Since 2188 people were employed, the average volume of production per employee was only 1.5 MT (1.7 ST). This indicates the need for close supervision and the high value of the fish per unit of sale. In 1974 there were 10,274,000 fry stocked. The total supply of food fed was 37,277 MT (41,000 ST). Hence the feed conversion for the 3298 MT of production was 11.3 units of feed per unit of output. The diet is almost exclusively trash fish since manufactured food pellets are more expensive.

The three main areas of production are Nagasaki, Mie and Kumamoto Prefectures. These 3 areas accounted for 62% of production in 1974. See the map of Japan for the location of these three prefectures.

Red sea bream culturing is increasing rapidly. Production expanded from 454 MT (500 ST) in 1970 to 3298 MT (3634 ST) in 1974. However, the natural wild catch is 18,700 MT (20,607 ST). Hence culture production, while accounting for an increasing share of the total supply, is only about 15% of the total.

Horse Mackerel (Trachurus symmetricus)

Horse mackerel is the third most important marine finfish cultured in Japan. Prior to 1970 the volume of cultured production was listed under "other" cultured fish. Production must have been negligible because in 1970, when separate records were made, production was only 7 MT (less than 8 ST). Within 4 years production reached 619 MT (681 ST) (Table 21.19). Production per management is small. For example, in 1974 there were 125 separate managements producing 619 MT—less than 5 MT (5.5 ST) per management (Table 21.20). The 125 managements were using a total of 403 floating net marine enclosures (Table 21.21). These enclosures had a total surface area of only 39,000 m² (420,000 ft²). The average net enclosure covered 97 m² (1044 ft²) of surface area. Production was 1.5 MT (1.65 ST) per net enclosure for an average output of nearly 16 kg per m² of surface (3.2 lb per ft²).

A total of 86 people were employed, so production was only 7.2 MT (8.0 ST) per person. Hence it can be seen that stocking is intensive and labor requirements high.

Total volume of feed fed was 7221 MT (7958 ST), for a feed conversion of 11.7 units of fish feed per unit of fish produced. Production is mainly west of the Tokyo area on the Pacific side of the main island of Honshu. Over one-third of total production comes from Mie Prefecture.

Fishermen catch 50 to 100 g (2 to 4 oz) fry in marine waters for stocking. After only 2 to 3 months of feeding sardines and other trash fish, the horse mackerel are between 200 and 300 g (7 to 11 oz) and are sold as fresh iced fish in the round. In 1976 the wholesale values varied between $1.67 and $2.00 per kg ($0.76 to $0.91 per lb). It can thus be seen that horse mackerel is a relatively expensive fish.

Although production of horse mackerel is increasing rapidly and reached 619 MT (681 ST) in 1974, this volume is only a minor contributor to total supply. During the same year about 128,000 MT (141,056 ST) were caught in marine waters. Culturing accounts for less than 1% of total supply and is evidently going to continue to expand rapidly.

Other Cultured Marine Fish

Under this heading are: (1) yellowjack *(Caranx delicatissmus)*, (2) puffer *(Fugu rubripes)*, and (3) filefish *(Monacanthus cirrhifer)*.

Separate statistics were only started for yellowjack in 1974. In that year there were 48 MT (53 ST) of cultured production (Table 21.19), produced by 23 separate managements (Table 21.20). Nearly all culturing was in floating marine enclosures or pens. Eighty-six people were employed, making the output per person only about 0.5 MT. However, many of these managements produced other species, so the real output per person is higher than this information indicates. The major producing area is Kochi Prefecture on the island of Shikuko. Farmers stock 200 to 300 g (7 to 11 oz) wild fish caught by fishermen. These are fed minced sand eels, anchovies, sardines and other trash fish. They are raised to a size of 2 to 3 kg (4.4 to 6.6 lb) in a period of 12 to 18 months. The fish are considered a gourmet item and are sold to first-class restaurants at prices averaging $6.67 per kg ($3.03 per lb). Yellowjack production will undoubtedly increase under the double impetus of good prices and ability to continue expansion. In 1974, 54,700 MT (60,279 ST) of wild fish were marketed compared to only 48 MT of cultured production.

Puffer is also a cultured species. Puffer production reached 600 MT (661 ST) a year during the 1959-61 period along the coasts of the Inland Sea. Since then it has rapidly declined to only 8 MT (9 ST) in 1974. This was due to increases in natural production and imports. Most culturing is for restocking the sea waters. Spawning begins in May when adults are collected for eggs and fertilization. In 7 to 10 days the eggs hatch. Hatching waters range between 15° and 19° C (59° to 64° F). When the fry are 5 cm (2.5 in.), they are stocked in floating net enclosures or impoundments having sea water exchange.

They are fed 2 to 3 times daily at 17 to 25% of body weight. Food is sand eels, minced mackerel and other trash fish. At 350 g (12.5 oz) feeding rates drop to 0.6% of body weight. By the end of December the fish average 400 g (14 oz). In January growth stops because of low water temperatures. The fish are fed for 2½ years and are sold at 1.5 to 2.0 kg (3.3 to 4.4 lb). They are sold live to specialized gourmet restaurants at very high prices. Producer prices range between $10 and $20 per kg ($4.55 to $9.09 per lb). The skin, liver and ovaries of puffer are extremely poisonous and cooks in the specialized restaurants undergo intensive training and are licensed by the government. The Japanese have a saying that "to savor fugu [puffer] one courts death" and "Honey is sweet, but the bee stings." In spite of all precautions some 20 to 30 Japanese die each year from eating puffer. There are no available statistics on consumption, but a minimum of 1000 MT (1102 ST) are consumed annually.

Filefish are another cultured marine fish. They are cultured by only a few individuals and production varies between 25 and 40 MT (28 to 44 ST) annually. They are stocked with other marine cultured fish in small numbers. They eat only marine growth which adheres to the floating and submerged nets; the nets are thus kept cleaner and labor costs are reduced in cleaning and replacing nets. Filefish are usually stocked with horse mackerel and red sea bream. Wild fingerlings caught by fishermen are stocked at about 50 g (2 oz). In 10 to 15 months they reach market sizes of 500 to 600 g (18 to 21 oz) and sell for about $1.00 per kg ($0.45 per lb).

Shrimp (Penaeus japonicus)

This species ranges from southern Japan southward to northern and northeastern Australia, eastward to Fiji and westward to Africa. Shrimp, sometimes called prawns, are common in Asian waters. They dwell in coastal waters of moderate temperature and salinity. The shrimp are attractive in color with alternate cross bars of brown, blue and yellow and weigh between 20 and 25 g. They are highly valued for their delicacy. Weights of 130 g (4.6 oz) have been attained. Shrimp are a carnivorous species and live two years.

The northern limit of their range on the Japan Sea side is Akita Prefecture. On the Pacific side it is Miyagi Prefecture. The wild catch in Japanese coastal waters is about 2800 MT (3086 ST), while cultured production was 912 MT (1005 ST) in 1974. The wild catch decreased in the 1960s because of pollution. In the 1970s the various prefecture governments have been restocking coastal waters and the catch increased to near the old level of about 3000 MT (3306 ST). Cultured

FIG. 21.23. SHRIMP CULTURING POND SHOWING WATER GATE LEADING TO
THE SEA

production continues to increase.

The smaller, wild prawns are found in the sand beds of the inner parts of bays. As the prawns grow they move into deeper waters. In the summer they generally move into shallower waters and in winter move into deeper waters and bore themselves into the sand. The winter catch is small and the market price is correspondingly high.

Prawns are nocturnal and have a long spawning period. Female prawns gravid with mature ova can be seen as early as April 1 off southern Kyushu, and as late as September in the northern limits of their range. The females measure about 15 cm (4 in.) and males about 12 cm (3 in.). Mating is done during the night. Spawning also takes place at night and usually takes from 3 to 4 min. Fertilization is finished as soon as the ova leave the female body. The ova are slightly heavier than salt water and sink to the bottom.

In Japan experimental work began on culturing prawns as early as 1933. However, it was 1963 before techniques were adequately developed so that culturing was economically practical. In 1976 there were more than 100 prawn farms in Japan. They are mainly in the Seto Inland Sea area (Fig. 21.24). They are established on deserted salt beds or on sandy beaches of the inner bays.

The depth of the culture pond is shallow, being about 60 to 180 cm (24 to 46 in.) with a flat sandy bed. The turnover of the seawater is done by using floodgates, which permits partial exchange of the sea-

FIG. 21.24. MAJOR PRODUCTION AREAS FOR PRAWN CULTURING, 1976

water twice a day, as the tides ebb and flow. The bed of the pond should be a light-colored slightly reddish sand, since the color of the prawn is said to be affected by the color of the sand and the feed. Prawns with red coloring and clear stripes are most valuable in Japan.

Seed Production.—Commercial concrete tanks having capacities of 200 m³ (7000 ft³) or more are common.

To produce prawns within the year, the young prawns must be stocked by the end of May in the finishing culture pond. This means that heated water of about 28° C (82° F) may be required in the early stages of incubation.

In April or May the gravid female prawns must be obtained, and at this time must have a dark green ova cluster. They must be collected, shipped in cooled wet sawdust, and placed in the incubation tank before the evening of the day they are captured.

The gravid female prawns spawn during the night while swimming. After 14 to 15 hr in water temperatures of 25° to 28° C (77° to 82° F) the ova become nauplii. They do not require feeding, but absorb the yolk of the ova and go through repeated ecdyses. Between 20 and 36 hr after the sixth ecdysis they become zoeae.

The zoeae repeat ecdysis 3 times in 4 days and become myses. During the zoea stage cultured Diatomaceae are placed in the tank for food. Water quality and the quality and quantity of Diatomaceae affect the growth rates of the zoeae.

When the zoeae enter the mysis stage supplemental food can be added to the Diatomaceae. The ova and larvae of oysters are sometimes used. The mysis stage takes three days. The first legs develop and they become post-larvae.

In the post-larva stage they begin actively feeding on microorganisms and the nauplii of brine shrimp are fed along with minced short-necked clam meat. At this point about 25% of new water is exchanged daily in the tank. By P_{20} the body weight may be 10 to 20 mg. They are now ready for stocking in the finishing or grow-out culturing pond.

The P_{20} post-larvae must be placed in the culturing pond by early June to prepare them for the winter market season. An average stocking density of 20 per m² (2 per ft²) of surface area is recommended.

Raising.—Before stocking the P_{20} shrimp the pond must have been prepared. After the last crop of adult prawns are harvested the pond is drained and dried. The sand bottom is turned over to aid in decomposition of organic matter. When all odors have vanished, the sand is usable. New sand is often added to renew the proper culturing characteristics of the pond bottom. The pond is then filled with seawater and small foreign fish and predators are killed by using derris powder.

If the P_{20} shrimp are stocked in early June at a density of 20 per m^2 (2 per ft^2), they can reach market sizes of 20 g by December or January when the water becomes too cold for growth. The young shrimp are fed short-necked clams. The clams are run through crushers to crack the shells and are then dropped into the culturing pond. Other types of feed are sometimes fed, including other shrimp, crab and young fish. For the serious reader the authors recommend *Aquaculture in Japan*, by Akio Honma (1971) and *Shrimp Culture in Japan*, by Kunihiko Shigueno (1975).

Prawns spend most of their time in or on the sand bed at the bottom of the pond. Particular care must be given to avoid pollution. As feeding increases, the size of the prawns increase and hydrogen sulphide is formed. At 2 to 4 ppm the prawns lose their sense of balance and at 4 ppm they die. Pumping is sometimes necessary to assure good water quality. Care must also be exercised in the continuous suppression of foreign fish and predators. A 4 to 5% rotenone mixture is commonly used. Several other chemicals for use are discussed by Honma and Shigueno.

Harvesting and Selling.—Harvesting is done by several methods. Among these are the Pound net and the Pump net. Catching is done at night when the prawns are active. The Pound net is a simple trapping device which the prawns enter and then are lifted out of the water. The Pump net is used when the water is cold and the prawns have burrowed into the sand. Jets of water plough the sand bed, forcing the prawns to surface into a trailing catch net. Nearly 100% harvesting can be done by using these two methods.

After harvesting the prawns are transferred to a cooling tank. The prawns are cooled by slowly lowering the water temperature. When the prawns can barely move and the body color becomes reddish, cooling is stopped. This is usually at 12° to 14° C (54° to 57° F). They are then packed in wet sawdust by size. Each layer of prawns is separated by a layer of sawdust. Sometimes a lump of ice is packed in the middle of the shipping container. Between 2 and 3 kg (4.4 to 6.6 lb) of prawns are packed in each box. The number and weight is recorded on the outside of each box. The prawns are moved to airports or trains by refrigerated trucks. In these shipping containers they can live for 3 to 5 days. At the central auction the prawns are examined for condition and auctioned off to restaurants and fish-retailers.

Sales are usually at one of the six largest central fish markets. These are: (1) Tokyo, (2) Yokohama, (3) Nagoya, (4) Kyoto, (5) Osaka, and (6) Kobe. Average prices received by one farm are shown

TABLE 21.24

ELEVEN YEAR HISTORY OF THE UBE SHRIMP CULTURING FARM, UBE CITY,
YAMAGUCHI PREFECTURE, JAPAN, 1965-75

Year	Post-larvae Stocked (000s)	Survival Rate (%)	Avg Harvest Weight (g)	Total Harvest Weight (kg)	Avg Price ($ per kg)	Avg Price ($ per lb)
1965[1]	600	50	20.0	6,000	8.33	3.79
1966	870	61	11.0	5,836	6.51	2.96
1967	800	68	13.9	7,520	11.63	5.29
1968	800	71	17.5	8,509	11.41	5.19
1969	800	37[2]	22.1	7,033	15.62	7.10
1970	800	44[3]	18.2	6,268	14.31	6.50
1971	800	79[4]	20.9	10,115	12.30	5.59
1972	680	69	25.9	12,430	14.11	6.41
1973	880	39[5]	26.6	8,806	15.09	6.86
1974	800	76	27.7	16,103	15.66	7.12
1975	900	53	27.0	12,950	19.20	8.73
Averages	—	59.7	21.08	9,557	13.59	6.18

Source: Anon. (1976E), pp. 6-7.
[1] First year of operation.
[2] Oxygen depletion.
[3] Pollution problem and red tide.
[4] Aeration added.
[5] Sea grass infestation.

for the 11 years between 1965 and 1975 (Table 21.24). This data indicates the relationship between size of prawns and price, the increasing prices of cultured prawns, and the high value per unit of sale. In 1975 the average price received at this farm was $19.20 per kg ($8.73 per lb).

Production Costs.—While selling prices of cultured prawns are high, production costs are too. One example of this is shown in Table 21.25. The farm, which has been in operation since 1965, produced 16,103 kg (35,429 lb) of prawns in the June 1, 1974 to May 31, 1975 period. For this year, when the survival rate was 76% (Table 21.24), the cost of production was $14.69 per kg ($6.68 per lb). Wages, which were regarded as a fixed cost, were the major cost item. Feed was the second highest expense item. Feed at this farm consists of 40% pellets, 40% other shrimp and 20% clams. Nearly 10% of the total sales was profit. This was after all taxes, including the tax on profits. This indicates the relatively high return to a successfully operated prawn farm.

Restocking.—The Japanese have intensive restocking efforts for both freshwater and marine species. For example, in 1974 over 54,-

000,000 shrimp larvae were released in marine waters. In the same year a total of 841,411,000 fry were cultured and released. These were as follows:

Salmon and marine trout	47,393,000
Landlocked salmon and trout	19,552,000
Ayu	229,678,000
Common carp	42,898,000
Crucian carp	22,941,000
Eel (adults)	9,841,000
Others	469,108,000

Under "others", nearly every conceivable species found in Japanese fresh and marine waters was included (Anon. 1976 [Fifth Fisheries Census]).

TABLE 21.25

PRODUCTION COSTS FOR THE UBE SHRIMP CULTURING FARM, UBE CITY, YAMAGUCHI PREFECTURE, JAPAN, JUNE 1, 1974 to MAY 31, 1975[1]

Item	Cost ($)
Variable expenses	
Gravid female prawns	3,733
Feed	54,772
Wages	1,531[1]
Power	4,352
Repayment of principal	7,190
Miscellaneous	1,224
Transportation	7,419
Packaging	2,992
Storage	3,336
Repairs	12,653
Rent	1,483
Interest	1,427
Others	306
Taxes	26,667[2]
Year end bonuses	3,333[2]
Total variable expenses	132,418
Total fixed expenses	104,110
Total costs	236,528
Profit	23,580

Source: Anon. (1974-75), Volume of prawns produced was 16,103 kg (35,427 lb).
[1] Most wages are included under Fixed Expenses.
[2] Estimated amounts.

SPECIAL ACKNOWLEDGEMENTS

DR. MASARU FUJIYA, Staff of Freshwater Fisheries Research Laboratory, Hino-shi, Tokyo, Japan

MR. YOSHIHISA FUMUTA, (No address available)

MR. YOSHIHISA FURUTA, Freshwater Fisheries Research Laboratory, Hino-shi, Tokyo, Japan

MR. AKIRO HOMMA, Research Division, Fisheries Agency, Kasumigaseki, Tokyo, Japan

MR. TOSHIO HOMMA, 2-7-20 Himonya, Meguroku, Tokyo, Japan

MR. SHIGEO INABA, Director of Fuji Rainbow Trout Experiment Station, Fujimiya-shi, Shizuoka Pref., Japan

MR. RIKIZO ISHIDA, Freshwater Fisheries Research Laboratory, Hino-shi, Tokyo, Japan

MR. JUKICHI ISHII, Director of Koijyu Carp Co., Ltd., Ogo-machi, Gunma Pref., Japan

MR. YOSHITO IWAHASHI, Fisheries Experiment Station, Numazu-shi, Shizuoka Pref., Japan

DR. TAKEICHIRO KAFUKU, Staff of Freshwater Fisheries Research Laboratory, Hino-shi, Tokyo, Japan

MR. YUKIMASA KUWATANI, Staff of Freshwater Fisheries Research Laboratory, Hino-shi, Tokyo, Japan

MR. KUNISUKE MUDA, Staff of Freshwater Fisheries Research Laboratory, Hino-shi, Tokyo, Japan

MR. HAJIME NAKAGAWA, Director of Nakagawa Fish Culture Coop., Anan-shi, Tokushima Pref., Japan

MR. ICHIRO NAKAMURA, Japan Fisheries Association; Akasaka 1, Minatoku, Tokyo

MR. TOSHIICHI NAKAMURA, Director of Midori Shobo Publishing Co., 4-6-5 Iidabashi, Tokyo, Japan

MR. HIKARU NIIHAMA, Eel Division of Anan Agri. Coop., Anan-shi, Tokushima Pref., Japan

MR. KAZUTAMI NISHIO, Director of Yoshida Eel Laboratory, Yoshida-machi, Shizuoka Pref., Japan

MR. KENJI NOSE, Staff of Freshwater Fisheries Research Laboratory, Hino-shi, Tokyo, Japan

MR. TOKUZO SAMESHIMA, Fisheries Division, Tokushima Pref., Japan

MR. SHOZABURO SETO, Vice-chief of Fisheries Division, Yamaguchi Pref., Japan

MR. YAICHIRO SHINMA, Freshwater Fisheries Research Lab., Hino-shi, Tokyo, Japan

MR. EIMATSU TOKUNAGA, Staff of Freshwater Fisheries Research Laboratory, Hino-shi, Tokyo, Japan

MR. KUNIO WATANABE, Director of Fuji Rainbow Trout Coop., Fujimiya-shi, Shizuoka Pref., Japan

MR. MASAHIRO (FRANK) WATANABE, Director of Western Publications Distribution Agency, Nishiokubo, Shinjukuku, Tokyo

MR. YOSHIO WATANABE, Editor of Midori Shobo Publishing Co., 4-6-5 Iidabashi, Tokyo, Japan

DR. TAKESHIGE YAMAKAWA, Director of Freshwater Fisheries, Research Laboratory, Hino-shi, Tokyo, Japan

MR. AND MRS. YOKOTE, Staff of Freshwater Fisheries Research Laboratory, Hino-shi, Tokyo, Japan

REFERENCES

ANON. 1967-74. Monthly Reports of Japan Exports and Imports, Customs Bureau, Ministry of Finance.

ANON. 1972-74. Annual Reports of the Tokyo Central Wholesale Market, Fish Production Editions, Tokyo Metropolitan Government.

ANON. 1974-75. Business report. Ube Shrimp Culture Farm, Ube City, Yamaguchi Prefecture, Japan.

ANON. 1975A. Fresh water synthesis report for 1974. Hiyogo Prefecture, Japan.

ANON. 1975B. Kansiu monthly report. Kansiu Fish Culture Association, Kobe, Japan, Report No. 116.

ANON. 1976A. Fish farming in Japan, 1975. Japan Fisheries Association, Akasakal, Minato-Ku, Tokyo, Japan.

ANON. 1976B. Fisheries of Japan, 1975. Japan Fisheries Association, Akasakal, Minato-Ku, Tokyo, Japan.

ANON. 1976C. Yearbook of Japanese Fisheries and Culture Production. Ministry of Agriculture and Forestry, Tokyo, Japan.

ANON. 1976D. Fifth Fisheries Census, Report No. 10. Ministry of Agriculture and Forestry, Tokyo, Japan.

ANON. 1976E. 10 Year History of Ube Shrimp Culture Farm. Ube City, Yamaguchi Prefecture, Japan.

ANON. 1976F. Miscellaneous report on eel production costs. Fisheries Cooperative, Tokushima Prefecture, Japan.

CHIBA, KENJI. 1968. Fish Culture Seminar, Vol. I, II, III. Mydori Shobo Book Publishing Company, Tokyo, Japan.

FUJIYA, MASARU. 1976A. Yellowtail farming in Japan. J. Fish. Res. Board of Canada 33, No. 4, Pt. 2.

FUJIYA, MASARU. 1976B. Coastal aquaculture of fish in Japan. Experience paper, FAO Aquaculture Technical Conference, Kyoto, Japan.

HONMA, AKIO. 1971. Aquaculture in Japan. Japan FAO Association, Kita-ku, Tokyo.

SHIGUENO, KUNIHIKO. 1975. Shrimp Culture in Japan. Association for International Technical Promotion, Tokyo, Japan.

Chapter 22

People's Republic of China

Clinton E. Atkinson

From the very earliest times, fish provided a basic source of food for the Chinese people. Archeological records dating back some 15,000 to 50,000 years provide evidence that grass carp *(Cteno-pharyngolon idellus)* were regularly eaten by the "mountain cave dwellers" of that period and skull bones of this fish were used as primitive ornaments. In the next 5000 to 10,000 years, hooks and spears made of bone and net weights of stone and clay were commonly found among the artifacts of this primitive civilization. There is also evidence that the method of preserving fish in salt, for use when food became scarce, was developed during this period.

By the time of the Shang Dynasty (1766 to 1122 BC) fishing was a well-developed skill; various types of metal fish hooks and remnants of other fishing gear are frequently seen in museums today.

Shortly thereafter the Chinese rulers of the Chou Dynasty (1122 to 255 BC) recognized the need to protect their fishery resources by special closures and other regulations, and established an office within the government to exercise control over their fisheries. This was the beginning of our modern concept of fisheries conservation.

Chinese literature is filled with references to the almost sacred position that carp, "the King of Fishes," occupied among the early Chinese. The carp was primarily known for its stamina and persistence in overcoming areas of fast water. The carp was said to be the only fish that was able to ascend the famous rapids of the Yangtze River—the Dragon's Gate. Thus when a student ultimately overcame all the obstacles of obtaining the final *Han Lin* degree at school, it was said that he had passed "through the Dragon's Gate." The Chinese ideograph for fish is said to represent a carp with nose pointed upwards in the ascent of a falls. Similarly the Chinese character for wealth and abundance is symbolic of two carp swimming upstream side-by-side. The flying of the carp streamers during boys' day festivities, the "nobori-koi" of Japan, signifies the strength of a carp (koi) swimming upstream and is an ideal for the sons to follow in overcoming the adversities of life.

Over the centuries carp have been closely associated with religious beliefs and ritual. The place of carp in the oriental religions probably originated in the teachings of Confucius and developed from the knowledge of carp and carp rearing of the Chou Dynasty. However, it is significant that even at the present time Buddhists of China and other nations in eastern Asia celebrate Buddha's birthday (April 8th) by purchasing young carp and releasing them. Further, the Buddhist goddess of compassion, Kwan-yin, is usually shown as a middle-aged woman carrying a common carp in a basket. Carp also are the subject of worship and ritual in the Shinto faith, especially in its preparation for cooking.

Although some scholars believe that fish culture actually may have originated in one of the more southern Asian countries, there is little question that the high position occupied by the common carp in diet, folklore and religion, and the advanced civilization of the Chinese people had a strong influence on the development of the fisheries and methods for culturing carp and other freshwater fish.

The origin of aquaculture in China is attributed to Wen Wang, the founder of the Chou Dynasty. Twenty years after his accession to the throne until his death, the last ruler of the Shang Dynasty had Wen Wang confined to an estate in Honan Province (1135 to 1122 BC). It was there that Wen Wang built a pond, filled it with fish and made the first recorded reference to their behavior and growth. The fish pond was apparently the source of great pleasure for Wen Wang and much of the early skill in the rearing of fish is said to have originated from his efforts and those of the people working with him.

Although a number of references are found in the early Chinese writings that appeared in the next several centuries, it was not until 460 BC that Fan Li, a land-holding official in the ancient kingdom of Yüeh, wrote his "Fish Culture Classic," which describes in some detail the results of numerous experiments made by Fan Li and others at that time. His observations on the location and size of ponds and principles of stocking and harvest served as a guide for fish culturists for many years and is even applicable today.

It was also during the latter part of the Chou Dynasty that the keeping of carp for pleasure, such as described by Wen Wang, changed to the rearing of carp for food. The size of ponds were expanded and earnings from fish culture proved to be exceptionally profitable.

Over the next 70 years the methods established by Fan Li were further refined and expanded. Carp culture in "large water areas" was fully developed during the Han Dynasty (206 BC to 220 AD) and laws were enacted to protect the fisheries in the larger areas and to govern the use of large fish nets. The use of rice paddy fields for carp culture began during the Three Kingdom Era (221 to 277 AD), colored carp appeared during the Tsin Dynasty (265 to 419 AD), and methods to transport live fish were developed during the Era of Division between North and South (420 to 500 AD).

Laws passed during the T'ang Dynasty (618 to 917 AD) were of special significance in the culture of freshwater fish in China. During this period the carp was made a symbol of the royal family and, in effect, became a national treasure. Thus, if carp were taken, either intentionally or accidentally, they had to be released, and if they were taken and sold, the offenders were punished. Because of this action, commercial carp culture was no longer possible and growers in the coastal areas of the Yangtze and Pearl Rivers turned to the culture of grass carp, black carp *(Mylopharyngodon piceus)* and bighead *(Aristichthys nobilis)* for their income—probably the beginning of "polyculture" in China.

By the time of the Sung Dynasty (960 to 1280 AD), methods of collecting fish fry had already been developed in Kiangsi Province and the young fish were being regularly transported into the interior of Kiangsi as well as the neighboring provinces of Chekiang and Fukien. It was also during this period that culturists began experimenting with feeding and hybridization.

The system of polyculture widely practiced in China today was actually developed during the Ming Dynasty (1386 to 1644 AD) and much attention was given at that time to problems of sex and species composition, density of fish in a pond, and to methods of pond fertilization. However, for all intents and purposes, this completed the active development of the major aspects of freshwater fish culture in China. Authors have described the next 300 years, or until the end of the Civil War in the 1940s, as a period of stagnation, with some effort made to refine existing technology but without significant new discovery or radical change in culture methods.

Since 1949 a new interest in aquaculture has developed in China. Most critical was the need to provide adequate food for the people, and aquaculture, with all of its history and tradition, played a natural

role in such a program. The extensive development of water-use projects on all the major rivers was a most significant development and
increased by many times the productive capacity of freshwater fisheries in China. Domestic subsidies of funds and labor, foreign grants and
technical assistance have all contributed much to the reactivation and
success of the present aquaculture programs.

In China today numerous fish breeding centers and scientific laboratories have been established to assist the fish culturist in obtaining
a higher yield. Hormone-induced spawning has become an integral
part of the fish culture program; production is no longer dependent
upon the abundance of naturally occurring fry. New species have
been introduced (such as *Tilapia*) to form an important supplement
to aquaculture production. The aquaculture program has been expanded into the more remote areas of China to provide a welcome
addition to the diets of the people in those areas where fish have not
been regularly eaten. Perhaps most exciting has been the increase in
production of brackish water species, such as milkfish *(Chanos
chanos)* and mullet *(Mugilidae)*, and in the culture of the truly marine
species. It is quite possible that the greatest increase in aquaculture
production in the future will come from marine aquaculture.

THE AREA

China has a large number of rivers and more than 1500 of them
have watersheds of 1000 km² (380 mi.²) or more. Three of the
rivers (the Yangtze, Yellow and Amur rivers) are ranked among the
ten longest rivers in the world. In addition, there are 130 lakes that
exceed 100 km² (38 mi.²) in area. The control and use of water
has been an integral part of Chinese civilization from very ancient
times. Extensive dykes, irrigation systems, canals and other projects
have been constructed over the past 4000 years, contributing greatly
to the natural water resources of the country and providing added incentive for the development of freshwater fisheries and aquaculture
in China.

Similarly, China has a long and sinuous coast, extending for a total
of about 20,000 km (12,500 mi.) and bordered by 4 seas (Po'hai,
Yellow, East China and South China Seas). China's coastline may
be roughly divided into two types: north of Hangchow Bay the coast
is generally flat and is composed mainly of sand, mud and silt from
the rivers; south of Hangchow Bay, the mountains run close to the
shore and the coast is characterized by outcroppings of rock, islands
and large deep-water harbors. Although coastal fishing existed on a
fairly large scale as early as 770 BC, China's economy has histori-

cally and traditionally been closely linked with the development of agriculture and freshwater fisheries. Only recently has emphasis been placed on modernization and expansion of the coastal and offshore fisheries and marine aquaculture.

FRESHWATER AQUACULTURE[1]

For centuries Chinese fish farmers have learned to rear a variety of freshwater fish that naturally occur in their waters. The common carp, of course, still retains a position of "King of Fishes" in China and is a favored item for festive occasions. In terms of production, however, many of the other carp-like fishes grow more rapidly, are less susceptible to mortalities and cost less to rear than the common carp. For these reasons, the amount produced of grass carp, black carp, silver carp *(Hypophthalmichthys molitrix)* and bighead far exceeds that of the common carp. A summary of the characteristics of the more common freshwater fish cultured in China is given in Table 22.1

With the exception of the common and crucian carp *(Carassius carassius)*, the fish farmer in China has had to depend upon a supply of eggs and fry collected annually from the rivers and lakes. Because of natural fluctuations in the environment and perhaps other factors, the supply of eggs and the species composition varied widely both within the season and from year to year. The instability of supply has led to the development of a system of polyculture, mixing different groups of fish together according to feeding habits in order to make the best use of whatever fish were available and, if properly managed, utilizing the total productivity of the pond at all trophic levels.

The fish farmer would also receive from time to time young of the rarer species, either mixed with the other fry or for some reason or other especially abundant in the fry collections. Although some are highly predacious, they are frequently mixed with the other fish if their size and growth rates are compatible. Or, if the market value is high, they may be reared in separate ponds.

Pond Rearing

Although there is great effort being made at the present time to

---

[1] According to the *Yearbook of Fisheries Statistics, Catches and Landings, 1973* (United Nations, Food and Agricultural Organization, Vol. *36*, 1974) the estimated total inland, freshwater fish catch was 4.6 million MT. Estimates indicate that two-thirds of this amount or 3.0 million MT (3.3 million ST) were cultured. Including the large, but unknown, volume of brackish and marine cultured fish, China produces about two-thirds of the world's cultured fish.

Courtesy of K. Kondo, Tokai-ku
Fisheries Research Institute, Tokyo

FIG. 22.1.   HARVESTING FISH (SILVER AND BLACK CARP AND BIGHEAD) FROM
A LAKE FARMED BY THE WUHAN FRESHWATER FISHERIES COOPERATIVE

The lake has an area of 325,000 m^2 and a depth in the center of about 5 m. Much of the
fertilization is by cattle manure from nearby farms. Production in 1974 was 780,000 kg.

modernize fish culture practices in China, there is still much to be
done in the training of the traditional fish farmer in new techniques of
pond design, stocking rates, species balance, feeding and disease. There
are still many simple ponds, fertilized by an adjoining pig pen or with
the waste from other animals or ducks, trash from kitchen and gar-
den, and nightsoil. From time to time the ponds are cleaned, the rich
bottom soil placed on the neighboring fields, the pond sterilized by
leaves from the tea bush or other native plants, and restocked. Although these methods have produced fish for centuries, production
per pond area is low and generally inefficient by present standards.

Change is not easy but through the work of a series of newly established scientific laboratories, experimental stations and fish breeding centers, new methods of rearing freshwater fish have been developed and are gradually being adopted by the various communes, cooperatives and even the traditional farmer. The methods of pond
management reviewed here represent some of the aspects of the
"new look" in Chinese aquaculture.

Eggs and Fry

Traditionally Chinese freshwater fish culturists have obtained their

TABLE 22.1
FRESHWATER FISH COMMONLY CULTURED IN CHINA[1]

Species	Food Habits	Growth Age	Growth Length (cm)	Maturity Age	Fecundity (No.)	Egg Size (mm)	Egg Type	Hatching Time Hr	Hatching Time °C
Grass carp (Ctenopharyngodon idellus)	Herbivore (grass, aquatic plants, etc.)	1 5 7 10 max	20-25 56 75 84 122	3-4	100,000-800,000	2.2	Nonadhesive	32-40	26-40
Black carp (Mylopharyngodon piceus)	Shell animals (snails, mollusks, crustaceans, etc.)	1 2 3 max	22-32 45-50 5 kg 120 36 kg	6-7	400,000 or more		Nonadhesive		
Silver carp (Hypophthalmichthys molitrix)	Plankton (phytoplankton, algae, etc.)	2 3 4 5 7 max	30 47 52 55 64 100	3-4	400,000-600,000		Nonadhesive		
Bighead (Aristichthys nobilis)	Plankton (macroplankton, algae, etc.)						Nonadhesive		
Common carp (Cyprinus carpio)	Omnivorous (plant and animal forms, organic debris, etc.)	1 2 3 4 5 7 max	10-17 22 30 40 45 50 150	3-4	96,000-1,810,000	1.4	Adhesive	75	20
Crucian carp (Carassius carassius)	Omnivorous (zooplankton, algae and diatoms, insect larvae, organic debris, mud, etc.)	1 2 3 5 10 max	5 10-12 18 22-25 35 40	2-3	160,000-380,000	1.5	Adhesive	85-95 75	20-21 22-23

TABLE 22.1 (Continued)

Species	Food Habits	Growth Age	Length (cm)	Maturity Age	Fecundity (No.)	Egg Size (mm)	Type	Hatching Time Hr	°C
Bream (Wuchan fish) (*Parabramis pekinensis*)	Omnivorous (mainly grass and aquatic plants, algae, crustaceans, mollusks, etc.)	1 2 4 5 7 10 max max	8 14 22 25 33 37 55 60	4-5			Adhesive		
Black amur bream (*Megalobrama terminalis*)	Omnivorous (mainly algae and other aquatic plant forms, etc.)								
Mud carp (Dace) (*Cirrhina molitorella*)	Omnivorous (diatoms, algae, organic debris, mud, etc.)	1 2 3	8-9 20 25-30						
(*Elopichthys bambusa*)	Predacious (small fish, etc.)	1 2 3 4 5 7 max	25 35-40 42 55 71 75 200	4-5	1,800,000	7			
Walking catfish (*Clarias fuscus*)	Omnivorous	1	120 g	1.5-2		0.2	Adhesive	30	27-29
Japan eel[5] (*Anguilla japonica*)	Predacious (mainly aquatic animal material, dead or alive, etc.)					1.1	Nonadhesive	38	23
Taiwan dojo[2] (*Ophiocephalus maculata*)	Predacious (small fish, insects, macrozooplankton, etc.)	max	60				[6]	45	25

Species	Food	Age	Length/Weight		No. of eggs	Egg diameter (mm)	Egg type		
Snakehead (Ophiocephalus argus war-pachowskii)	Predacious (small fish, insects, macrozoo-plankton, etc.)	1 2 4 max	25 35 55 100	2-3	50,000-60,000	1.0	Nonadhesive	48	25
Mandarin fish (Siniperca chuatsi)	Predacious (small fish, insects, macrozoo-plankton, etc.)	1 2 3 max	12 20 25-30 5-6 kg	2-3	500,000	2.0	Nonadhesive	72	21-25
Milkfish[2,3] (Chanos chanos)	Vegetation (phytoplanton, algae, diatoms, some zooplankton, etc.)	1 2 max	20-40 40-50 180		3,000,000-7,000,000				
Grey or striped mullet[3] (Mugil cephalus)	Vegetation (phytoplankton, algae, diatoms, some zooplankton, organic detritus, mud, etc.)	1 2 5 max	18 35 55 80	2	50,000-60,000	0.7	Nonadhesive	48-120	
Northern Asian mullet[3] (Mugil so-iuy)	Vegetation (phytoplankton, algae, diatoms, some zooplankton, organic detritus, mud, etc.)	1 max	13-18 60	2-3					
Tilapia[2,3] (Tilapia mossambica)	Omnivorous (mainly aquatic plants, algae, diatoms, also macro-zooplankton, etc.)	⅓	17-22	¼	80-500+		[4]	72-120	

[1] Various measurements are best averages available and could differ significantly by environmental conditions (temperature, food, population density, etc.).
[2] Tropical fish and distribution generally limited to the warmer waters of South China (and Taiwan).
[3] Euryhaline.
[4] Tilapia are mouthbreeders, the male incubates the eggs in his mouth. The eggs are ovoid, 1.8 × 2.3 mm.
[5] Eels are catadromous, the young elvers captured for rearing when they enter fresh water to feed.
[6] Airbreathers, male builds floating nest of air bubbles and keeps the eggs in the nest and guards until young are free-swimming.

supply of eggs and fry from two sources: (1) from brood stock kept in ponds on the farm (common and colored carp, goldfish, tilapia, snakehead [*Ophiocephalus argus*], etc.) and (2) from wild stock naturally spawning in rivers and lakes (grass carp, black carp, silver carp, bighead, etc.).

Whether from pond or wild stock, spawning is triggered by rain, water temperature, stream flow, water depth, turbidity and other environmental conditions. The uncertainty of the weather has always been a critical problem for the fish farmer since it controls the frequency of spawning and the total number of eggs and fry available during the season, as well as the time of spawning in relation to the length of the growing season. Modern techniques of hormone-induced spawning have revolutionized fish farming in China by stabilizing the supply of eggs and fry.

Until recently the common carp (and the colored carp and goldfish) were the only species that could be spawned in artificial ponds. The breeder would first prepare a wintering pond 1 to 1.5 m in depth (up to 57 in.) by fertilizing with animal manure up to rates of 300 to 350 kg per 100 m² (6 to 7 lb per ft²) to produce and maintain a healthy plankton bloom. At the appropriate time (perhaps December) the brood fish would be placed in the pond at a stocking rate of about 150 kg per 100 m² (e.g., 100 fish averaging 500 g per fish). The fish are fed an artificial diet in addition to the natural food, adjusting the amount to produce a healthy fish, but not too fat a fish, since fat reduces the number of eggs that are developed in a female.

When the water begins to warm in the spring and the fish show signs of maturity, the brood stock is carefully selected for size, sex and condition and transferred to special spawning ponds 50 to 100 m² in area and 0.5 to 1.0 m in depth (538 to 1076 ft² and 19 to 39 in. deep). The sex ratio used by the breeders in China varies from 1 female to 1 to 3 males, but there are exceptions depending upon the breeder's experience and preference.

The eggs of the common carp are adhesive and, when deposited by the female, cling to aquatic plants or other materials. To collect the spawn the breeder simply suspends bunches of aquatic plants or other fibrous material in the pond, either from poles arranged along the bank or from pieces of wood or rafts floating on the surface and anchored to the bottom. After spawning, the spawn collectors with the eggs are transferred to special containers for hatching and rearing (frequently 10 to 50 m² in area and 15 to 60 cm in depth, or 108 to 540 ft² and 6 to 23 in. deep). During this period the fry are fed a variety of very fine foods (egg yolk, soybean liquor and curd, etc.), or even better, whatever plankton organisms might be available. This is a critical stage in the rearing of young carp and a mortality of 50% or higher is common. When the fry reach a length of about 2.5 or 3.0 cm (1 to 1.25 in.) they are transferred to rearing ponds or shipped to other areas.

Other fish are suitable for pond culture because of their unique breeding habits. One example is tilapia, which are mouth-breeders —the males collect the eggs, incubate them in their mouths and shelter the young until they are free-swimming and able to fend for themselves. Another example is the snakehead, which is an air-breather—the males build a nest of air bubbles on the surface of the water among the aquatic vegetation, place the eggs in the nest and guard them until the young are free-swimming, and the nests can be easily collected in a dipper and transferred to nursery containers for rearing.

Unlike the common carp, for many years the only source of supply for fry of grass carp and the other carp-like fish for culture had to be obtained from natural spawning that occurred in the lower reaches of most of the rivers in China. These spawning areas are located in the river channels where the currents and other characteristics are such as to attract the spawning fish. The areas are well-known to the professional fry collectors and, at the first indication of eggs or fry in the river, the traps are placed in the water below the spawning area to collect the fry. There are thousands upon thousands of traps that operate along the banks of the rivers and the operations of the fry fishermen, through generations of training and experience, are both practical and efficient.

The most common gear for collecting the young fry are floating traps, triangular in shape, with a round, semi-round or rectangular mouth, and discharging into a square pen of fine-meshed netting. The nets were formerly made of ramie fiber but are undoubtedly made of synthetic twine at the present time. As might be expected, there is a great variation in size and design of these traps. Generally the length is 4 or 5 m (13 to 16 ft), the mouth 1 to 4 m (3 to 13 ft) in width and depth, and a trap opening 10 to 15 cm (4 to 6 in.) depending upon the size of fry. The traps may be fished singly, in gangs of 4 to 7, with or without wings, and placed at the surface or at depths up to 2 m (6 ft). The traps are operated in the river either from anchored, floating bamboo poles or from poles driven into the bottom.

There are a variety of other types of gear used in the collection of fry, ranging from the common dip nets and push nets to beach seines and permanently installed traps. The species of fish and the spawning usually dictate the best gear to use. For example, mullet fry are usually collected from the estuaries and river mouths where there is little current and the conventional type of trap would be useless.

The nets take a variety of fry that are spawned in the river above. Although the fry can be identified by certain body characteristics, no attempt is usually made to sort the catches by species because of the small size and fragile nature of the fry. Fortunately, however, there are slight differences in the times of spawning for the various species which provides some species dominance in the catch. The size of the fry and the rate of growth also allow the catches to be partially separated by screening. The resulting collections, whether screened or not, are sold on the basis of a sampling of each lot. The farmer will balance the mixture in his ponds with older fish, if necessary, at a later date and when the fish are hardier and easier to handle.

Formerly, when a sufficient number of fry had been collected, the fry were sold to a middleman (a fry merchant) who would readily travel up and down the rivers contacting the various local fishermen. The middleman operated his business as a monopoly, more or less fixing the price he wished to extort from the culturists, and even on occasion releasing some of the fry when there was a surplus and the price was too low. The fry were usually sold in two basic units: a *wan* of about 10,000 fry or a *bowl* of 800,000 fry. Prices, of course, would fluctuate widely from season to season and within the season in accordance with the numbers available and the demand.

The middleman usually operated a series of nursery ponds in connection with his business, placing the fry in the ponds as soon as possible after collection. These ponds were about 6 or 7 m (13 to 16 ft) in length and breadth, relatively shallow and shaded in the summer to prevent overheating. The location was preferably near the seashore where brackish water could be mixed with the fresh water in order to reduce the mortality of the young fish.

In the vicinity of Canton the middlemen have specialized in rearing the fry and young fish to a length of about 15 cm (6 in.) before distribution. Over the years fish of this size have proven to be much hardier than the younger fish and can be easily shipped to fish farms throughout China and even distant countries, such as Malaysia, Singapore, etc. The ponds were usually 5 to 6 m long and 30 cm deep (16 to 20 ft long and 12 in. deep) arranged in a series separated by only a low dyke and with enough slope to allow some flow of water. If no gradient was available, the farmers would frequently have to resort to the use of a treadle pump or other means creating some circulation of water in the ponds.

These middlemen/growers sold the young fish in two basic sizes: 7.5 to 12 cm in length (3 to 5 in.) and 12.5 to 18 cm in length (5 to 7 in.). To separate the fish into different size groups, a screen made of bamboo slats is placed in the pond—the smaller fish can escape through the slats and the larger ones are retained for sale. The middle-man/grower frequently controlled the growth of the fish by feeding or crowding to obtain the size of best price in the market. Normally all fish were sold within 2 or 3 months of collection.

The shipment of the fry from the middleman to the farmer is made in a number of ways depending upon the distance involved, weather conditions, species, etc. Originally the containers were of rattan or bamboo, closely knit and treated to hold water. More recently, kerosene tins have been substituted for the knit baskets and have been used where the distances are short. For longer hauls the fry were frequently transported in boats, or "live barges," with holds so con-structed that water could circulate continuously through the tanks and without pumps. Because of the extensive network of canals and waterways in the plains area of China, transportation of fry by "live barge" was the standard method used for many years. The mortality, however, was still high and a loss of 50% was often anticipated by the middleman and the cost of such loss included in the delivered price.

At the present time, the distributors have adopted the familiar method of using plastic bags for transporting fry and young fish, fill-ing the bags with oxygen and placing them in a cardboard carton or other container for shipment. This method is commonly used by cul-turists throughout the world and has reduced markedly the fry mor-tality that was formerly experienced by the middlemen.

The most significant advance in the culture of freshwater fishes in China in recent years has been the development of hormone-induced spawning in grass carp, black carp, silver carp and bighead. The first success in the artificial propagation of carp was in 1958 when, with the assistance of Russian experts, silver carp and bighead were in-duced to spawn by injection of a pituitary extract from carp. Since then, the use of hormone-induced spawning for the cultured carp species has rapidly expanded to fish breeding centers in all the prov-inces and has gone far to increase and stabilize production of fresh-water fish farms.

A recent visit by a group of fisheries experts provides a good des-cription of the artificial breeding techniques now practiced at the Fish Breeding Farm near Shanghai (Anon. 1976).

> Brood stock are kept in separate ponds of 3 to 5 mu area[2]; the species are mixed in these ponds which contain only 10 to 15 brook fish per mu and they receive special care. Spawning of the family fish in this region occurs in May and June when water temperatures are increasing from 18° to 25° C (64° to 77° F). Silver carp spawn earliest followed by bighead, then grass carp and finally black carp. As a species matures, 5 or 6 females and 7 to 9 males

---

[2] 1 mu = 99.15 m^2 (approx 1066 ft^2).

are placed in a special cement-lined spawning pond, oblong in shape, 25 m (82 ft) in length, 10 m (33 ft) in width, 1.3 m (51 in.) in depth and with a good flow of water to simulate conditions of natural river spawning sites. To induce spawning, these fish are injected in mid-afternoon with pituitary extract from common carp, or (except for grass carp), a commercial preparation of gonadotropin. (The dosages used range from 2 to 8 mg of dried pituitary extract or 500 to 1000 IU of gonadotropin per kilogram of spawner). Spawning takes place the next morning about 4:00 or 5:00 AM. The fertilized eggs are collected in traps at the outlet end of the spawning pond and transferred immediately to special circular incubating tanks about 8 m (26 ft) in diameter. These tanks, constructed of concrete, consist of 2 concentric circular channels which are about 1.5 m wide and 1.5 m deep (57 in.). A good flow of water is maintained in the circular channels by injecting the inflow of water through a series of flanged nozzles protruding along the bottom and directed horizontally around the circle. An adequate flow (0.3 to 0.5 m per sec) is maintained to keep the eggs and developing fry in suspension. The capacity of this size of incubating tank is 30 million eggs (20 million in the outer channel and 10 million in the inner channel). The eggs hatch after one day in the incubation tank but are held there for an additional 6 or 7 days. By this time they are about 6 mm in length and are ready to begin feeding.

The modern fish farm in China consists of a series of specially designed ponds, each serving a certain function. The site of the farm is important and, where there is a choice, it is important that the land has sufficient slope to provide for a natural flow of water from pond to pond. Much attention is also given to water quality (mineral content, alkalinity, etc.) and to exposure to the weather (both sun and winter winds). Because of natural association with water, the selection of a site safe from inundation from floods is perhaps most difficult.

Most of the larger ponds are still made of dirt, with banks sloped to minimize erosion and to provide a surface for production of natural food for the fish. Where necessary, the banks of the ponds are strengthened with rock-work. At the present time an increasing number of ponds, particularly those used for spawning, incubation of eggs and the rearing of the very young fry, are being constructed of concrete.

Preparation of the ponds generally follows the traditional methods practiced by fish farmers for centuries. Ponds which have been in use for several years are cleaned during the winter months. Accumulation of organic material is removed from the bottom, and the pond is treated with lime, rotenone or one of the natural organic poisons (Camellia, Croton, etc.) to eliminate undesirable predators and other pests. The pond is then left exposed to the sun to "cure" before filling and stocking.

Probably the most difficult phase of freshwater pond culture concerns the proper application of fertilizers, in order to produce the

Courtesy of K. Kondo, Tokai-ku
Fisheries Research Institute, Tokyo

FIG. 22.2.   SERIES OF DIRT-BANKED PONDS AT CHINGPU PROVINCIAL
FRESHWATER FISH FARM

optimum concentration and species composition of plankton organisms to satisfy the food preferences of the various kinds of fish in the pond. Each pond differs in water quality, mineral balance and naturally occurring nutrient chemicals and materials. The kinds of fertilizers applied are related to the type of natural food eaten by the fish. The amount of fertilizer needed depends not only on the stocking rate and kinds of fish in the pond, but on weather conditions (i.e., temperature, sunshine, amount of rain, etc.). These same weather factors, in turn, affect the growth and feeding habits of the fish. Too much fertilizer can easily produce a super-abundance of plankton which will deplete the oxygen supply and kill all the fish in the pond. There have been endless experiments conducted in China on the proper rates of fertilization of the ponds and the ultimate yield in terms of fish. There is still much to be learned.

Through trial and error the Chinese fish farmer has developed a feel for pond fertilization, based on color and transparency of the water and the growth and behavior of his fish. Most fish farmers still prefer to use organic materials, including pig and cow manure, droppings from ducks or other fowl, nightsoil, rice bran, soybean and peanut meal, and organic compost. Ponds rich in nutrients (or heavily fertilized) are suitable for such fish as silver carp, bighead, common carp, crucian carp and tilapia, and especially for duck-*cum*-fish and hog-*cum*-fish farms. Ponds low in nutrients, and an abundance of grass and weeds from nearby fields, should be used predominately for the culture of grass carp. Predator fish (snakehead, mandarin fish, *Elopchthys bambusa*, etc.) should be used sparingly in a mixed culture pond and provided with suitable forage fish (*Xenocypris* sp., etc.) (Chen 1973).

In the last ten years, much attention has been given to the use of superphosphates and other inorganic fertilizers to supplement the organic fertilizers or to serve as the sole source of added nutrients. Although the results of experiments show the added production that can be obtained from a pond by using inorganic fertilizers, the Chinese fish farmer still prefers to use organic fertilizers, except in those areas where the supply is limited.

Most of the fish farmers in China plan to harvest their fish by the end of the second year, allowing time to drain their ponds and prepare them for the next season's fry. The very fast growing fish, such as tilapia, are marketed the year round. Silver carp are marketed as soon as they reach the weight of about 500 g (18 oz). The remaining species are usually not ready for marketing until fall, or are even held until the New Year to take advantage of the higher demand and price.

With very few exceptions, polyculture of freshwater pond fish is almost universally practiced in China at the present time and there is almost an infinite "mix" of species available to the fish farmer. There are many factors that influence his choice of "mixes," generally very practical. For example, his decision is based on the kinds of fry available, past successes and failures, projected market demand, various inducements from the commune or the government, etc. Basically, of course, the farmer is concerned with creating the right balance in his ponds, both between species competing for similar food and groups of fish with different food habits, so that he will utilize the entire productive capacity of a pond. Examples of a variety of mixes are given in Table 22.2.

The same is true in Taiwan. For example, the stocking rate (fry per hectare) for a freshwater pond in central Taiwan is given as: 800 silver carp (7 to 12 cm or 3 to 5 in.); 100 bighead (7 to 12 cm or 3 to 5 in.); 50 grass carp (7 to 12 cm or 3 to 5 in.); 1000 mud carp (5 cm or 2 in.); 2000 mullet (5 cm or 2 in.); and 1000 common carp (2.5 cm or 1 in.). All fry were put into the pond in March or April. In southern Taiwan the stocking rate (fry per hectare) was: 1000 silver carp (10 to 13 cm or 4 to 5 in.); 400 bighead (10 to 13 cm or 4 to 5 in.); 200 grass carp (12 to 15 cm or 5 to 6 in.); 500 mud carp (7 to 10 cm or 3 to 4 in.); 2000 common carp (3 to 4 cm or 1 to 1.5 in.); 2000 crucian carp (3 cm or 1 in.); 2000 mullet (5 cm or 2 in.); 500 walking catfish (5 cm or 2 in.); and 500 snakehead (10 cm or 4 in.). Most of the fish

TABLE 22.2

SPECIES, SIZE AND RATE OF STOCKING
PER 100 M² OF A POND OF MIXED SPECIES

Grass Carp		Black Carp		Silver Carp		Bighead	
Size (cm)	No.	Size (cm)	No.	Size (cm)	No.	Size (cm)	No.
15-18	2,000-4,000			11-12	8,000-12,000		
15-18	4,000-8,000			8-9	20,000-25,000		
8-12	10,000-25,000			11-12	4,000-5,000		
15-18	2,000-4,000					11-12	8,000-12,000
12-14	4,000-6,000					8-9	15,000-20,000
8-12	10,000-25,000					11-12	4,000-5,000
		12-14	5,000-6,000	12	4,000-5,000		
		14-15	5,000-6,000			[1]	200-250
		12-14	5,000-6,000			11-12	4,000-5,000
		9-12	10,000-15,000			[1]	200-250
15-18	2,000-4,000	12-14	400-600	11-12	8,000-12,000		
12-14	2,000-4,000	12-14	400-600			11-12	8,000-12,000
8-12	10,000-25,000			[1]	50-60	[2]	150-200

Source: Anon. (1973).
[1] Number not given, weight 500 to 750 g per fish.
[2] Number not given, weight 500 to 625 g per fish.

were introduced in February and March, walking catfish in May and snakehead in June (Chen 1973).

Rice Paddy Culture

Since the early 1950s, the government has encouraged cooperatives and communes to culture fish in rice fields, pointing out the multiple benefits of such ventures. For example, in addition to the added yield from the fields, the fish would eliminate mosquitoes and other insect pests, cultivate the plants by their constant digging, and provide fertilization from their feces. Originally common carp and the related golden carp were used in the paddies, but more recently silver carp, bighead and snakehead are proving profitable. Grass carp is not recommended until after harvest and the paddy is lying fallow.

The deep water method of culture is used in the northern provinces of Szechwan, Kweichow, Kwangsi and Hupeh, where the depth of water in the rice fields may be as much as 0.5 to 0.8 m (19 to 31 in.). The growth of fish in the paddy fields is good; fry about 8 cm in length (3 in.) will weigh 250 to 500 g (9 to 18 oz) within 3 months. The yield of fish from a paddy is reported to be 35 to 75 kg per 100 m² (0.8 lb per ft²).

The shallow water method is used in Kwangtung and other southern provinces, where the depth of water in the rice fields may be only 6 to 8 cm (3 in.). To allow the fish space to swim and hide, several ditches are dug in a criss-cross pattern in the field, about 20 cm in depth (8 in.) and with pockets 0.6 to 1.0 m deep (23 to 39 in.). Stocking at the recommended rate of 1000 to 1500 fish per 100 m² (0.9 to 1.4 fish per ft²) will produce fish 10 to 15 cm in length (4 to 6 in.) by the time the rice is ready for harvest (Solecki 1966).

Statistics on yield of fish from rice paddies are not known but the amount must be considerable for mainland China. In 1958 one report stated that about 10% of all rice paddies in China were also being used for the culture of fish. After the pressures of the Great Leap Forward and the Cultural Revolution, the amount is certain to have increased.

Open Water Culture

Production of fish in lakes and rivers, or "fish ranching," has been conducted in China for at least 100 years. Many of the rivers and lakes were owned by the local land lords and held for their own use and development. The yield in 1936, a peak year, was 25 to 30% (about 370,000 MT) of the total landings. Production from fish culture would probably account for 8 to 10% of the amount and the remainder would be from conventional fisheries on the rivers and lakes.

There was little attempt in the earlier years to farm these natural bodies of water. There were regulations, of course, either by the private owners or in a few instances by the government. Unique, perhaps, was the utilization of parts of the canals for fish culture. Weirs were installed in the channels, of bamboo or other material that could be opened to let the boats pass up and down the canal. These fenced areas were stocked with fish which in due time were caught and marketed. The canals (as well as the rivers and flowing lakes) could not be fertilized because of rapid dilution. With the exception of grass carp, the use of artificial foods was limited.

However, the Water Conservation Program of the Great Leap Forward (1958-60) placed renewed emphasis on the fishery production of the inland waters. During the first 2 years, 1958-59, some 47 large reservoirs, 3000 large irrigation projects, and 26,000 small and medium-sized projects were completed—many by local cooperatives and communes and without the use of heavy equipment.

There were many problems, however, in the realization of full reservoir use. Many of the reservoirs were in mountainous areas where little fishing or fish culture had taken place in the past. Also, during construction little thought was given to the needs of the fishing in-

Courtesy of K. Kondo, Tokai-ku
Fisheries Research Institute, Tokyo

FIG. 22.3.   A BAMBOO FISH FENCE IN A CANAL SHOWING A CENTRAL
OPENING, FLEXIBLE AT THE TOP, WHICH ALLOWS BOATS TO SLIP THROUGH

dustry and the bottoms were left with stumps and other debris, restricting the choice of gear that could be used. The first years after
completion of the program were disappointing for fisheries.

However, in time small areas were cleared in some of the reservoirs to allow fishing with seines and traps. More important were the
attempts to stock the reservoirs with fry in order to increase production from the inland waters. The recommended stocking rate was
100 to 150 fish (12 to 14 cm in length, or 5 to 6 in.) per 100 m² (1076
ft²). Grass carp, silver carp and black carp so introduced were expected to reach an average weight of about 5 kg (11 lb). The yield
from these early attempts was disappointingly low, on the average
about 8 kg per ha (7 lb per acre) (Solecki 1966).

It would appear from recent reports that many of the difficulties
have been overcome. For example, present annual rates of production
are: 900 to 1800 kg per ha (800 to 1600 lb per acre) in screened-off
sections of rivers and canals; 300 to 450 kg per ha (267 to 400 lb per
acre) in natural lakes; and 180 to 300 kg per ha (160 to 267 lb per acre)
in reservoirs (Anon. 1974).

Other Freshwater Culture

During the past 20 years the Chinese have conducted a number
of experiments with the introduction and culture of various freshwater

Courtesy of K. Kondo, Tokai-ku
Fisheries Research Institute, Tokyo

FIG. 22.4.   HARVESTING FISH FROM ONE OF THE PONDS AT CHINGPU
PROVINCIAL FRESHWATER FISH FARM

species. Tilapia were introduced into mainland China from Vietnam
in 1958 and are now extensively cultured. Although elvers have been
trapped and exported to Japan since 1968, eel culture on a commercial
scale was established in mainland China about 1970. Rainbow trout
from North Korea and bullfrogs from Cuba have been introduced into
mainland China but the success of the transplants is not known. There
no doubt have been other attempts to introduce new species into
China (various kinds of oysters, shrimp, etc.) that have escaped
attention in this review. Of interest to culturists are the experimental
studies on the rearing of freshwater clams at Chinphu Fish Breeding
Farm near Shanghai for freshwater pearl production and to supply
food for eels and black carp.

MARINE AQUACULTURE

Development of marine aquaculture in China has not progressed
as rapidly as the program for freshwater fisheries. There are three
basic reasons for this: the long history and experience of the Chinese
fish culturists in freshwater fisheries; the close association with the
Water Conservation Programs of 1958-60 and 1967-69; and the com-

mon interests of the agriculture cooperatives and communes in farming and fish culture. The history of marine aquaculture began about 300 years ago in the Pohai Sea area but expansion was limited by the availability of suitable land.

The potential of marine aquaculture in China is enormous, with some 20,000 km (32,200 mi.) of coastline, waters relatively rich in nutrients, and a combination of protected bays and channels to the south and low, flat coastal shorelands to the north. The total marine area suitable for culture has been estimated by several sources to be between 450,000 and 900,000 ha (182,000 to 364,000 acres), of which 60,000 ha (24,300 acres) were utilized in 1957, 120,000 ha (48,600 acres) in 1958 and 250,000 ha (617,000 acres) by 1959 (Kenji 1962).

Although considerable study by various research laboratories on the life histories and methods of culture of various marine seaweeds, fish and shellfish was undertaken in the early 1950s, it was not until the Changchiang Conference of 1959 that a definitive program of marine aquaculture was adopted. The conclusions and recommendations reached at this conference included six main points. (1) Some important species can be cultured only in salt water. (2) Marine culture does not infringe upon the use of agricultural land. (3) Production from marine aquaculture is more stable than that from freshwater farming and provides a more stable supply of food to the consumer. (4) Seawater is rich in natural food and simplifies feeding with artificial foods. (5) Marine fish fry are abundant and can be easily collected and shipped. (6) Seawater is continually being moved by action of the tides and wind and the supply of nutrients and oxygen is constantly being replenished. As a result, 10 to 15 large culture farms were built by the government for growing seaweed, oysters, sea cucumbers, mullet, shrimp, scallops, crabs, octopus and mussels (Solecki 1966).

Although the descriptions are not complete, there are apparently four types of marine culture: dyked ponds, fixed net enclosures, floating pens or rafts, and open shore culture.

The fish farms built along the Pohai Sea and in Kwangtung Province provide examples of dyked ponds. These farms consist of a series of ditches built in shallow, natural basins or in the intertidal zone of a suitable beach. The ponds are arranged with a main ditch about 6 m wide and 1.2 m deep (20 ft wide by 4 ft deep) and a series of side ditches separated only by low banks or dykes. The water exchange is controlled by a main gate, 2.5 to 3 m wide (8 to 9 ft) and 1.5 m (5 ft) high, and smaller flood gates at the entrance of each side ditch. The seaward bank of the farm is usually 65 to 90 cm (2 to 3 ft) higher than the maximum tide level. The ponds are usually rebuilt every spring and before stocking, seawater is allowed to circulate freely through the system to thoroughly clean the ponds and stabilize the salt content of soil. After several cleanings, the ponds are stocked with gravid shrimp, egg-carrying crabs or fish fry, which are allowed to grow until harvesting in the fall.

Raft culture is basically the same as used in Japan, consisting of a series of bamboo poles, tied together by rope or wire in a rectangular shape, and supported by floats of wood, styrofoam or oil drums. For culture of the various species, ropes are suspended from the bamboo poles with spat or spore collectors fixed to each rope at regular intervals.

Of course, the use of either natural or supplemental food is limited to ponds or pens where the fish are confined and where the food or nutrients are not rapidly dissipated by tidal currents. Organic fertilization (including a wide array of manures, vegetable and animal meals, etc.) is an integral part of the successful culture of milkfish and other marine species and must be used to provide an unidentified "life factor" that helps maintain the natural qualities of seawater. If fresh seawater or organic fertilizers are not added from time to time, the ponds soon become sterile and nonproductive (Lin 1968).

Care must also be taken in the use of inorganic fertilizers. Surprisingly, the application of superphosphates or N-P-K fertilizers has little effect, if any, on increasing the production of milkfish. In fact, the application of these types of inorganic fertilizers may have a detrimental effect upon the production of plankton and bottom algae in saltwater ponds by stimulating the growth of nano-plankton organisms. These organisms are too small to be utilized by milkfish and multiply rapidly to cause turbidity, an oxygen deficiency and high mortality to the fish (i.e., the so-called "yellow water").

On the other hand, it appears that silica is a limiting factor in milkfish production and the application of zeolite (72.95% SiO_2) will help to keep the water clear and maintain the bottom "algal pastures" during the period that milkfish are cultured in ponds.[3]

The practice of flooding the milkfish and other marine culture ponds with seawater three times a year, allowing the water to evaporate and then filling and leaving stagnant to preserve the accumulated nutrients is a very effective way to increase natural productivity of the culture ponds.

There is little information available on the composition of diets

---

[3] Lin (1968) states: "A preliminary experiment carried out in 1966 indicates the possibility that silica may be required for the maintenance of bottom algal pasture throughout the culture period and to keep the water clear to allow a better chance for algae to grow. Because of the clearing effect, the plague of "yellow water" may be reduced to the minimum.

and methods of supplemental feeding practiced in mainland China. Several references (e.g., Kenji 1962) note the use of supplemental feeding in the culture of marine species but the details are lacking. In Taiwan, milkfish have been found to grow faster on supplementary food than on the natural bottom algae. Furthermore the best growth was obtained from diets high in flax seed and peanut cake, and not rice bran usually used by milkfish farmers (Lin 1968).

Unfortunately there are no national production statistics for fisheries from mainland China during recent years. In 1957 the production from mariculture was reported as 430,000 MT (473,860 ST) and for 1959, 1,000,000 MT (1,102,000 ST) (Kenji 1962). There are also numerous references to the relatively slow growth of production from marine farms, certainly significantly less than the growth of production from the freshwater fish culture[4]; kelp and certain shellfish are the exceptions. Although impossible to confirm, a good estimate of mariculture production at present might be 1.5 million MT (1,653,000 ST), or possibly 2 million MT (2,204,000 ST) at the most. However, it should be kept in mind that much progress has been made by scientists and culturists at the several research laboratories and stations in developing an understanding of the marine species and the necessary technology for an expanded mariculture program. It is in this field of aquaculture that we can expect the greatest advance by China in the future.

SPECIAL ACKNOWLEDGEMENTS

DR. T. ABE, Tokai Regional Laboratory, Japan

DR. T. KONDO, Tokai Regional Laboratory, Japan

DR. T. NISHIYAMA, University of Alaska, Fairbanks, Alaska

REFERENCES

ANON. 1960. Chukoku Tansui Gyorui Yoshoku Gaku (Freshwater Fish Culture of China). China Freshwater Fish Culture Joint Committee Compilation (translated into Japanese by T.Shuh). (Japanese)

ANON. 1973. Chungkuo Tanshui Yülei Yangshi (Freshwater Fish Culture of China). China Freshwater Fish Culture Joint Committee Compilation (2nd Edition). (Chinese)

ANON. 1974. China—A Geographical Sketch. Foreign Language Press, Peking.

---

[4] The rapid advance in freshwater aquaculture during the first post-civil war years is attributed to the long history and tradition of freshwater fish culture, the natural association with agricultural cooperatives and communes, and the drive to fulfill the goals of the Water Conservation Programs of 1957-59 and 1968-69.

ANON. 1976. Final Report—Visit of the Canadian Fisheries Mission to the People's Republic of China, November 22-December 9, 1974. Fisheries and Marine Science, Ottawa.

CHEN, T.P. 1973. II. Polyculture of Chinese Carps in Freshwater Ponds. Chinese-American Joint Commission on Rural Reconstruction.

DREW, R.A. 1951. The cultivation of food fish in China and Japan: A study disclosing contrasting national patterns for rearing fish consistent with the differing cultural histories of China and Japan. Ph.D. Dissertation. University of Michigan.

KENJI, ASAKAWA. 1962. Fishery production and policy in Communist China. US Joint Publications Research Service, February 1, 1962 (JPRS: 12252).

KONDO, K. 1975. Chulolu Ryokoki (An Account of Travel in China). Tokai Regional Research Laboratory, Japan Fisheries Agency, Contributions C Separate 15, pp. 51-58.

LIN, S.Y. 1968. Milkfish farming in Taiwan—A review of practice and problems. Taiwan Fisheries Research Institute, Fish Culture Report No. 3, February 1968.

LIN, S.Y. 1968. Pond fish culture and the economy of inorganic fertilizer application. Chinese-American Joint Commission on Rural Reconstruction. Fisheries Series No. 6, June 1968.

PRYBYLA, J.S. 1970. The Political Economy of Communist China. International Textbook Company, Scranton, Pa.

SASAKI, TADAYOSHI. 1975. Japan oceanography, fisheries, ocean development exchange visit to China group. La Mer (Bulletin de la Societe franco-japonaise d'oceanographie) 13, No.2 (May), 91-108.

SOLECKI, J.J. 1966. Economic Aspects of the Fisheries Industry in Mainland China. Institute of Fisheries, University of British Columbia.

Chapter 23

Taiwan

Dr. Hsi-Huang Chen

Taiwan, with a group of over 70 small islands, lies between the East China and South China seas. The main island is subtropical and extends nearly 390 km (240 mi.) north and south across the Tropic of Cancer. The maximum width is less than 150 km (about 90 mi.).

Its total area is 35,980 km² (13,890 mi.²), but about two-thirds is highly mountainous. It has a coastline of over 1600 km (994 mi.). Along the eastern coast the deep waters abutting precipitous cliffs form a favorite highway for migratory fish from both the north and south. The gradually inclined shelf on the west abounds in marine resources and provides excellent grounds for the habitation and propagation of many species of fish.

Taiwan is roughly comparable in size to Belgium or the Netherlands, but with less arable land and more people. This small island supports a population of more than 16 million people, increasing at an annual rate of 1.8%. However, Taiwan's agricultural and fisheries production has increased rapidly enough to stay comfortably ahead of the domestic demand for food while at the same time steadily increasing exports of surplus commodities.

Fisheries production rose from a record catch of 119,520 MT (131,-472 ST) in 1940 to 779,825 MT (857,808 ST) in 1975. This increase amounted to 552%. This rate of increase is second only to that of Peru.

Fisheries in Taiwan are classified, for statistical purposes, into four categories, on the traditional basis of type of fishing craft and gear used and the relative distance of fishing areas. These categories are deep-sea, inshore, coastal and culture. According to 1975 statis-

TABLE 23.1

FISHERIES PRODUCTION, TAIWAN, 1975

Category of Fisheries	Production (MT)	(ST)	Value ($000)	Quantity (%)	Value (%)
Deep-sea	326,707	360,031	131,664	41.9	28.6
Inshore	295,920	326,104	137,949	37.9	29.9
Coastal	29,644	32,668	15,917	3.8	3.5
Culture	127,554	140,505	175,384	16.4	38.0
Totals	779.825	859,367	460,914	100.0	100.0

Source: Taiwan Fisheries Bureau, Provincal Government of Taiwan.

tics, the total catch of 779,825 MT was: (1) deep-sea, 41.9%; (2) inshore, 37.9%; (3) coastal, 3.8%; and (4) culture, 16.4%.

Fish culture ranked third in volume but first in value. Culturing, with 16.4% of production, accounted for 38.0% of value, deep-sea for 28.6% of value, inshore for 29.9% and coastal for 3.5%. The value of cultured production in relation to volume indicates the culture of highly valued species (Table 23.1).

Cultured production in 1975 totaled 127,554 MT (140,565 ST), which was an increase of 11.4% over 1974. With respect to value, fish culture production totaled $175,384,000 in 1975, an increase of 13.6% from 1974 (Table 23.2).

In 1975 culturing included: (1) brackish water ponds, about 18,-798 ha (46,431 acres), chiefly for milkfish but including some mullet and shrimp; (2) shallow sea waters for oysters and clams, about 13,480 ha (33,296 acres); (3) freshwater ponds, about 12,005 ha (29,-652 acres), and pools and reservoirs about 9163 ha (22,633 acres), for raising carp, tilapia, eel and mixed culture; (4) paddy fields in central and southern Taiwan for raising tilapia. These paddy fields covered about 115 ha (284 acres). Total water area devoted to culture in 1975 was 53,561 ha (132,296 acres) (Table 23.3).

The western coast of Taiwan, extending from Tanshui in the north to Tungkang in the south with a total length of 400 km (240 mi.), is characterized by tidal lands having sandy bottoms exposed above water at low tide. This area is utilized by coastal villagers for culture of shellfish and seaweeds.

The brackish water ponds are constructed on the tidal lands for monoculture of milkfish, mullet and shrimp. They represent an ecosystem which is basically different from that of freshwater ponds.

Shallow water culture is usually conducted by coastal households on a part-time basis. Thus the average size of each unit is small. The average cultured acreage per household in 1972 was 0.78 ha

TABLE 23.2

FISH CATCH, CULTURED PRODUCTION AND VALUES, TAIWAN

Year	Totals (MT)	(ST)	Value ($000)	Cultured Fish Industry (MT)	(ST)	Value ($000)	Cultured Fish % of Total Quantity	Value
1941	85,336	94,040	1,241	12,338	13,596	148	14.46	11.89
1946	51,474	56,724	33,492	9,970	10,987	9,838	19.37	29.39
1951	104,180	114,806	11,650	24,966	27,513	3,489	23.96	29.95
1956	193,410	213,138	31,917	42,480	46,813	8,136	21.96	25.49
1961	312,439	344,318	66,957	57,354	63,204	18,535	18.36	27.68
1966	425,326	468,709	102,549	58,515	64,484	20,855	13.76	20.34
1971	650,188	716,507	221,495	77,789	85,723	46,448	11.96	20.97
1972	694,330	765,152	282,379	81,236	89,522	69,644	11.70	24.66
1973	758,484	835,849	377,475	107,489	118,453	108,550	14.17	28.76
1974	697,871	769,054	405,772	114,472	126,148	128,290	16.40	31.62
1975	779,825	859,367	460,914	127,554	140,565	175,384	16.36	38.05

Source: Taiwan Fisheries Bureau, Provincial Government of Taiwan.

TABLE 23.3
AREA[1] USED FOR FISH CULTURE BY TYPE OF CULTURE, TAIWAN

Year	Brackish Water Pond	Shallow Sea Culture	Fresh-water Pond	Paddy Field	Reservoirs and Others	Total
1956	14,178	5,704	4,938	7,328	7,400	39.547
1961	17,095	9,743	4,938	927	7,552	40,254
1966	15,587	9,822	5,336	123	7,261	38,129
1971	16,461	11,877	8,094	55	6,851	43,338
1972	16,744	12,943	10,275	26	7,180	47,167
1974	17,137	13,151	11,686	128	7,818	49,920
1975	18,798	13,480	12,005	115	9,163	53,561

Source: Taiwan Fisheries Bureau, Provincial Government of Taiwan.
[1] Figures indicate hectares. To determine acres multiply by 2.47. For example, 115 ha is 284 acres.

(1.9 acres). In general, 80% of the shallow water culturing units are less than 0.5 ha (1.25 acres) and are part-time enterprises. Those households with over 0.5 ha are either full-time enterprises or consider culturing a major sideline.

The freshwater ponds on the coastal plain are richer in minerals and are generally more productive than those on the higher levels in the interior. Many freshwater ponds are polluted by sewage from villages and cities. *Tilapia mossambica* constitutes the principal biological unit of the ecosystem in coastal plain ponds. In the same ponds silver carp, grass carp, goldfish, common carp, gray mullet, eel and milkfish are also stocked.[1]

The pools and reservoirs lie some distance from the coast at altitudes of 20 to 500 m (65 to 1620 ft) above sea level. They are chiefly for carp production. Also grown are silver carp, bighead, goldfish, grass carp, gray mullet, catfish and perch.

The eel ponds lie either along the coastal plain or in the interior close to the hill area.

The ecosystem of individual ponds never remains the same. Because of this, fish production varies considerably from year to year and from pond to pond. For example, ponds or reservoirs on the high land in the interior may produce only 100 kg per ha per year (89 lb per acre) while those constructed on alluvial soil on the coastal plain may yield more than 500 kg per ha (445 lb per acre) annually without fertilization or supplementary feeding. A coastal pond under

[1] Silver carp *(Hypothalmichthys molitrix)*, grass carp *(Ctenopharyngodon idellus)*, common carp *(Cyprinus carpio)*, gray mullet *(Mugil cephalus)*, eel *(Anguilla japonica)*, milkfish *(Chanos chanos)*, bighead *(Aristichthys nobilis)*, and tilapia *(Tilapia mossambica)*.

identical treatment could yield more than 2000 kg per ha per year (1781 lb per acre).

In 1975 the yield from fish culture amounted to 127,554 MT (140,-565 ST) valued at $175,384,000. Milkfish was predominant with production of 33,309 MT (36,707 ST) valued at $33,778,000. Tilapia ranked second with 18,260 MT (20,123 ST) valued at $6,262,600; then carp with 17,419 MT (19,196 ST) valued at $10,633,952; and eel with 13,575 MT (14,960 ST) valued at $83,753,315 (Table 23.4).

Also cultured were 13,850 MT (15,263 ST) of oysters, valued at $20,843,501, and 12,481 MT (13,754 ST) of clams valued at $7,267,-639.

MILKFISH (Chanos chanos)

It is generally believed that the culture of milkfish in Taiwan dates back to the Ching Dynasty about 300 years ago. The Chinese fish farmers who migrated to Taiwan built dykes on the low land along the coast of the southern prefecture of Tainan and stocked these ponds with milkfish fry obtained from the littoral waters. About 1910 a fish culture station was established in Tainan, Taiwan to conduct experiments in milkfish culture. At present milkfish farming is very important in the cultured fish industry. With the use of chemical

Courtesy of Dr. Hsi-Huang Chen

FIG. 23.1.   HARVESTING CULTURED MILKFISH

TABLE 23.4

VOLUMES¹ AND VALUES² OF FISH CULTURE PRODUCTION BY MAJOR SPECIES, TAIWAN, 1975

Species	Total		Brackish Water Pond		Freshwater Pond		Shallow Sea Culture		Paddy Field		Reservoirs and Others	
	Quantity	Value	Quantity	Value	Quantity	Value	Quantity	Value	Quantity	Value	Quantity	Value
Carp	17,419	10,633	—	—	12,803	8,010	—	—	95	51	4,521	2,572
Tilapia	18,260	6,263	4,355	1,279	12,565	4,552	95	29	110	38	1,044	365
Eel	13,575	83,754	17	95	13,516	83,400	—	—	1	2	41	257
Other fresh-water fish	3,466	2,323	8	4	2,784	1,960	—	—	60	41	614	318
Milkfish	33,309	33,778	33,164	33,631	145	147	—	—	—	—	—	—
Mullet	1,355	1,932	185	313	1,141	1,576	29	43	—	—	—	—
Other brack-ish water fish	1,346	207	1,189	182	38	7	115	18	—	—	4	0.3
Lobster, prawn and shrimp	775	1,672	684	1,564	85	104	—	—	—	—	6	4
Oyster	13,850	20,844	3	4	—	—	13,847	20,839	—	—	—	—
Hard clam	12,481	7,268	1,570	836	—	—	10,911	6,432	—	—	—	—
Freshwater clam	1,375	577	—	—	604	305	—	—	—	—	771	252
Other shell-fish	3,238	3,143	197	806	22	11	2,905	2,301	—	—	114	25
Total	120,449³	172,374³	41,372	38,714	43,794	100,072	27,902	29,662	266	132	7,115	3,793

Source: Anon. (1976).
¹ Quantities given in metric tons. For short tons multiply by 1.1023.
² Quantities given in thousands of U.S. dollars, computed at the rate of 37.7 Taiwanese dollars to each U.S. dollar.
³ Differs slightly in volume and value from data given in Tables 23.1 and 23.2.

fertilizers and pest control measures yields as high as 5700 kg per ha (5096 lb per acre) have been reported. Average yields are over 2000 kg per ha (1781 lb per acre).

Milkfish ponds are located in areas where: (1) the temperature is above 15° C (59° F) for eight months of the year; (2) there is little likelihood of flooding; (3) the pond water is not likely to be seriously diluted during the rainy season; and (4) fresh water is available to adjust the pond water to a salinity of not over 50% during the dry season.

The main problem in milkfish farming in Taiwan is high mortality in the wintering ponds. The causes of high mortality are (1) low temperatures and (2) oxygen deficiency. To avoid chilling of the waters by the strong winter winds, milkfish farmers construct wintering ponds with narrow ditches between 1 and 2 m (39 to 78 in.) in depth and protected on the windward side by windbreaks. The milkfish are placed in the wintering ponds in October and held there until March, when they are transferred to the rearing ponds.

Harvesting begins at the end of May. About eight partial harvests of market-size fish are made before the middle of November. Then the rearing ponds are drained of water after netting and any fish left over are picked up.

The fish are packed in bamboo baskets, loaded onto trucks and taken to the market. Crushed ice is spread on top of each basket to maintain quality. Milkfish are sold in the domestic market only. Farm prices in 1975 average $1.01 per kg ($0.46 per lb). Total production was 33,309 MT (36,707 ST) (Table 23.5).

TABLE 23.5

VOLUMES AND VALUES OF FISH CULTURE PRODUCTION
BY MAJOR SPECIES BY KILOGRAMS AND POUNDS, TAIWAN, 1975

Species	Quantity Harvested (000s)		Value	
	(MT)	(ST)	($ per kg)	($ per lb)
Carp	17,419	19,196	0.61	0.28
Tilapia	18,260	20,122	0.34	0.15
Eel	13,575	14,961	6.17	2.80
Other freshwater fish	3,466	3,820	0.67	0.30
Milkfish	33,309	36,707	1.01	0.46
Mullet	1,355	1,493	1.42	0.65
Other brackish water fish	1,346	1,483	0.15	0.07
Lobster, prawn shrimp	775	854	2.16	0.98
Oyster	13,850	15,263	1.50	0.68
Hard clam	12,481	13,754	0.58	0.26
Freshwater clam	1,375	1.515	0.41	0.19
Other shellfish	3,238	3,568	0.97	0.44
Total	120,449	132,736	—	—

Source: Computed from Table 23.4.

TILAPIA *(Tilapia mossambica)*

Tilapia were first introduced to Taiwan from Singapore in 1974. They are raised in brackish and freshwater ponds as well as paddy fields. They have become one of the most important cultured fish in Taiwan, ranking second in volume only to milkfish in 1975. They are, however, a low-valued fish. Farmer prices in 1975 were only $0.34 per kg ($0.15 per lb) (Table 23.5).

In Taiwan tilapia start to spawn at four months. The number of eggs from each spawning increases with the age and size of the fish, varying from 100 to over 1000. The number of spawnings by one fish in one year ranges between 6 and 11 in southern Taiwan. The interval between spawnings is generally 22 days. Optimum water temperatures for spawning range from 20° to 35° C (68° to 95° F).

One of the most interesting developments is the utilization of wastes for tilapia farming. For example, there are about 30 ha (74 acres) of fish ponds situated along drainage canals in Taiwan which use sewage water from the canals to supply nutrients and food for the fish. During the winter tilapia are moved into wintering ponds. At this time rearing ponds are thoroughly dried. Then the sewage water is let into the ponds and allowed to evaporate. After the pond bottoms are thoroughly dried the process is repeated. This process is done 3 to 4 times before the fish are stocked from wintering ponds in March. Generally no supplementary feeds are given. Selective harvesting begins about 40 days after stocking and is continued at intervals of 10 to 15 days. The annual yield is 6500 to 7800 kg per ha (5789 to 6947 lb per acre).

Beginning in 1972 tilapia culture in combination with hogs or ducks became popular. Many farmers found rice farming to be unprofitable due to low prices and high labor costs. They converted their rice paddies into fish ponds and built pigsties or duck houses beside the ponds. The excretions of the hogs or ducks are diverted into the ponds, with or without fermentation, to serve as fertilizers and/ or feeds. Since 1972 more than 5000 ha (12,350 acres) of paddy fields have been converted into fish ponds. It is estimated that 50 to 70 hogs or 2000 ducks can supply sufficient fertilizer and feed for fish in a 1 ha (2.47 acres) pond.

Fish farmers harvest tilapia of marketable size many times during a rearing season to avoid overcrowding and to obtain money for farm and household expenses. Tilapia are harvested and marketed nearly every day of the year. In some fish stalls in rural areas of southern Taiwan, only tilapia and mullet can be found in December and January.

Tilapia are generally sold fresh iced; however, live fish bring a better price.

JAPANESE EEL *(Anguilla japonica)*

The climate of Taiwan is ideal for eel farming. However, the eel industry did not really grow until 1970, mainly due to the lack of demand. Since 1970 increased demand from Japan has resulted in higher prices for seed eel (elvers) and market-size eel. The total area used for eel farming increased from 60 ha (148 acres) in 1966 to about 830 ha (2050 acres) in January, 1972. In 1976 it was in excess of 1600 ha (3952 acres). Although eel culturing in Taiwan has a short history, it appears to still offer great potential.

Eel ponds require plenty of fresh water and should be located in areas of good water supply, both quantitatively and qualitatively. The water should be free from pollution and have a pH between 6.5 and 8.0. Most eel ponds in Taiwan use underground water from deep wells but irrigation water is also used in some parts of northeastern Taiwan. Most eel-rearing ponds have concrete or brick walls and sandy bottoms, although some are mud ponds with steep earthen embankments. It is claimed that if the water quality is good and the

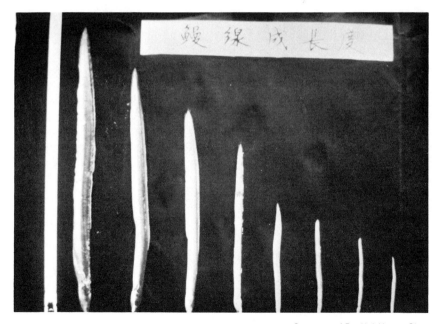

Courtesy of Dr. Hsi-Huang Chen

FIG. 23.2.   ELVERS (EEL) AT DIFFERENT STAGES

TABLE 23.6
EEL EXPORTS FROM TAIWAN TO JAPAN

Year	(MT)	(ST)	Eel Total $ (000s)	($ per kg)	($ per lb)	(MT)	(ST)	Eel Fry Total $ (000s)	($ per kg)	($ per lb)
1970	950	1047	3,325	3.50	1.59	120	132	3,475	290	132
1971	2,300	2535	9,200	4.00	1.82	220	242	4,300	195	87
1972	4,500	4959	22,500	5.00	2.27	180	198	2,500	139	63
1973	6,750	7439	54,000	8.00	3.64	—	—	—	—	—
1974	6,872	7573	42,269	6.15	2.80	—	—	—	—	—
1975	7,924	8732	50,700	6.40	2.91	—	—	—	—	—

Source: Taiwan Fisheries Bureau, Provincial Government of Taiwan.

eels are well fed they will not try to escape, even from ordinary fish ponds with earthen banks.

From June to September, when some of the eels have reached marketable size, they are selectively harvested daily or every few days at feeding times. A net placed below the feeding platform is used.

Eels are sold either in the domestic market or exported to Japan. For transportation over short distances, the eels are first chilled in ice water to put them in a state of suspended animation and then put in bamboo baskets over which ice water trickles. For export to Japan, the previously chilled eels are put into a plastic bag with a small quantity of water. The bag is then filled with oxygen. Two such bags are encased in a strong carton for shipment by air.

The average farm price in 1975 was $6.17 per kg ($2.80 per lb). Total production was 13,575 MT or 14,960 ST (Table 23.6).

Exports increased from 950 MT (1047 ST) in 1970 to 7924 MT (8732 ST) in 1975. In 1970 foreign exchange earnings were $3,325,-000 while in 1975 they amounted to $50,700,000 (Table 23.6). The export price in 1970 average $3.50 per kg ($1.59 per lb) and averaged $6.40 per kg ($2.91 per lb) in 1975. The domestic market in 1975 accounted for 42% of the market and the export market claimed 58%.

GRAY MULLET *(Mugil cephalus)*

Traditionally the dried roe of mullet is considered a gourmet food by Chinese people. Thus the gray mullet is one of the important commercial fishes in Taiwan.

Mullets migrate southward in many schools for spawning. As they enter the Taiwan Strait they head towards the central part of the west coast of Taiwan and then proceed southward along the coastline. During the period of December to January mullets are captured and the roe is dried for local consumption and export to Japan.

For mullet culture the fingerlings are normally taken from estuaries along the west coast of Taiwan from December to March, for stocking in fresh and brackish water ponds. Recently the catch of fingerlings from natural waters has been poor; thus a shortage of supply has occurred. In order to produce fingerlings by artificial propagation, Tungkang Marine Laboratory (also known as Tungkang Shrimp Culture Center) has carried out experiments for six years during the mullet spawning season. Significant success was achieved from 1969 to 1973. According to the last report from Tungkang, a total of 431 hatched larvae were obtained, measuring 3.28 cm (1.5 in.) and weighing 0.34 g on the 45th day. They were about 1.5 times larger than the fingerlings normally collected from natural waters and were robust and strong. Some of them were stocked in fish ponds and grew to 28.1 cm (11 in.) and 217.4 g (8 oz) by the 200th day.

In 1975 total mullet culture production was 1355 MT (1493 ST) (Table 23.5) of which 1141 MT were raised in freshwater ponds. Generally speaking, mullet production is not important from a volume standpoint. However, its market prices are high both for the fish and roe. In 1975 the average farm price of fish body was $1.42 per kg ($0.65 per lb) while the market price of dried roe was more than $44 per kg ($20 per lb).

SHRIMP

It is not known when shrimp culture began in Taiwan. For more than 100 years, farmers have captured juvenile grass and sand shrimp (*Penacus monodon* and *Metapenaeus monoceros*, respectively) from the coastal waters and stocked them in milkfish ponds, where they were given no special care and were harvested as an extra crop. It is only in recent years that shrimp culture has become intensified and appeared to offer economic potential.

In association with milkfish, the number of shrimp planted per hectare varies from 5000 to 8000 (2024 to 3239 per acre). They are stocked from February to early April. They grow to about 40 g each in 3 months and 2 crops can be raised per year. The survival rate is usually over 80%. No special feeds are fed. The shrimp live off the natural foods produced in the shrimp-milkfish ponds.

Taiwan has shrimp hatcheries for spawning, hatching and rearing the larvae (see shrimp section, Chapter 21 (Japan) for details). These hatcheries are located along the sea coast where clean seawater as well as fresh water can be found. Circular plastic tanks of 0.5 to 1.0 MT capacity and concrete tanks of 30 to 70 MT (33 to 71 ST) capacity are used. The number and size of tanks is determined by the scale of operation desired.

The major problem is to acquire spawners for seed production. Tungkang Marine Laboratory (located in southern Taiwan) has been conducting research on this challenging problem and has been successful. A shrimp culture demonstration station has been built to provide young shrimp to growers on a large scale and to demonstrate the economic feasibility of production.

Harvesting marketable size shrimp is relatively easy. A net is placed at the sluice gate when water is being discharged at low tide. The shrimp attempt to escape by swimming with the current. After capture they are iced and shipped live to domestic markets by truck. In 1975 farm value was $1,672,000 for the 775 MT (854 ST) of

Courtesy of Dr. Hsi-Huang Chen

FIG. 23.3.   SPAWNING CARP, TAIWAN

shrimp harvested. Market values are high, amounting to $2.16 per kg, or $0.98 per lb (Table 23.5).

CARP[2]

Polyculture is usually used for carp production. More than one species may be cultured in the same pond. Production is chiefly in freshwater ponds or large reservoirs (Table 23.4). The carp subsist on natural foods found in the ponds and very little supplemental food is fed.

Total carp production in 1975 was 17,419 MT (19,196 ST). Hence carp production is very important from a volume standpoint. However, market prices are relatively low compared to other cultured species. In 1975 the average farm price was $0.61 per kg, or less than $0.28 per lb (Table 23.5).

OUTLOOK

It is expected that the cultured fish industry in Taiwan will con-

---

[2] Silver *(Hypothalmichythys molitrix)*, grass *(Ctenopharyngodon idellus)*, common *(Cyprinus carpio)*, and crucian *(Carassius carassius)*.

tinue to expand production. This will be due to results of research, which will increase the supply of seed fish, and to newer scientific methods of rearing and feeding. Nearly every species now raised should continue to increase in volume. This is particularly true for eel and shrimp.

REFERENCES

ANON. 1976. Taiwan Fisheries Yearbook. Taiwan Fisheries Bureau, Provincial Government of Taiwan.

Chapter 24

Israel

Dr. G.W. Wohlfarth

The annual consumption of fish in Israel is about 10 kg (22 lb) per capita, about 4 kg (9 lb) of which is produced in ponds. The proportion of pond-raised fish in the total supply of animal protein is likely to increase in the future as a result of rising prices of meat and fowl and recent advances in fish farming. In 1975, 13,000 MT (14,326 ST) of fish were produced at 89 farms in a total pond area of about 4500 ha (11,100 acres). Two-thirds of this yield consisted of common carp, the rest being tilapia, silver carp and gray mullets.

HISTORY

The first successful attempts at fish farming in Israel (then Palestine) were made by the Central European immigrants, Messrs. Schwartz and Sitzer, in ponds dug at Kurdaneh in the coastal marches south of Acre (Hornell 1935). They introduced the common carp to this area from Central Europe. Yields attained were 2 or 3 times as high as those commonly attained in Central Europe, clearly a result of higher temperatures and a longer growing season. This encouraged the establishment of commercial fish farming in communal farms.

The first commercial ponds were constructed in 1938 in the Beisan Valley, a semi-arid region with a number of brackish water springs too saline for irrigation. Further developments of fish farming have been documented by Bertram (1946), Tal and Shelubski (1952), Reich (1952), Hofstede (1955) and Jones (1956). Since 1961

TABLE 24.1

DEVELOPMENT OF FISH FARMING IN ISRAEL

Year	Net Pond Area[1] (ha)	Total Yield[2] (MT)	Average Yield[3] (MT per ha)	Common Carp (%)	Tilapia (%)	Mullets (%)	Silver Carp (%)
1939	15	14	0.93				
1941	120	128	1.07				
1943	560	689	1.23				
1945	993	1,260	1.26				
1947	1380	2,250	1.64				
1949	2100	3,700	1.76				
1951	2580	3,850	1.49				
1953	2950	4,650	1.58				
1955	3630	7,320	2.01	98.3	1.6	0.1	
1957	3640	7,530	2.07	98.6	0.5	0.9	
1959	3890	7,990	2.03	97.2	0.2	2.6	
1961	4520	8,870	1.96	96.2	2.6	1.1	
1963	4900	10,050	2.04	92.2	6.0	1.8	
1965	5100	10,180	2.00	94.3	3.3	2.4	
1967	4960	8,680	1.76	88.1	8.0	3.9	
1969	4780	10,260	2.15	81.8	11.6	6.6	
1971	4870	12,530	2.57	85.4	8.6	4.1	1.9
1973	4790	13,780	2.88	83.3	8.1	3.5	5.1
1974	4820	12,170	2.52	75.7	12.6	3.3	8.4
1975	4540	12,910	2.84	68.6	14.2	3.9	13.2

[1] Net pond area includes operative production ponds, auxilliary ponds used for spawning, nursing, storage etc., as well as any ponds not in use during a particular year due to pond repairs, etc.

[2] Total yield is in fact total amount marketed during a given year. This may not be identical with yield during that year, if a surplus of fish was left from the previous year, or if a surplus of fish is left to the following year.

[3] Average yield is total yield divided by net pond area, and is lower than the yield of operative production ponds.

Sarig has published yearly reports entitled: "Fisheries and Fish Culture in Israel in [appropriate year]". These reports, published in "Bamidgeh, Bulletin of Fish Culture in Israel," enable us to follow the yearly development of fish farming.

The total pond area increased from 10 ha (25 acres) in 1938 to 5100 ha (12,600 acres) in 1965. Since then it has decreased to 4500 ha (11,100 acres). Largely due to chronic water shortages, the construction of new ponds is stringently limited. The total yield of fish ponds increased from 14 MT in 1939 to 14,000 MT in 1973. This thousand-fold increase resulted from the larger pond area and the increase in average yield from 1 ton per ha in the early forties to 2 tons per ha between 1955 to 1967. Since 1969 the average yield has tended to increase again and now is close to 3 tons per ha. The data for pond areas and yields are shown in Table 24.1. The ease with which it is possible to obtain such accurate data is due to the organization of fish farming in Israel. All fish are marketed through a central marketing board from which the annual total amounts of fish marketed may be obtained. Licenses are required for constructing ponds, so the exact pond area during each year is noted at the Ministry of Agriculture.

PONDS

Production ponds in Israel are relatively small and vary in area from 2 to 10 ha (5 to 25 acres). The small size is due to the high yields and the usual practice of harvesting the fish by draining the pond. The size of the pond is dictated by the amount of fish the farm can conveniently handle in one operation. Each farm also has a number of smaller ponds from 0.1 to 1 ha (0.25 to 2. acres) for spawning, nursing, storage and various manipulations. The layout of two fairly typical farms is shown by Jones (1956). As a rule each pond has a separate water supply through an iron pipe and a separate water outlet to a drainage ditch via a concrete construction with sluice boards— the so-called monk. The early ponds were shallow with an average depth of 80 cm to 1 m (31 to 39 in.), but later deeper ponds were constructed and some existing ponds deepened to average depths of 3 m (10 ft) for storing runoff water. All ponds are earthen, sometimes with stone reinforcement to prevent erosion by wave action. Some ponds have been constructed in marshy areas and on sand dunes while others were constructed on good agricultural land.

Water is the main factor limiting the area devoted to fish farming. The decrease in pond area since 1965 is largely a result of water shortage. The amount of water supplied to each farm is strictly rationed, so that the alternative value of water is the real cost criterion. In many cases this is evaluated by comparison to cotton, i.e., the expected profit attainable by using a given amount of water for cotton. On this basis water is the main cost factor in fish farming, together with feed costs. Since water always has been in short supply, every effort is made to utilize sources of water not suitable for irrigation. This is the reason why the Beisan Valley with its brackish springs was the first area of fish farming. In the forties close to 60% of the fish pond area was in the Beisan Valley (Bertram 1946). The drop in this proportion to 30% in the seventies, as fish culture spread to other regions[1], indicates the use of other sources of water. In some cases

---

[1] Proportion of the total pond area in different regions in 1975 was: Beisan and Jordan Valleys, 31%; Galilee, 28%; Coastal Plain, 17%; Zvulun Valley, 14%; Yizrael Valley, 10%.

this is runoff water, trapped during the short rainy season in deep ponds and reservoirs. But in other cases fish culture is in direct competition with irrigated agriculture for its water supply.

PRODUCTION METHODS

Fish culture in Israel is geared mainly to supplying the market with live carp of about 600 g (21 oz) all year round. Schematic figures per hectare of production pond for carp are: 3000 fish growing at 5 g per day give a daily production of 15 kg. During a growing period of 250 days per year, this results in a yearly production of about 3.5 MT of carp per ha (3125 lb per acre). As a rule the carp reach market size in a period of 100 to 120 days, so that generally each pond produces two crops per year. Typical production ponds are stocked, in addition to common carp, with 500 to 1500 tilapia and 500 to 1000 silver carp per ha, as well as varying numbers of gray mullet which produce about 1 ton per ha.

PROPAGATION

Carp are spawned in ponds in their natural spawning season in late spring, and often this is the only source of carp fry for the whole year. Since carp attain sexual maturity in one year in warm climates, the year-old fingerlings stocked into ponds in early spring often spawn "wild spawnings." The harm caused by these uncontrolled spawnings is two-fold—loss of control over the number of fish in the pond and loss of weight of the spawning fish. Wild spawning has been avoided by stocking immature fish, less than a year old, from either "early" or "late" spawnings. Early spawning is accomplished by spawning carp in very early spring, about two months before their natural season, in the water of warm springs, mainly in the Beisan Valley. Late spawning consists of spawning the carp out of season, in late summer with the aid of carp pituitary extracts.

Tilapia are also spawned "naturally" in ponds. As a rule they attain sexual maturity during their first year and spawn in the production ponds at the end of the summer. However, the main harm of wild tilapia spawnings comes from mature fish restocked into ponds in spring. Attempts to avoid this wild spawning by stocking only males has not been completely successful.

Silver carp are propagated by induced spawning, with the aid of carp pituitary extracts, at a number of fish hatcheries located at different fish farms.

FEEDING

The increased growth and yields of fish resulting from supplemental feeding were noticed in the very first observations of fish farming (Hornell 1935). The feeds used were lupine seed, maize and cottonseed cake, largely depending on availability and price. For many years the standard practice of feeding consisted of feeding cereal grains once a day, six days a week. The amount of feed was calculated at 4 to 5% of the biomass of the fish or according to the conversion ratio. Feed conversion ratio improved considerably during the years from 6 to 7 in the early years to about 3 in the fifties (Jones 1956) and to 2 to 2.5 in recent years. Feeding experiments starting in the sixties showed that a more balanced diet (i.e., addition of protein to the grain feed) improved growth and yields (Hepher et al. 1971). As a result high protein feed pellets containing 25% protein from fish meal and soy meal came into use. As a rule cereal grains, mainly sorghum are fed at the beginning of the season when the fish biomass is low. With increasing biomass varying proportions of high protein pellets are added to the grain, until at the end of the season, at high biomasses of fish, feed pellets only are fed. The increase in yields per unit area starting in the late sixties is presumably largely due to this improved feed. However, the rising costs of fish meal and soy meal require an examination of whether this level of protein in the feed is economical. Both 25 and 18% protein fish feeds are now commercially produced. The feed was standardly applied to the ponds by tractor-driven feed blowers before the advent of fish feeders. The first to come into use in Israel, demand feeders, were found to increase growth, but often at the expense of a higher feed conversion ratio. More complex automatic feeders, now commonly used, enable a stricter control over the amount of feed and the rate of feeding. The use of these feeders appears to have improved yields. It is also relatively easy to seine fish in the area around the feeder for intermittent cropping and spot checks.

FERTILIZING AND MANURING

The standard method of fertilizing fish ponds in Israel consists of biweekly applications of 50 kg (110 lb) ammonium sulphate and 50 kg of regular superphosphate per ha (2.47 acres). Potassium fertilizers are not used. Larger amounts and more frequent applications were found ineffective in increasing fish yields (Hepher 1963).

Variations in this method are:
(1) Use of liquid ammonia, which is a cheaper nitrogen source but more complicated to apply.

(2) Applying smaller amounts of nitrogen during the summer months when nitrogen-fixing blue-green algae supply some of the nitrogen needs.

(3) Occasional heavy doses of ammonium sulphate to combat the ichthyotoxin-producing alga *Prymnesium parvum.*

(4) Occasional heavy doses of regular superphosphate to combat lack of oxygen.

Chicken manure is often applied to the ponds at a rate 0.5 m³ per ha (330 lb per acre). The early European practice of applying 3 to 5 m³ of cattle manure per ha (1 to 2 MT per acre) was tried at the beginning of fish culture in Israel but soon discontinued. The effect of this manuring on growth and yields of carp could not be demonstrated and shortage of oxygen sometimes occurred in the manured carp ponds. Lately, however, manuring is coming into use in polyculture fish ponds, but in an entirely different method of application (see Future Trends).

FISH STOCKED IN PRODUCTION PONDS

Originally the only fish stocked was the common carp. The idea of culturing supplemental fish in carp ponds has been considered almost from the inception of fish farming in Israel. Catla, grass carp, rainbow trout, tench and largemouth buffalo fish were introduced years ago and attempts made to evaluate their performance. All these early attempts failed. At present four main fish are grown in Israel: the common carp, tilapia, gray mullet and silver carp. Two other fish, grass carp and bighead carp, have been spawned and some attempts are being made to evaluate them.

Common Carp (Cyprinus carpio)

The original introduction of common carp was from Yugoslavia. Subsequent further introductions were from Europe and the Far East. The so-called "Israeli mirror carp," sometimes mentioned in the literature, is by and large a European carp, though lately crossbreeds between European and Chinese common carp have been evaluated (see Genetic Improvement). Frequently carp of two sizes are grown in production ponds: fish to be grown to market size, and fry for the next season. Nursing carp fry to fingerling size in production ponds requires a large weight difference between the two weight classes. Reich (1952) and Jones (1956) describe 2 such methods, i.e., stocking 2 to 5 g fry with 100 g fish or 10 g fry with 300 g fish. This "mixed nursing" method is thought to increase pond yields, obviates the need for special nursery ponds and spreads the fry all over the pond area, so that a disaster in any one pond cannot kill off all the available fry.

Tilapia

The species mainly grown is *T. aurea* of local origin. Problems besetting tilapia culture in Israel are similar to those in other parts of the world, i.e., high fecundity and sensitivity to cold. Originally young of the year were stocked into ponds in early summer and were marketed in the fall. This limited the supply of tilapia in the market to a short season. In order to increase the marketing period some of these fish are over-wintered, the survivors manually sexed and only the males stocked into production ponds in spring. The females are discarded, when possible, to prevent spawning and because their growth is slower than that of males. Some attempts have been made to produce all male broods by crossing female *T. nilotica* of African origin with local *T. aurea* males, but at present this has only a limited commercial application. *T. aurea* is regarded as a highly desirable species because its reported cold resistance is higher than that of other mouth-breeding tilapia and because of its wide spectrum of feeding. The total yield of pond-grown tilapia has been increasing from year to year and its proportion in the total yield reached 14% in 1975.

Gray Mullet

Gray mullet, of the species *Mugil cephalus* and *M. capito*, are the most highly priced of all fish grown in ponds in Israel. Their price is often twice that of common carp and close to that of trout. The main factor limiting mullet production is the supply of fry, seined from river estuaries since all attempts at artificial propagation of mullet in Israel have failed. This supply has been decreasing, apparently due to either overfishing or pollution or both. The proportion of mullets in the total yield increased to a maximum of 6.6% in 1969, but has decreased to between 3 and 4% since.

Silver Carp

Silver carp *(Hypophthalmichthys molitrix)* were introduced to Israel from Japan in 1965. Commercial production started in 1970 after preliminary experiments had shown the great potential of this fish and its spawning problems were solved. Yields increased from 1.9 to 13% of the total production between 1971 and 1975. It has been estimated that yearly yields of up to 1 MT per ha (892 lb per acre) in polyculture do not detract from the yield of other fish, meaning that the 1975 yield could probably be doubled. The silver carp is not a popular fish in Israel; it demands the lowest price of all pond fish and at present about half the yield is exported. However, excellent fish products have been prepared from silver carp in the form of fish sausage and smoked and canned fish.

Grass Carp

Grass carp *(Ctenopharyngodon idella)* were introduced along with silver carp but their spread has been much slower. Observations to date indicate that even at low stocking rates they are capable of keeping the ponds clean of all higher plants, but they compete with common carp for supplemental feed. Small amounts of grass carp are likely to be marketed for the first time in 1976.

GENETIC IMPROVEMENT

Genetic improvement of the common carp has been under investigation since 1958. The results of these investigations show:
 (1) Mass selection of the largest individuals is apparently ineffective in improving growth of carp (Moav and Wohlfarth 1976). This was the traditional method of attempting genetic improvement in Europe, together with mass selection for a high-backed and "scarcely scaled mirror carp," and was transferred to Israel along with other European practices (Hofstede 1955).
 (2) Family selection, though more tedious than mass selection, appears to be more effective.
 (3) Crossbreeding between common carp with different origins was found to improve growth rate, viability and disease resistance (Hines *et al.* 1974; Moav *et al.* 1975).
 (4) A particularly successful crossbreed was that between common carp of Chinese and European origins. Its growth rate was as good as that of the best commercial crossbreeds under conditions of supplemental feeding. In ponds whose only inputs were manures and fertilizers, these Chinese and European crossbreeds showed the best relative performance (Moav *et al.* 1975; Wohlfarth *et al.* 1975).
Crossbred common carp are standardly stocked in production ponds in Israel (Wohlfarth *et al.* 1965).

MARKETING

All fish are marketed through a central marketing board (Tnuva), the amount marketed daily and the farms supplying the fish being organized by the Fish Breeders Association. The full market demands for fish are usually met, i.e., all four kinds of fish are marketed during the full calendar year, in spite of the fact that pond fish in Israel do not grow during winter (approximately four months from November to March). Common carp are live-hauled to the market, sold live to retailers and again to customers. The price of common carp is government controlled. All other fish are marketed dead and cooled, and price fluctuates with supply and demand.

DISEASES AND PARASITES

The first "disease" to strike fish farming in Israel was the ichthyo-toxin-producing alga *Prymnesium parvum*. Later it was found that blue-green algae of the genera *Microcystis* and *Anabaena* also release fish toxins into the water. These algae are now partially controlled by application of ammonium sulphate and copper sulphate. External parasites are a chronic nuisance due to their adverse effects on the fish and religious food laws, which forbid eating fish infested by external parasites. These parasites belong to the genera *Lernea, Argulus, Ichthyophtherius, Dactylogyrus, Gyrolodactylus, Costia* and others. Sarig (1971) has described the parasite fauna of pond fish in Israel, as well as control measures for these parasites. No serious bacterial or viral diseases of fish have been detected.

EQUIPMENT

From its inception efforts have been made to mechanize fish farming in Israel. This is the result of the concentration of fish farming in kibbutz farms, with their chronic labor shortage. Tractors are widely used for hauling nets and transport in the pond areas. Feeding has been described. Fish for market are raised from ponds to trucks with mechanical fish elevators. Intricate grading and sorting devices are in use, especially for harvesting fish from polyculture ponds. Carp are live-hauled to market in tanks with mechanical aeration. Lately intensive growing systems have been introduced, utilizing high stocking rates and mechanical aeration of the pond water.

EXTENSION SERVICE

Extension service of the Ministry of Agriculture provides an instructor in fish farming for each fish growing area. These instructors, who are fish farmers or former fish farmers, pay periodic visits to the farms in their areas and are members of the professional committees of the Fish Breeders Association. They also organize yearly courses of several weeks duration on fish farming and participate in occasional courses on fish diseases, etc.

RESEARCH

Research on many aspects of fish farming is carried out at the Fish and Aquaculture Research Station, Dor, which had its forerunner at Sdeh Nahum. Feeding, fertilizing and manuring, genetic improvement, polyculture, induced spawning, etc. are or have been under investigation. Lately integrated research plans have enabled the simultaneous investigation of several factors, e.g., manuring and breeding of common carp, and the interactions between these factors. The development of fish farming in Israel to its present level is a result of the contributions made by research as well as the initiative of the fish farmers. Due to the close contact between fish farmers and research personnel, the results of investigations at the research station are applied to practical farming in a short time.

FUTURE TRENDS

Since fish farming was established in Israel by immigrants from Europe, the European "feed lot" approach of fish farming was adopted. Feed inputs were always larger than fish yields on a calorie basis. Since the introduction of high protein feeds the protein yields of the ponds are smaller than the protein inputs. So long as the unit of input was much cheaper than the unit of yield, converting cheap feed stuffs into relatively expensive fish was economically feasible. Rising feed costs make this approach less and less profitable, unless pond fish become a luxury food as in the USA. One possible future trend is the intensive culture of fish at high stocking rates, in monoculture on a protein-rich feed, and in running water or with mechanical aeration. This can in the long run only produce a product at a luxury price. Some intensification of fish farming of this type is already being practiced in Israel, but we do not believe that this will be economically viable in the future.

A different future trend is to adopt the Chinese method of balanced polyculture, where the inputs are largely agricultural by-products such as manures. The applicability of this Chinese type of fish farming to conditions in Israel has been demonstrated recently in a series of experimental ponds at the Dor Fishculture Research Station. Yields as high as 8 MT per ha per year (3.6 ST per acre) resulted from daily applications of liquid cow manure without any supplemental feed. The ponds were stocked with common carp, tilapia, silver carp and grass carp. Moreover this practice of frequent manuring of polyculture ponds has already spread to commercial fish farming, an example of the rapid commercial application of experimental results. Preliminary reports of the fisheries extension people indicate increased yields or improved feed conversion ratios in many of these manured ponds.

A demonstration of what may be obtained by polyculture with supplemental feeding was given at Gan Shmuel, one of the best fish farms in Israel (Anon. 1976). This farm is beset by particularly difficult water problems, as shown by the fact that a majority of its ponds are deep ponds for storing runoff water and most of the shallow ponds are to be deepened too. The results of 3 years of fish culture in 93 ha (230 acres) of ponds are shown in Table 24.2. During this period the feed conversion ratio decreased from 1.86 in 1974 to 1.73 in 1975. The table shows:

(1) Yields rose from 4 MT per ha in 1973 to 7 MT per ha in 1975 (1.8 to 3.1 ST per acre).

(2) The absolute yield of common carp increased from 2.7 MT per ha to 3.7 MT per ha (1.2 to 1.65 ST per acre), but its proportional yield decreased from 66 to 51%.

(3) Both absolute and proportional yields of tilapia and silver carp increased, from 17 to 20% and from 10 to 23%, respectively.

It seems likely that the better yield and feed conversion result from the decreased proportion of common carp, that is, from a closer approximation of a balanced polyculture. Results similar to those at Gan Shmuel in 1975 have been attained for years in a number of 2 ha (5 acre) ponds at Dor.

TABLE 24.2

ABSOLUTE AND PROPORTIONAL YIELDS AT GAN SHMUEL

Fish	Average Yield (MT per ha)			Proportional Yield (%)		
	1973	1974	1975	1973	1974	1975
Common carp	2.74	3.52	3.70	66	60	51
Tilapia	0.69	0.99	1.42	17	17	20
Mullets	0.26	0.25	0.40	6	4	6
Silver carp	0.43	1.06	1.66	10	18	23
Total	4.12	5.82	7.18			

SPECIAL ACKNOWLEDGEMENTS

DR. B. HEPHER, Fish and Aquaculture Research Station, Dor, Israel
DR. G. HULATA, Fish and Aquaculture Research Station, Dor, Israel
PROFESSOR R. MOAV, Institute of Animal Science, Agricultural Research Organization, Beth Dagan, Israel

REFERENCES

ANON. 1976. Fish farming in Gan Shmuel. Gan Shmuel's fish farm staff. Daig Vemidgeh Be-Israel *11*, No. 2, 53-62. (Hebrew)

BERTRAM, G. 1946. Carp farming in Palestine. Emp. J. Exp. Agric. *14*, 187-194.

HEPHER, B. 1963. Ten years research in fish pond fertilization. II. Fertilizer dose and frequency of fertilization. Bamidgeh. *15*, No. 4, 78-92.

HEPHER, B., CHERVINSKI, J., and TAGARI, H. 1971. Studies on carp nutrition. III. Experiments on the effect on fish yields of dietary protein source and concentration. Bamidgeh *23*, No. 1, 11-37.

HINES, R., WOHLFARTH, G., MOAV, R., and HULATA, G. 1974. Genetic differences in susceptibility to two diseases among strains of the common carp. Aquaculture *3*, No. 2, 187-197.

HOFSTEDE, A. 1955. Report to the government of Israel on inland fisheries. FAO Rep. 327.

HORNELL, J. 1935. Report on the fisheries of Palestine. Crown Agents for the Colonies, London.

JONES, S. 1956. Pond fish culture in Israel. Bamidgeh *8*, No. 4, 57-67.

MOAV, R., HULATA, G., and WOHLFARTH, G. 1975. Genetic differences between the Chinese and European races of the common carp. I. Analysis of genotype environment interactions for growth rate. Heredity *34*, No. 3, 323-340.

MOAV, R., and WOHLFARTH, G. 1976. Two way selection for growth rate in the common carp (*Cyprinus carpio* L.) Genetics *82*, No. 1, 83-101.

REICH, K. 1952. Some observations on the present state of fish culture in Israel. Proc. Gen. Fish. Counc. Mediterranean *1*, 71-78.

SARIG, S. 1971. Diseases of fishes. Book III. The prevention and treatment of diseases of warmwater fishes under subtropical conditions, with special emphasis on intensive fish farming. T.F.H. Publications, Jersey City, N.J.

TAL, S., and SHELUBSKI, M. 1952. Review of the fish farming industry in Israel. Trans. Am. Fish. Soc. *81*, 218-223.

WOHLFARTH, G., LAHMAN, M., MOAV, R., and ANKORION, Y. 1965. Activities of the Carp Breeders Union in 1964. Bamidgeh *17*, No. 1, 9-15.

WOHLFARTH, G., MOAV, R., and HULATA, G. 1975. Genetic differences between the Chinese and European races of the common carp. II. Multicharacter variation—a response to the diverse methods of fish cultivation in Europe and China. Heredity *34*, No. 3, 341-350.

Chapter 25

Philippines

In 1973 total national fish production was 1,204,837 MT (1,337,-722 ST). Production of cultured fish, essentially from ponds, amounted to 99,600 MT (109,759 ST), which was 8.3% of total fish production. This amount, however, represents a significant increase in cultured fish production. In the 10-year period from 1963 to 1973 the production of fish farmers increased from 62,044 MT (68,372 ST) to 99,600 MT (109,759 ST), or more than 60% (Table 25.1). Fish ponds in production in 1963 covered 131,850 ha (325,669 acres) (Table 25.2). By 1963 the area in production had increased 34% to 176,184 ha (435,174 acres). Investment had increased from $36,123,287 to $62,-974,109, or 74%. With an average of one person employed per hectare of production about 176,000 people were engaged in fish farming in 1973. The value of the crop increased from $14,193,198 to $59,486,-986, or 319% in 10 years. In 1963, 1 kg of cultured fish was worth $0.23 wholesale ($0.105 per lb). In 1973, 1 kg of cultured fish was worth $0.60 ($0.27 per lb).

Of the 176,184 ha in fish culture ponds in 1973, 158,565 ha (391,-656 acres) or 90% were devoted to the culture of milkfish *(Chanos chanos)*, called bangus in the Philippines. Since the fish farming industry is predominately milkfish, this report deals with this species only.

MILKFISH *(Chanos chanos)*

The average production from fish ponds in 1973 was 565 kg per ha (503 lb per acre). This figure is low compared to other milkfish-producing countries. For example, Taiwan averages about 2000 kg per ha per year (1903 lb per acre per year). Efforts are being made to

371

TABLE 25.1

FISH PONDS PRODUCTION AND VALUE, PHILIPPINES

Year	Quantity (MT)	(ST)	Value[1] (cents per kg)	(cents per lb)
1963	62,044	68,372	22.9	10.4
1964	62,680	69,073	22.7	10.3
1965	63,198	69,644	23.0	10.5
1966	63,654	70,147	27.9	12.7
1967	63,912	70,431	29.0	13.2
1968	86,811	95,666	28.8	13.1
1969	94,573	104,216	27.7	12.6
1970	96,461	106,299	35.9	16.3
1971	97,915	107,798	45.2	20.5
1972	98,923	109,010	46.0	20.9
1973	99,600	109,759	59.7	27.1

Source: Anon. (1963-73).
[1] Values for 1965 to 1973 computed at estimated wholesale prices.

increase the productivity of milkfish per unit of water area in the Philippines to this figure.

The milkfish industry has been and is becoming more critically dependent upon the natural supply of fry from natural sources. Recent shortages in fry may be attributed to: (1) increase in acreage farmed; (2) increases in stocking rates; and (3) consistent high mortalities in collecting, sorting, counting, storage and transportation of fry.

The fry collection grounds are the sandy, shallow coasts, tidal creeks and mouths of rivers. Generally the milkfish fry season starts in March and extends to June. However, in some areas it occurs

TABLE 25.2

AREA AND INVESTMENT IN FISH PONDS, PHILIPPINES

Year	Area (ha)	Investment ($000)
1963	131,850	36,123
1964	134,242	36,779
1965	137,251	37,603
1966	138,968	38,073
1967	140,055	38,371
1968	160,807	44,605
1969	164,414	45,045
1970	168,118	46,060
1971	171,446	46,972
1972	174,101	47,699
1973	176,184	62,974

Source: Anon. (1963-73).

throughout the year with the peak season in April and May. Various collection gears are used. These range from set bamboo and net traps to seining nets. After being caught, the fragile fry are counted and stored in wide-mouth earthen jars of approximately 20 liters capacity (5 gal.). Normally 2000 to 2500 fry are stored in a jar half filled with water. Water used in storage can vary from fresh to salt water providing the change in water salinity is not too abrupt. When sufficient fry are stored they are placed in oxygenated plastic bags for transportation to ponds.

The ponds must have a good source of water. In coastal areas this is the tide which exchanges either salt or brackish water. Freshwater ponds can be used if there is a dependable supply of water during the summer months. The ponds must be capable of being drained to get rid of undesirable fish and water plants. In a production facility between 2 and 3% of the water area is for nursery ponds, 30 to 35% is for transition ponds and 60% is for rearing ponds. Around the entire facility is a main dike which must be at least 1 m (39 in.) above the highest tide. There are one or more gates to permit entry of tidal waters and to maintain depth. Inside the main dike are secondary dikes, with secondary gates.

Before stocking, when the ponds are drained, agricultural lime ($CaCO_3$) is spread over the bottom. This is to correct for acidity and prevent pH fluctuations, prevent excessive buildup of magnesium, sodium and potassium ions, promote the release of nutrients, and increase the breakdown of organic matter. While drained different chemicals can be added to the moist bottom to eradicate polychaete worms and snails.

Fertilizers are added to pond waters or soil to stimulate and maintain growth of fish food, such as phytoplankton, microbenthic algae or filamentous green algae. Fertilizers may be either organic or inorganic. The most limiting nutrient is phosphorus (P) and nitrogen (N). Potash (K) is normally adequate in brackish waters. The most common organic fertilizer used is chicken droppings. Cattle and pig manures are sometimes used. Organic fertilizers are not spread because they may absorb excessive oxygen. They are placed in small piles at the rate of 20 to 30 kg per ha (18 to 27 lb per acre). Organic fertilizers may be broadcast over the drained bottom prior to stocking or placed on platforms in the water during culturing. The pond water for culturing is between 75 and 100 cm (2.5 to 3 ft). Success in culturing depends upon proper maintenance of water quality and production of natural foods. The number of fry or fingerlings to be stocked in a pond is dependent on the type and amount of food raised, the carrying capacity of the ponds and the size of fish desired at harvest time. When filamentous green algae are grown the stocking rate is 1000 to 1500 fingerlings per ha (400 to 600 per acre). With microbenthic algae stocking can be increased to 1500 to 3000 fingerlings per ha (600 to 1200 per acre). With plankton production stocking can be at the rate of 3000 to 5000 fingerlings per ha (1200 to 2000 per acre).

After stocking the farmer sometimes manipulates the fish stocks. Sometimes the fingerlings are stocked in the same pond for a three month rearing period. With this practice, there may be 2 or 3 different stocking and rearing periods per year.

Another practice which gives more yield is the stocking of fish in one pond with good growth of food, followed by transfer after a month to a larger adjacent pond with similar growth of food and so on until the fish has reached the desired size. The previous ponds are then prepared for the succeeding rearing periods. Six to eight harvests are probable by using this method.

As soon as milkfish reach the desired size for the market, they are harvested. The newest method utilizes the tendency of the fish to swim against the current. The rearing pond is partially drained of its water during low tide. At the next incoming high tide, new tidal water is allowed to flow into the rearing pond. The fish then will swim towards the inflowing water passing through the gate to the catching pond where they congregate. After a sufficient number of fish have been confined, the gate is closed. The confined fish are either seined or scooped, depending on the size of the catching pond. This method is used for total and selective harvesting or thinning of the stock. The remaining fish in the pond are harvested by totally draining the pond water.

In addition to the raising of milkfish in ponds, fish pens began to be used in 1971 in Laguna de Bay. In this method a pen is constructed enclosing an area of the bay. By 1973 there was a total of 993 fish pens ranging in size from less than 1 ha (2.47 acres) to more than 100 ha, with a total of 4802 ha (11,861 acres) in Laguna de Bay. Up to 4 MT are harvested per ha per year (3570 lb per acre).

After harvesting, the milkfish are washed with pond water, sorted by size and dipped in iced or chilled water to harden the scales and prevent further loss of scales. They are then packed in a variety of shipping containers with about equal quantities of crushed ice added to the fish. They are shipped to market by truck, air and water. As production continues to grow and seasonal volumes are reflected in seasonal low prices, freezing of deboned fish, canning and smoking are likely to be used to stabilize prices.

TABLE 25.3

AVERAGE GROSS INCOME, EXPENSES AND
NET INCOME, 93 FISH FARMS, PHILIPPINES, 1972

Item	$	%
Income		
20,704 kg of fish @ $0.392 per kg	8116	100.0
Expenses		
Fry and fingerlings	652	10.2
Labor	2010	31.4
Interest	2362	37.0
Depreciation	146	2.3
Taxes	61	1.0
Repairs	33	0.5
Fertilizers	568	9.0
Manure	115	1.8
Rent	252	3.9
Chemicals	34	0.5
Feed	63	1.0
Others	92	1.4
Total	6388	78.7
Net Income	1728	21.3

Source: Carandang and Darrah (1973).

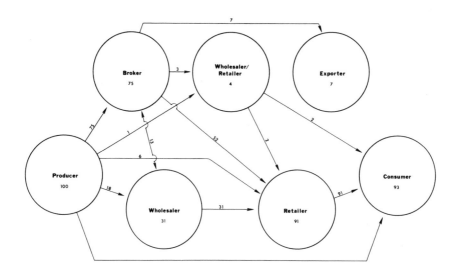

From Guerrero and Darrah (1975)

FIG. 25.1. MARKETING CHANNELS FOR MILKFISH, LUZON, 1974

Income and Expenses

A study of 93 milkfish farms in 1973 indicated an average size of 30.6 ha (76 acres) of water, an average production of 20,704 kg, or 677 kg per ha (603 lb per acre), and a selling price of $0.392 per kg ($0.178 per lb). Gross income was $8116 with expenses of $6388. Net income, not including marketing costs, was $1728 per producer. Net income was over 21% of total income, a relatively high figure. Interest was the largest expense item, followed by labor cost (Table 25.3).

Marketing

Big supplies of milkfish are sold at population centers like Manila because of higher prices relative to smaller population centers.

The marketing system is relatively simple. The main flow of fish is from producer to broker to retailer to consumer (Fig. 25.1). Of the consumers' fish dollar, the producer received $0.64, the retailer $0.10 and the other intermediaries $0.26.

REFERENCES

ANON. 1963-73. Fisheries Statistics of the Philippines. Bureau of Fisheries and Aquatic Resources, Manila.

ANON. 1976. The Philippines Recommends for Bangus, 1976. Philippine Council for Agricultural Research, Manila.

CARANDANG, F.L., and DARRAH, L.B. 1973. Bangus production cost. Marketing Research Unit, National Food and Agricultural Council, Department of Agriculture, Manila.

GUERRERO, C.V., and DARRAH, L.B. 1975. Bangus marketing. Special Studies Division, National Food and Agricultural Council, Department of Agriculture, Manila.

Chapter 26

Indonesia

Indonesia is a nation containing about 3000 islands along the equator. The nation extends for about 4700 km (2900 mi.) from Papua New Guinea to the Malay Peninsula. The 120 million people live on a land area of about 2 million km² (about the size of Texas). Because of the crowded land area the Indonesians also rely on food from water resources. Per capita consumption of fish is about 10 kg (22 lb) per year. About two-thirds of this comes from marine fisheries and one-third from inland fisheries. Total consumption is about 1,268,-000 MT (1,398,000 ST). About 838,000 MT (924,476 ST) come from marine waters and 430,000 MT (473,860 ST) from inland fisheries.

There is little or no culturing of fish in marine waters. The total inland water area, including fresh and brackish, is about 9.4 million ha (23.2 million acres) with an estimated potential yield of 1.6 million MT annually. Present production is only 430,000 MT (474,000 ST) or less than 30% of the potential. About 73% of the catch is caught in natural, open waters such as rivers and lakes, 12% in brackish waters and 15% or 65,000 MT are cultured in ponds or rice paddies. Most of the brackish water production of about 52,000 MT (57,304 ST) is cultured. Hence cultured production is over 100,000 MT (110,200 ST).

The area committed to fish culture in 1971 was 290,403 ha (717,-295 acres), including 36,167 ha (89,332 acres) of freshwater ponds, 70,437 ha (173,979 acres) of paddy fields and 183,799 ha (453,984 acres) of brackish water ponds (Folsom and Reilly 1974). The principal species cultured is milkfish *(Chanos chanos)*. Milkfish are essentially cultured in brackish water. Other fish cultured include common carp *(Cyprinus carpio)*, grass carp *(Ctenopharyngodon idella)*, silver carp *(Hypophthalmichthys molitrix)*, kissing gourami *(Helostrama temminchi)*, gourami *(Osphronemus goramy)*, nilem *(Osteochilus hasseltii)*, tawes *(Puntius gonionotus)*, sepat siam *(Trichogaster pectoralis)*, catfish (species unknown), and tilapia *(mugil engeli)*. Polyculture is practiced. Annual production ranges between 200 and 400 kg per ha (89 to 178 lb per acre) which must be considered low.

FRESHWATER CULTURE

Fish farming in Indonesia's fresh water has been practiced for centuries. Farming in rice fields and brackish water is more recent. The cultivation of fish is in both family ponds and commercial ponds on farms.

Family ponds generally aim for fish production for family use. However, some ponds are in cultivation by hobbyists. These ponds may be as small as 10 m² (100 ft²). They are usually located near the living quarters. All kinds of fry are stocked. The fry are bought from middlemen who buy them from commercial fish farmers. The fish are fed various foods that are available from the table or from the farm itself. As a result of fishing out, the density varies and a few fish may reach relatively large sizes.

The commercial farms may be for producing fry or for production of food-size fish. The latter farms buy fry for stocking. Artificial feeding is not generally practiced except for Chinese carp which are fed plants and grass cut nearby. Manuring is commonly practiced. This is in the form of compost or animal feces. Sometimes the latrines or stables overhang the ponds or are close enough that rainwater washes the feces into the ponds. Foods of low economic value are fed, such as rice bran and other agricultural products. The ponds may be stocked with more than one species to increase production. In fresh water ponds kissing gourami, nilem, common carp, silver and grass carp and tawes may be all stocked together. In brackish water ponds milkfish may be the only stocked species. In general, small fish of 50 to 100 g (2 to 4 oz) are produced. This size results in maximum production and in some cases corresponds to the purchasing power of the buyers.

Nearly all the fish cultured live off phyto- and zooplankton or other natural foods and are tolerant of low dissolved oxygen content in the waters.

Most of the cultured fish is sold fresh, in the round in local markets. This is in marked contrast to marine-caught fish which are salted and dried.

BRACKISH WATER CULTURE

Culture of brackish water fish, such as milkfish and shrimp (species unknown), is widely practiced wherever there are marshlands and inlets.

The milkfish are caught and stocked at about 1.5 cm (0.5 in.) and reared in culture ponds of about 1 ha (2.5 acres). Depth of the ponds is about 1 m (39 in.). Tides exchange the water twice daily and no artificial foods are fed. After about 4 months the fish are marketed at about 250 g (0.5 lb). If density is low, fish of over 500 g (1 lb) may be harvested. The survival rate is between 40 and 50%.

Sometimes around the beginning and ending of the wet season large amounts of shrimp are found near shore. Gates to the brackish water ponds are then opened to permit the shrimp to enter with incoming seawater. No artificial foods are fed. Management consists only of removing predator fish from the pond. After periods of about two months the shrimp are harvested.

REFERENCES

AKAI, M., and SHUNICHI, H. 1974. Fishing Industry in Indonesia, 17th Issue. Japan Fisheries Resource Association, Tokyo.

FOLSOM, W.B., and REILLY, J. 1974. Fisheries of Indonesia, 1972. Office of International Fisheries, National Marine Fisheries Service, Foreign Fisheries Leaflet 74-4, Washington, D.C.

HUET, M. Textbook of Fish Culture, Breeding and Cultivation of Fish, Fishery News. Books, Ltd., London, England.

Chapter 27

Papua New Guinea

J.L. MacLean

RAINBOW TROUT *(Salmo gairdneri)*
and BROWN TROUT *(Salmo trutta)*

The highlands of Papua New Guinea have been supporting trout populations since their first recorded introduction in 1949. Regular stockings began in 1971 with the establishment of a government hatchery. Eyed eggs for the hatchery are imported from Australian trout farms. Both rainbow and brown trout have become established in several small rivers above 1200 m (4000 ft) altitude.

The successful acclimatization in stocking led to the commencement of the country's first and only private culturing farm. This is the Kotuni trout farm near Goroka which was established in 1973. This farm imports eyed rainbow trout eggs from several private disease-free hatcheries in Australia. The eggs are hatched at the farm.

The hatchery, circular, and rectangular ponds are gravity fed by pipes and channels which convey 6400 liters (1500 gal.) of water per minute. Fry and fingerlings are fed on standard fish pellet food which is imported. An equivalent product may be produced locally in the near future.

Marketing began in 1975 when 9 MT (10 ST) of dressed trout were produced from 446 m² (4800 ft²). In 1976 production was anticipated to be 10 MT (11 ST) from 566 m² (6090 ft²). The capacity of the farm is estimated to be about 15 MT (16.5 ST). This volume is seen as sufficient to fill the total demand in Papua New Guinea. In addition to this volume of food fish, some fingerlings are also sold for river stocking. While the present capacity of the farm is 15 MT, there is ample cold water in the district for a projected expansion to 100 MT (110 ST) if export markets can be commercially exported.

A number of problems are associated with the venture. The most difficult are the logistics of marketing and distribution of the product to a widely scattered population. On the technical side, the major problem is water temperature. On a sunny day the temperature of the water reaching the farm climbs from an overnight low of around 8° C (46° F) to 18° C (64° F). This plays havoc with fish feeding and affects feed conversion rates. Growing time of the fish is 18 months.

So far the wild and farmed stocks of trout in Papua New Guinea are disease-free. Legislation is in hand to ban import of salmonoids from countries known to have disease problems.

There are no other aquaculture projects in the country at present. Growing international trade in small freshwater crayfish is creating interest, however. Ten species of *Cherax* (Parastacidae), some of which are suitable for export, inhabit a wide range of habitats throughout Papua New Guinea, and there are large tracts of land ideal for extensive farming.

SPECIAL ACKNOWLEDGEMENTS

MR. D. HUNTER, Kotuni Trout Farm, Goroka, P.N.G.

Chapter 28

Australia

J.L. MacLean

The Australian aquaculture scene is dominated by oyster producers —10,000 MT of Sydney rock oysters are produced annually in New South Wales. The year 1960 can be taken as the starting point of finfish aquaculture. A small trout farm began then as well as a research station to study reproduction in native fish. Commercial-scale trout production commenced in the mid-sixties while native fish are still not reared to market size commercially. The seventies have seen research begin on several freshwater crustaceans with plenty of promise, but as yet there are no signs of a farmed crayfish industry.

Australia is the dryest continent. The few native freshwater fish the early settlers found in the rivers were not very appealing and trout for stream stocking began to be imported as early as 1864, feeding highland streams and now supporting a vigorous sport fishery.

RAINBOW TROUT *(Salmo gairdneri)*

Rainbow trout were the first finfish species used for farming in Australia. A small farm in Victoria produced small quantities of table trout in 1960. The first commercial-scale development began in 1964 in Tasmania. Rainbow trout production commenced there in 1965 on a site containing 4 ha (10 acres) of ponds and earthen raceways. Production costs seemed likely to outweigh the price of imported trout until 1970, when a tariff board enquiry found the embryonic industry worthy of protection and placed a tariff on trout imports. By 1971 there were five trout farms in operation. The future of the industry was further enhanced by a ban placed on the importation of all fresh and frozen salmonoids in 1975. The reason was valid enough. Australia had found itself still free of the highly infectious diseases which were currently devastating trout farms in some parts of the world. Quarantine and fisheries authorities in Australia wish to preserve this happy situation.

There are now 12 trout farms in Australia spread through southern New South Wales, Victoria, South Australia and Tasmania. Total production of dressed trout from these farms for the year ending June 30, 1976, was 177 MT (195 ST), almost all of it being rainbow trout from 3 farms totaling 6.2 ha (15 acres) in pond and raceway surface area. The remainder of farms supply fingerlings to local farmers, and there are a few "fish-out" operations where anglers pay to catch the trout from farm ponds.

Most of the farms produce eggs and/or fingerlings for sale to other farms, acclimatization societies and for export. The total quantities are unknown, but a minimum estimate of private hatchery production is 100,000 fingerlings and 20 million eggs per year. In addition government hatcheries supply various farms, rivers and impoundments with both brown *(Salmo trutta)* and rainbow trout.

Current producers anticipate increasing their production of table trout to a total of 482 MT (531 ST) of dressed fish from their present pondage area of 6.5 ha (16 acres). Only cultured trout can be sold. Wild trout are reserved for amateur anglers.

The problems faced by trout farmers in Australia are mainly of "teething" nature—the lack of expertise in trout husbandry and local feed formulations. Australian banks are still skeptical of aquaculture. Demand far exceeds supply as yet but farmers see a need for consumer education to maintain sales. On the other hand, much of the land suitable for trout farming is in national parks and the remainder, in which developments such as fish farming are permitted, is at a premium. In Tasmania at least, intermittent closing down of hydroelectric stations prevents use of the majority of waterways.

NATIVE FISH

In contrast to the mystique of introduced trout, native freshwater Australian fish have dubious reputations as far as edible qualities are concerned. Limited in their distribution to muddy inland river systems, they have not found ready acceptance by Australians. These species in order of preference are Murray cod *(Maccullochella macquariensis)*, golden perch *(Plectroplites ambiguus)*, silver perch *(Bidyanus bidyanus)* and catfish *(Tandanus tandanus)*.

Of these fish only the catfish, not related to the American farmed species but an eel-tailed fish, will breed in enclosed waters. Lake (1967), at the New South Wales Inland Fisheries Research Station, Narrandera discovered the environmental trigger that induces the other species to spawn—the spring flooding of rivers—and reproduced this phenomenon with miniature floodplains around tiny spawning ponds at the research station.

That station, which began in 1960, produces fry and fingerlings for farm dams in addition to its research functions. There are 37 ponds totaling 4 ha (10 acres) used for fingerling production.

The floodplain spawning technique was adopted by a private hatchery and farm in New South Wales in 1969. The farm currently produces Murray cod and golden perch fingerlings. Some 30 ha (74 acres) of grow-out ponds are being prepared for production of table fish.

A second New South Wales hatchery began using the floodplain spawning technique in 1971 and now produces golden perch by this method. Catfish, requiring no floodplain, are also produced.

Spawning of native fish occurs in October and November, and fingerlings can be sold about three months later.

Total production from government and private hatcheries for the year ending 30 June, 1976 was 10,000 Murray cod, 39,000 golden perch, 45,000 silver perch and 11,500 catfish. Total pond area is 9 ha (22 acres).

The private hatcheries envisage increasing production in the near future to some 50,000 Murray cod, 91,000 golden perch and 6000 catfish, from ponds which will then total 10.4 ha (26 acres).

Farmers of native fish face many problems. The floodplain technique is not fully understood and there have been cases where failures have occurred. Both government and private hatcheries are using hormone injections on an experimental basis, which may enable fingerling production to become regularized. However, the critical point is the search for adequate live food, which is required in the early stages, particularly by newly emerged fry. This is proving the main obstacle to production of sufficient quantities of fingerlings for subsequent commercial-scale farming.

A private hatchery in Queensland has also been producing for some years limited quantities of fingerlings of catfish, golden perch, Murray cod, sleepy cod *(Oxyeleotris lineolatus)* and saratoga *(Scleropages leichardti)*, the last two being indigenous to northern Australia.

The native fish industry is in its infancy. Hatchery operators could sell many times their present production of fingerlings alone. That demand plus the prospects of growing fish to marketable size assures the industry of a viable future.

Courtesy of J.L. MacLean

FIG. 28.1.   SPAWNING POND FOR TWO PAIRS OF "NATIVE" FISH,
SPAWNING INDUCED BY FLOODING ADJACENT AREA

CRAYFISH

There are in Australia the largest freshwater crayfish species in the world. They are carefully guarded. Live specimens are prohibited from export, and one, the marron *(Cherax tenuimanus)* from south Western Australia has been the subject of experimental aquaculture for several years under a Commonwealth government grant. There is only an amateur fishery for marron, so limited are natural stocks.

The main interest has been in yabbies *(Cherax destructor)*, small freshwater crayfish about the size of the prized European kraftor and consequently exported to Europe both live and processed. Trade is promising enough to warrant research into their aquaculture by the South Australian government, under a Commonwealth government grant commencing 1976. That state is the largest producer with about 300 MT (330 ST) per year. There are also farming enterprises in Victoria and New South Wales but these are not yet commercial. The resource is widespread and still abundant, facts which have discouraged concentrated efforts on yabbie aquaculture to date.

Macrobrachium rosenbergii inhabits northern rivers of Australia. A private hatchery has begun producing juveniles but none have yet been reared to market size. Depending on funding, it seems only a matter of time before these shrimp are marketed in Australia.

SPECIAL ACKNOWLEDGEMENTS

MR. J. BARKLEY, Snowy Mountains Trout Ltd., Tumut, New South Wales

MR. A. BROWN, Triton Trout Farm, Tumut, New South Wales

MR. N. DOUGLAS, Hume Weir Trout Farm, Albury, New South Wales

MR. P. MCLAREN, Murray Cod Hatcheries of Australia, Wagga Wagga, New South Wales

MR. L. PARKER, "Fairfield", Tabbita, New South Wales

MR. A. PURVES, Sevrup Fisheries Pty. Ltd., Bridport, Tasmania

DR. K. SHEARER, Inland Fisheries Research Station, Narrandera, New South Wales

REFERENCES

LAKE, J.S. 1967. Rearing experiments with five species of Australian freshwater fishes. I. Inducement to spawning. Aust. J. Mar. Freshw. Res. *18*, 137-53.

Outlook

In spite of a worldwide expansion in fish farming, only limited progress has been made in the compilation of statistical data to indicate trends and the rate of growth. The lack of specialized enumerators, the unwillingness or inability of producers to provide detailed information, the widely scattered location of production units and areas, and the lack of interest and unwillingness to finance the collection of usable data by some countries have all contributed to this scarcity.

In 1970 a partial estimate, made on the basis of data gathered from 36 countries, indicated total fish production through aquaculture of 2.6 million MT (2.9 million ST) (Pillay 1972). A more comprehensive but still rough estimate for 1975 indicated nearly 4.0 million MT (4.4 million ST) of finfish production, 16,000 MT (17,632 ST), of shrimps and prawns, 608,000 MT (670,016 ST) of oysters, 239,000 MT (236,378 ST) of mussels, 39,000 MT (42,978 ST) of clams, 63,000 MT (69,426 ST) of scallops, 30,000 MT (33,060 ST) of cockles and other mollusks and 1,055,000 MT (1,162,610 ST) of seaweeds. The total was over 6 million MT (6.6 million ST) (Pillay 1976). The largest producer was the People's Republic of China with 2,200,000 MT (2,420,000 ST) of finfish. This volume is undoubtedly a rough and perhaps low estimate. Similar estimates by China watchers whom the author contacted ranged up to 3 million MT (3.3 million ST). If it is reasonable to assume that these two estimates are the minimum and maximum, then finfish culture ranged between 4 and 5 million MT (4.4 to 5.5 million ST) in 1975. This would indicate a world increase of 80 to 127% in 5 years. These increases occurred largely through application of technology which existed in 1970 and by the intensification of production per unit of area engaged in fish farming.

Pillay (1976) projected a doubling of output for the decade between 1975 and 1985, and a 5-fold increase by the year 2000 if the rate of increase is maintained. The author believes that a doubling of output to the area of 9 million MT (9.9 million ST) is possible by 1985. This can be accomplished through application of present-day technology which involves more efficient use of each unit of water through greater concentrations of numbers of fish, polyculture, more area devoted to culture, feeding efficiency and breeding. However, the author does not believe that production will increase by 5-fold to 20 to 25 million MT (22 to 27.5 million ST) by the year 2000. This belief is based on the assumption that the rate of increase will not be sustained. In every industry initial gains are large, but as the industry increases in size and output the percentage rate of increase declines. Thus the author predicts that output will increase only 3 to 3.5 times the 1975 level by the year 2000. This would put output at a 12 to 18 million MT (13.2 to 19.8 million ST) level. To achieve even this lower level will require a greater proportion of the world's supply of fish meal to be directed to fish feeds, or the development of substitutes. The ability of the fish farming industry to use a greater share of the available supply of fish meal would indicate that fish feed conversions must improve or the price of cultured fish increase relative to other competing animal uses as the poultry industry. There are no existing data that indicate that fish is preferred to other meats by many people. The major determinant of how fast the volume of production of cultured fish will increase lies with forage fish which can be produced in fertilized waters and rely on natural foods. To a large extent, this type and volume of production will depend upon the use of more water area in those parts of the world where fish culture is not commonly practiced today. Hence if the African, Central and South American countries intensify production, then a rate of increase of 3 to 3.5 times the 1975 world level can be achieved by 2000.

Individual segments of the aquaculture industry may have phenomenal growth during the next 25 years. For example, some countries may turn to the production of one or more highly valued species for export to developed countries. This is particularly true for shrimp and prawns wherein the developing country may wish to increase the availability of foreign exchange earnings.

The ability of any one country to increase production significantly is related not only to marketing and price structures but to political and legal considerations. For example, in the United States most public waters are presently part of the public domain. Fish farmers, on the other hand, because of the nature of investment and the length of commitment, must have long-term rights to a water area. How this and similar situations are resolved will affect the aquaculture industry.

The types and amounts of development assistance by governments to fish farmers will also have considerable significance on the rate of development and the volume of production in any given time period. The ability to finance development, the repayment provisions, and governmental infrastructure such as research applied to local conditions, extension service efforts, roads and a host of other factors will affect aquaculture.

The author is confident that most of the restrictions will be removed as the need for more fish production increases. There is little or no disagreement on whether fish farming will increase, but only on the rate of increase in any given time period.

REFERENCES

PILLAY, T.V.R. 1972. Problems and priorities in aquaculture development *In* Progress in Fishery and Food Science. Univ. Washington College of Fisheries Fiftieth Anniv. Celebration Symp. Univ. Wash. Publ. Fish. (New Ser.) 5, 203-8.

PILLAY, T.V.R. 1976. The status of aquaculture 1975. Paper presented at FAO Technical Conference on Aquaculture, Kyoto, Japan, 26 May - 2 June, 1976.

Appendix

CONVERSION FACTORS

Metric and American Systems

10 millimeters (mm)	=	1 centimeter (cm)
1 centimeter (cm)	=	0.3937 inches (in.)
1 meter (m)	=	39.37 inches (in.)
1 kilometer (km)	=	0.62137 mile (mi.)
1 square meter (m²)	=	10.76 square feet (ft²)
10,000 square meters (m²)	=	1 hectare (ha)
1 hectare (ha)	=	2.471 acres
1 acre	=	4,047 square meters (m²)
1 liter	=	1.0567 liquid quarts (liq qt)
1 quart (qt)	=	0.25 gallon (gal.)
28.4 grams (g)	=	1 ounce (oz)
16 ounces (oz)	=	1 pound (lb)
454 grams (g)	=	1 pound (lb)
1 kilogram (kg)	=	2.20 pounds (lb)
1 metric ton (MT)	=	2204.6 pounds (lb)
1 short ton (ST)	=	2000 pounds (lb)
1 short ton (ST)	=	0.907 metric tons (MT)

TABLE A.2

CURRENCY CONVERSIONS

Norway	$1 U.S.	=	5.44 kroner
Denmark	$1 U.S.	=	5.94 kroner
West Germany	$1 U.S.	=	2.57 deutsche marks
Netherlands	$1 U.S.	=	2.64 gulden
Belgium	$1 U.S.	=	39.43 francs
Switzerland	$1 U.S.	=	2.66 francs
Italy	$1 U.S.	=	667.0 lire
France	$1 U.S.	=	4.36 francs
United Kingdom	$1 U.S.	=	0.47 pound
Ireland	$1 U.S.	=	0.47 pound
South Korea	$1 U.S.	=	500 won
Japan	$1 U.S.	=	300.0 yen
Taiwan	$1 U.S.	=	37.7 yuan
Philippines	$1 U.S.	=	7.3 pesos
Hungary	$1 U.S.	=	23.2 forints
USSR	$1 U.S.	=	3.45 rubles

Index